Graphene-Polymer Composites II

Graphene-Polymer Composites II

Editors

Artur Pinto
Fernão D. Magalhães

MDPI • Basel • Beijing • Wuhan • Barcelona • Belgrade • Manchester • Tokyo • Cluj • Tianjin

Editors
Artur Pinto
University of Porto
Portugal

Fernão D. Magalhães
University of Porto
Portugal

Editorial Office
MDPI
St. Alban-Anlage 66
4052 Basel, Switzerland

This is a reprint of articles from the Special Issue published online in the open access journal *Polymers* (ISSN 2073-4360) (available at: www.mdpi.com/journal/polymers/special_issues/Graphene_Polym_Compos_II).

For citation purposes, cite each article independently as indicated on the article page online and as indicated below:

LastName, A.A.; LastName, B.B.; LastName, C.C. Article Title. *Journal Name* **Year**, *Volume Number*, Page Range.

ISBN 978-3-0365-1678-3 (Hbk)
ISBN 978-3-0365-1677-6 (PDF)

Contents

About the Editors

Artur Pinto graduated in Pharmaceutical Sciences, obtaining his PhD in Biomedical Engineering from Faculty of Engineering - University of Porto (FEUP), visiting U. Washington. Afterwards, Artur has been appointed as a Post-Doctoral Researcher at Eindhoven University of Technology, Chemical Engineering Department, the Netherlands. Following his Post-doc, Artur has been awarded a long-term Researcher position at LEPABE-FEUP. He is the Principal Investigator of a collaborative project between FEUP and i3S (Instituto de Investigação e Inovação em Saúde) focused on developing graphene-based materials and adjusting their properties for biomedical applications, such as phototherapy of skin disease and cancer and immunotherapy.

Fernão D. Magalhães obtained his PhD in Chemical Engineering from University of Massachusetts at Amherst, USA, in 1997. He is currently Associate Professor at FEUP, and senior member of LEPABE, where he leads a research group on Materials Science (polymers, nanomaterials, and composites). His research interests include synthetic and natural adhesives for wood and cork based composites, graphene-based materials for biomedical applications, synthesis, and modification of waterborne polymer dispersions and high-performance industrial coatings.

Preface to "Graphene-Polymer Composites II"

Graphene-polymer nanocomposites continue to gain interest in diverse scientific and technological fields. Graphene-based nanomaterials present the advantages of other carbon nanofillers, like electrical and thermal conductivity, while having significantly lower production costs when compared to materials such as carbon nanotubes, for instance. In addition, in the oxidized forms of graphene, the large specific area combined with a large quantity of functionalizable chemical groups available for physical or chemical interaction with polymers, allow for good dispersion and tunable binding with the surrounding matrix. Other features are noteworthy in graphene-based nanomaterials, like their generally good biocompatibility and the ability to absorb near-infrared radiation, allowing for the use in biomedical applications, such as drug delivery and photothermal therapy.

This Special Issue provides an encompassing view on the state of the art of graphene-polymer composites, showing how current research is dealing with new and exciting challenges. The published papers cover topics ranging from novel production methods and insights on mechanisms of mechanical reinforcement of composites, to applications as diverse as automotive and aeronautics, cancer treatment, anticorrosive coatings, thermally conductive fabrics and foams, and oil-adsorbent aerogels.

Artur Pinto, Fernão D. Magalhães
Editors

polymers

Editorial

Graphene-Polymer Composites *

Artur M. Pinto and Fernão D. Magalhães

LEPABE—Laboratory for Process Engineering, Environment, Biotechnology and Energy, Faculty of Engineering, University of Porto, 4200-465 Porto, Portugal; arturp@i3s.up.pt
* Correspondence: fdmagalh@fe.up.pt

Graphene-polymer nanocomposites continue to gain interest in diverse scientific and technological fields. Graphene-based nanomaterials present the advantages of other carbon nanofillers, like electrical and thermal conductivity, while having significantly lower production costs when compared to materials such as carbon nanotubes, for instance. In addition, in the oxidized forms of graphene, the large specific area combined with a large quantity of functionalizable chemical groups available for physical and/or chemical interaction with polymers, allow for good dispersion and tunable binding with the surrounding matrix. Other features are noteworthy in graphene-based nanomaterials, like their generally good biocompatibility and the ability to absorb NIR radiation, allowing for the use in biomedical applications, such as drug delivery and photothermal therapy.

This Special Issue provides an encompassing view on the state of the art of graphene-polymer composites, showing how current research is dealing with new and exciting challenges. The published papers cover topics ranging from novel production methods and insights on mechanisms of mechanical reinforcement of composites, to applications as diverse as automotive and aeronautics, cancer treatment, anticorrosive coatings, thermally conductive fabrics and foams, and oil-adsorbent aerogels. The most impactful works are summarized in this editorial.

Citation: Pinto, A.M.; Magalhães, F.D. Graphene-Polymer Composites. *Polymers* **2021**, *13*, 685. https://doi.org/10.3390/polym13050685

Received: 19 January 2021
Accepted: 20 February 2021
Published: 25 February 2021

Taking advantage of graphene's unique mechanical properties, Irez et al. newly designed lightweight and cost-efficient composite materials for the aeronautic industry using recycled fresh scrap rubber, epoxy resin, and graphene nanoplatelets (GnPs). After manufacturing the composites, their bending strength and fracture characteristics were investigated by three-point bending (3PB) tests. Halpin–Tsai homogenization adapted to composites containing GnPs was used to estimate the moduli of the composites, and satisfactory agreement with the 3PB test results was observed. In addition, 3PB tests were simulated by the finite element method incorporating the Halpin–Tsai homogenization, and the resulting stress–strain curves were compared with the experimental results. Mechanical test results showed that the reinforcement with GnPs generally increased the modulus of elasticity as well as the fracture toughness of these novel composites. The typical toughening mechanisms observed were crack deflection and cavity formation. Considering the advantageous effects of GnPs on these novel composites and the cost efficiency gained by the use of recycled rubber, these composites have the potential to be used to manufacture various components in the automotive and aeronautic industries as well as smart building materials in civil engineering applications [1]. The mechanical behavior of graphene/polymer interfaces in the graphene-reinforced epoxy nanocomposite is one of the factors that dictates the deformation and damage response of the nanocomposites. Koloor et al. performed hybrid molecular dynamic (MD) and finite element (FE) simulations of a graphene/polymer nanocomposite to characterize the elastic-damage behavior of graphene/polymer interfaces under a tensile separation condition. The MD results show that the graphene/epoxy interface behaves in the form of an elastic-softening exponential regressive law. The FE results verified the adequacy of the cohesive zone model in accurate prediction of the interface damage behavior. The graphene/epoxy cohesive interface was characterized by normal stiffness, tensile strength, and fracture energy of

5×10^{-8} (aPa nm^{-1}), 9.75×10^{-10} (nm), 2.1×10^{-10} (N nm^{-1}), respectively, followed by an exponential regressive law with the exponent, $\alpha = 7.74$. It was shown that the commonly assumed bilinear softening law of the cohesive interface could lead up to a 55% error in the predicted separation of the interface [2].

Using a different resin, Liu et al. prepared graphene-reinforced tung oil (TO)-based unsaturated polyester nanocomposites via in situ melt polycondensation integrated with Diels–Alder addition. Functionalized graphene sheets derived from graphene oxide (GO) were then extracted from the obtained nanocomposites and carefully characterized. The mechanical and thermal properties of the TO-based unsaturated polyester resin (UPR) were greatly improved by the incorporation of GO. At the optimal GO content (0.10 wt.%) the nanocomposite showed a tensile strength and a tensile modulus of 43.2 MPa and 2.62 GPa, and T_g of 105.2 °C, which were 159%, 191%, and 49.4% higher than those of the unreinforced UPR/TO resin, respectively. Compared to neat UPR, the biobased UPR nanocomposite with 0.10 wt.% of GO even demonstrated superior comprehensive properties (comparable stiffness and T_g, while better toughness and thermal stability). Therefore, the developed biobased UPR nanocomposites are very promising to be applied in structural plastics [3].

Functional fabrics have gained attention as an environmentally-friendly synthesis route. Yang et al. developed novel bamboo pulp fabrics with thermal conductivity properties by coating the fabric with graphene and cellulose nanocrystal (G/CNC) solutions. The influences of G and CNC concentrations on the properties of fabrics were studied. It was found that the viscosities of the G/CNC solutions increased with an increase of G contents. G had an obvious thickening effect. Furthermore, compounded fabrics with different G and CNC contents (GCBPFs) were prepared and extensively characterized in terms of their thermal and mechanical properties as well as morphology. The ultimate thermal conductivity, bursting strength, and tensile strength of the GCBPF were 0.136 W/m K, 1.514 MPa, and 25.8 MPa, with 4 wt.% CNC and 3 wt.% G contents. The results demonstrated that the as-fabricated GCBPFs with favorable thermal conductivity could be applied as a novel fast cooling textile for the clothing industry [4].

Under the current situation of frequent oil spills, the development of green and recyclable high-efficiency oil-absorbing aerogel materials has attracted wide attention from researchers. Zhou et al. reported a high-strength, three-dimensional hydrophobic cellulose nanofiber (CNF)/polyvinyl alcohol (PVA)/graphene oxide (GO) composite aerogel with an anisotropic porous structure, which was fabricated by directional freeze-drying technology using anisotropically grown ice crystals as a template, followed by hydrophobic treatment with a simple dip-coating process. The prepared composite aerogel presented anisotropic multi-level pore microstructures, low density (17.95 mg/cm^3), high porosity (98.8 %), good hydrophobicity (water contact angle of 142°), and great adsorption capacity (oil absorption reaching 96 times its own weight). More importantly, the oriented aerogel had high strength, whose compressive stress at 80% strain reached 0.22 MPa, and could bear more than 22 times its own weight without deformation. Therefore, a CNF/PVA/GO composite aerogel could be prepared by a simple and easy-to-operate directional freeze-drying method, being a promising absorbent for oil-water separation [5].

Besides the graphene-based materials intrinsic properties and the interactions with the polymer matrix, the amounts of filler used and the composite production methods can have an impact on its final properties. Huang et al. studied the effects of increasing graphene nanosheet (GNS) concentrations on variations in the structure and properties of electrospun GNS-filled poly(trimethylene terephthalate) (PTT/GNS) composite fibers, such as in its morphology, crystallization behavior, mechanical properties, and electrical conductivity. The effects of GNS addition on solution rheology and conductivity were also investigated. GNSs were embedded in the fibers and formed protrusions. The PTT cold crystallization rate of PTT/GNS composite fibers increased with the gradual addition of GNSs. A PTT mesomorphic phase was formed during electrospinning, and GNS could induce the PTT mesomorphic phase significantly during PTT/GNS composite fiber electrospinning. The PTT/GNS composite fiber mats (CFMs) became ductile with the addition of GNS. The elastic recoveries of the PTT/GNS CFMs with 170 °C annealing were better than those of the as-spun PTT/GNS CFMs. Percolation scaling laws were applied to the magnitude of conductivity to reveal the percolation network of electrospun PTT/GNS CFMs. The electrical conductivity mechanism of the PTT/GNS CFMs differed from that of PTT/GNS composite films. Results showed that the porous structure of the PTT CFMs influenced the performance of the mats in terms of electrical conductivity [6]. Hendrix et al. developed a method exploiting high-speed elongational flow in a novel designed batch mixer, creating a distribution of pristine few to many layer graphene flakes. The method focuses on exfoliating in a molten polyamide 66 (PA66) matrix, creating a graphene reinforced polymer matrix composite (G-PMC). The process revealed that high-speed elongational flow was able to create few layer graphene. Graphite exfoliation was found driven in part by diffusion, leading to the intercalation of PA66 in graphite. The intercalated structure lead to increases in the hydrogen bonding domain, creating anisotropic crystal domains. The thermal stability of the G-PMC was found to be dependent on the degree of exfoliation, the PA66 crystal structure, and composite morphology. The aim of this research was to characterize uniquely produced graphene containing polymer matrix composites using a newly created elongational flow field. Using elongational flow, graphite can be directly exfoliated into graphene within a molten polymer [7].

Graphene impermeability against water, oxygen, and ions diffusion, together with its large surface area, make it an outstanding material for application in anticorrosion surface coatings. Lei et al. studied the corrosion behavior of zinc-rich epoxy primers or paints (ZRPs) with different conducting polyaniline-grafted graphene (PANI/Gr) contents. The conductivity of the formed PANI/Gr nanosheets was significantly improved by employing Gr as the inner template to synthesize the PANI. The protective properties and the electrochemical behavior of coatings with artificial defects were investigated by monitoring the free corrosion potential versus time and by using localized electrochemical impedance spectroscopy (LEIS). A synergetic enhancement of the physical barrier role of the coating and the zinc sacrificial cathodic protection was achieved in the case of ZRP including PANI/Gr nanosheets. In addition, the ZRP mixed with the PANI/Gr at a content of 0.6% exhibited the best anticorrosion performance across the range of investigated PANI/Gr contents [8].

Graphene and graphene polymer composites have been widely explored for the development of medical implants, antibacterial materials, biosensors, drug and gene delivery systems, as well as phototherapy platforms. Gurunathan et al. produced graphene oxide–green platinum nanoparticles (GO-PtNPs) using vanillin and studied its effect on human prostate cancer cells (LNCaP). GO-PtNP cytotoxicity in LNCaP cells was demonstrated by measuring cell viability and proliferation. Both decreased in a dose-dependent manner compared to that by GO or PtNPs alone. GO-PtNP cytotoxicity was confirmed by increased lactate dehydrogenase release and membrane integrity loss. Oxidative stress induced by GO-PtNPs increased malondialdehyde, nitric oxide, and protein carbonyl contents. The effective reactive oxygen species generation impaired the cellular redox balance and eventually impaired mitochondria by decreasing the membrane potential and ATP level. The cytotoxicity to LNCaP cells was correlated with increased expression of proapoptotic genes (p53, p21, Bax, Bak, caspase 9, and caspase 3) and decreased levels of antiapoptotic genes (Bcl2 and Bcl-xl). Activation of the key regulators p53 and p21 inhibited the cyclin-dependent kinases Cdk2 and Cdk4, suggesting that p53 and p21 activation in GO-PtNP-treated cells caused genotoxic stress and apoptosis. The increased expression of genes involved in cell cycle arrest and DNA damage and repair as well as the increased levels of 8-oxo-deoxyguanosine and 8-oxoguanine suggested that GO-PtNPs potentially induce oxidative damage to DNA. Thus, GO-PtNPs are both cytotoxic and genotoxic. LNCaP cells appear to be more susceptible to GO-PtNPs than to GO or PtNPs. Thus, GO-PtNPs have potential as an alternate and effective cancer therapeutic agent. In other words, this work shows that the combination of graphene oxide with platinum nanoparticles opens up new perspectives in cancer therapy. However, further detailed mechanistic studies are required to elucidate the molecular mechanism of GO-PtNPs induced cytotoxicity in prostate cancer [9]. Costa-Almeida et al. used a one-step thermal reduction and noncovalent chemical functionalization process to produce PEGylated reduced nanographene oxide (rGOn-PEG) from nanographene oxide (GOn). Nanomaterials were characterized in terms of particle size, dispersion stability, chemistry, and photothermal properties, in view of its use for photothermal therapy (PTT) of nonmelanoma skin cancer. GOn infrared spectrum presented more intense bands assigned to oxygen containing functional groups than observed for rGOn-PEG. GOn C/O ratio decreased more than 50% comparing with rGOn-PEG and nitrogen was present in the latter (N at.% = 20.6) due to introduction of PEG-NH_2. Thermogravimetric analysis allowed for estimating the amount of PEG in rGOn-PEG to be of about 56.1%. Simultaneous reduction and PEGylation increased the lateral dimensions from 287 $\pm$ 139 nm to 521 $\pm$ 397 nm, as observed by transmission electron microscopy and dynamic light scattering. It was found that rGOn-PEG exhibited roughly 13-fold higher absorbance in the near-infrared radiation (NIR) region, as compared to unmodified GOn. Low power (150 mW cm^{-2}) NIR irradiation using LEDs resulted in rGOn-PEG heating up to 47 °C, which is within the mild PTT temperature range. PEGylation strongly enhanced the dispersibility of rGOn in physiological media (phosphate buffered saline, fetal bovine serum, and cell culture medium) and also improved the biocompatibility of rGOn-PEG, in comparison to GOn (25–250 µg mL^{-1}). After a single NIR LED irradiation treatment of 30 min, a decrease of ≈38% in A-431 cells viability was observed for rGOn-PEG (250 µg mL^{-1}). Together, these results demonstrate the potential of irradiating rGOn-PEG using lower energy, cheaper, smaller, and safer LEDs as an alternative to high-power lasers for NIR mild hyperthermia therapy of cancer, namely nonmelanoma skin cancer [10].

Data Availability Statement: No new data were created or analyzed in this study. Data sharing is not applicable to this article.

Acknowledgments: The editors are thankful for the support from FEDER funds through the COMPETE 2020 Operational Programme for Competitiveness and Internationalisation (POCI), Portugal 2020, and by national funds (PIDDAC) through FCT/MCTES in the framework of the project POCI-01-0145-FEDER-031143, and Base Funding-UIDB/00511/2020 of the Laboratory for Process Engineering, Environment, Biotechnology and Energy (LEPABE). Artur Pinto thanks the Portuguese Foundation for Science and Technology (FCT) for the financial support of his work contract through the Scientific Employment Stimulus-Individual Call [CEECIND/03908/2017].

Conflicts of Interest: The authors declare no conflict of interest.

References

1. Irez, A.B.; Bayraktar, E.; Miskioglu, I. Fracture Toughness Analysis of Epoxy-Recycled Rubber-Based Composite Reinforced with Graphene Nanoplatelets for Structural Applications in Automotive and Aeronautics. *Polymers* **2020**, *12*, 448. [CrossRef] [PubMed]
2. Rahimian-Koloor, S.M.; Karimzadeharani, A.; Hamdi, M.; Petrů, M.; Tamin, M.N. Nano-Level Damage Characterization of Graphene/Polymer Cohesive Interface under Tensile Separation. *Polymers* **2019**, *11*, 1435. [CrossRef] [PubMed]
3. Liu, C.; Wang, C.; Tang, J.; Zhang, J.; Shang, Q.; Hu, Y.; Wang, H.; Wu, Q.; Zhou, Y.; Lei, W.; et al. High-Performance Biobased Unsaturated Polyester Nanocomposites with Very Low Loadings of Graphene. *Polymers* **2018**, *10*, 1288. [CrossRef] [PubMed]
4. Yang, F.; Lan, C.; Zhang, H.; Guan, J.; Zhang, F.; Fei, B.; Zhang, J. Study on Graphene/CNC-Coated Bamboo Pulp Fabric Preparation of Fabrics with Thermal Conductivity. *Polymers* **2019**, *11*, 1265. [CrossRef] [PubMed]
5. Zhou, L.; Zhai, S.; Chen, Y.; Xu, Z. Anisotropic Cellulose Nanofibers/Polyvinyl Alcohol/Graphene Aerogels Fabricated by Directional Freeze-drying as Effective Oil Adsorbents. *Polymers* **2019**, *11*, 712. [CrossRef] [PubMed]
6. Huang, C.-L.; Wu, H.-H.; Jeng, Y.-C.; Liang, W.-Z. Electrospun Graphene Nanosheet-Filled Poly(Trimethylene Terephthalate) Composite Fibers: Effects of the Graphene Nanosheet Content on Morphologies, Electrical Conductivity, Crystallization Behavior, and Mechanical Properties. *Polymers* **2019**, *11*, 164. [CrossRef] [PubMed]
7. Hendrix, J.W.; Szeto, R.; Nosker, T.; Lynch-Branzoi, J.; Emge, T.J. Evaluation of Exfoliated Graphite to Graphene in Polyamide 66 Using Novel High Shear Elongational Flow. *Polymers* **2018**, *10*, 1399. [CrossRef] [PubMed]
8. Lei, Y.; Qiu, Z.; Liu, J.; Li, D.; Tan, N.; Liu, T.; Zhang, Y.; Chang, X.; Gu, Y.; Yin, Y. Effect of Conducting Polyaniline/Graphene Nanosheet Content on the Corrosion Behavior of Zinc-Rich Epoxy Primers in 3.5% NaCl Solution. *Polymers* **2019**, *11*, 850. [CrossRef] [PubMed]

9. Gurunathan, S.; Jeyaraj, M.; Kang, M.-H.; Kim, J.-H. Graphene Oxide–Platinum Nanoparticle Nanocomposites: A Suitable Biocompatible Therapeutic Agent for Prostate Cancer. *Polymers* **2019**, *11*, 733. [CrossRef] [PubMed]
10. Costa-Almeida, R.; Bogas, D.; Fernandes, J.R.; Timochenco, L.; Silva, F.A.L.S.; Meneses, J.; Gonçalves, I.C.; Magalhães, F.D.; Pinto, A.M. Near-Infrared Radiation-Based Mild Photohyperthermia Therapy of Non-Melanoma Skin Cancer with PEGylated Reduced Nanographene Oxide. *Polymers* **2020**, *12*, 1840. [CrossRef] [PubMed]

polymers

MDPI

Article

Fracture Toughness Analysis of Epoxy-Recycled Rubber-Based Composite Reinforced with Graphene Nanoplatelets for Structural Applications in Automotive and Aeronautics

Alaeddin Burak Irez [1,*], Emin Bayraktar [2] and Ibrahim Miskioglu [3]

1 LMT—Laboratoire de Mécanique et Technologie, Université Paris-Saclay, ENS Paris-Saclay, CNRS, 94235 Cachan, France
2 School of Mechanical and Manufacturing Engineering, SUPMECA-Paris, 93400 Saint Ouen, France; bayraktar@supmeca.fr
3 ME-EM Department, Michigan Technological University, Houghton, MI 49931, USA; imiski@mtu.edu
* Correspondence: irez@lmt.ens-cachan.fr

Received: 31 December 2019; Accepted: 10 February 2020; Published: 14 February 2020

Abstract: This study proposes a new design of lightweight and cost-efficient composite materials for the aeronautic industry utilizing recycled fresh scrap rubber, epoxy resin, and graphene nanoplatelets (GnPs). After manufacturing the composites, their bending strength and fracture characteristics were investigated by three-point bending (3PB) tests. Halpin–Tsai homogenization adapted to composites containing GnPs was used to estimate the moduli of the composites, and satisfactory agreement with the 3PB test results was observed. In addition, 3PB tests were simulated by finite element method incorporating the Halpin–Tsai homogenization, and the resulting stress–strain curves were compared with the experimental results. Mechanical test results showed that the reinforcement with GnPs generally increased the modulus of elasticity as well as the fracture toughness of these novel composites. Toughening mechanisms were evaluated by SEM fractography. The typical toughening mechanisms observed were crack deflection and cavity formation. Considering the advantageous effects of GnPs on these novel composites and cost efficiency gained by the use of recycled rubber, these composites have the potential to be used to manufacture various components in the automotive and aeronautic industries as well as smart building materials in civil engineering applications.

Keywords: toughening mechanisms; graphene nanoplatelets; recycled rubber; Halpin–Tsai; SEM

1. Introduction

Over the past decades, aeronautic companies have been continuously trying to reduce the overall cost and mass of an aircraft to better compete with their rivals. Mass reduction leads to lower fuel consumption and CO_2 emissions. Therefore, development of low-cost and lightweight materials to be used in the manufacture of various aircraft parts constitutes an important task for engineers in aeronautic companies.

In aircrafts, polymer-based composites are used widely in the manufacturing of various structural and functional components including the wings, tail, and skin panels. In material selection, structural requirements are, of course, important. Also, the cost of the material should be minimized without compromising the structural requirements. To that end, proposing a low-cost, lightweight aircraft material was the main objective of this study.

Polymer-based composites are extensively used to manufacture lightweight structural components. Including recycled materials in the production may result in cost efficiencies as well as ecological solutions. Epoxy, a thermosetting polymer, which is relatively easy to process and has low cost, is often used as the matrix material in the polymer-based composites. Although epoxies have high stiffness and specific strength and are environmentally stable [1], they are brittle due to the fact of their highly cross-linked network structure. Therefore, to improve the toughness of epoxy, secondary phase particles, such as soft (thermoplastic particles, rubber) and rigid fillers, are added [2–4]. Hence, the use of recycled rubber in the proposed composites can meet the expectations on toughness and material cost.

The disposal of used or scrap rubber parts poses a technical, ecological, and economic challenge due to the fact of their vulcanized structure [5]. For instance, discarded tires in a landfill can hold water, creating habitats for mosquito larvae as well as other animals such as rodents and snakes [6]. These sites can potentially be sources for diseases such as malaria, cephalitis, dengue, and chikungunya. Besides, if the rubber piles in the landfills burst into flames, it is difficult to extinguish [7]. In addition, some additives in the rubber discarded at landfills, such as colorants, stabilizers, flame retardants, and plasticizers, may leach into the soil and cause further ecological problems [8–10]. Using recycled rubber in composite manufacturing can help to reduce the impact of discarded rubber on the environment with the added benefit of cost reduction. Various research groups have reported studies on recycled rubber-epoxy blends. As the first example, a research group used recycled rubber to modify epoxy resin to improve its toughness, with minimal change on strength and stiffness. In their study, it is proposed that the manufactured material can be used to manufacture railroad cross ties in high volumes [11].

Another research group argued that the blends of epoxy and recycled rubber can be used in agricultural areas for cementing the adhesive bonding of larger units. In addition, the same material can serve to fill larger cracks and to shape imbalances [12]. Also, due to the quite inhomogeneous cross-linked structure of rubbers and rubber-like materials, recycled rubber-modified epoxies can be utilized as sound and vibration dampers [13–16].

Apart from the positive outcomes, it is claimed that recycled rubber particles significantly decrease the shear strength of the overlapped adherent because the boundary of the particles and the resin promote the formation of cracks, decreasing the strength [17]. In order to compensate for the decrease in strength, hybridization of the composite by fillers, such as graphene nanoplatelets (GnPs), offer an optimum composition as a bracket material to be used in aircraft wings [10,18,19].

Nano graphene has unique properties as a result of its 2D honeycomb structure which makes it a promising nanoscale inclusion for polymer nanocomposites. It has an outstanding mechanical strength (130 GPa) and specific elasticity modulus (1 TPa) [20–22]. Besides, the high surface area of the nano fillers can improve the properties of the composites even at very low contents compared to the microscale fillers [23–25]. Also, as an added benefit, high electrical conductivity of graphene can reduce the risk of damage to aircrafts from lightning strikes.

In the frame of this research, after manufacturing these novel composites, three-point bending (3PB) tests were carried out to determine the fundamental mechanical properties, and the results were compared with FEM modelling and with a modified Halpin–Tsai homogenization adapted to GnP-consisting composites. In addition, composite fracture toughness was examined using notched specimens. Lastly, fracture surfaces were observed with a scanning electron microscope (SEM) to study the toughening and damage mechanisms.

2. Experimental Procedure

2.1. Materials

In this study, graphene nanoplatelets, recycled EPDM (ethylene propylene diene monomer) rubbers, and epoxy matrix were used to manufacture the specimens. Graphene nanoplatelets were procured from Alfa Aesar™ with the specific name of "Graphene nanoplatelets aggregates, sub-micron particles, S.A. 500 m^2/g". This product consists of sub-micron platelets which have a diameter of less than 2 microns and a thickness of around 5 nanometers. The tensile modulus and the density of the of the GnPs were listed as 1 TPa and 2.25 g/cm^3, respectively. For the matrix, Araldite DBF epoxy resin and its hardener Aradur HY 956 EN were obtained from Hunstman™. Araldite DBF has a tensile modulus of 2880 MPa and a density of 1.1 g/cm^3. The hardness of Araldite DBF is given as 80 in Shore D scale for 25 °C. Recycled EPDM rubber was supplied by a sports equipment manufacturer in Sofia, Bulgaria, as fresh scrap, i.e., they were collected directly from the production line as waste parts and pulverized. No contaminants were found in the rubber such as metallic particles which could oxidize and overheat the rubber or degrade the adhesion of rubber with the matrix. The average diameter of the rubber particles was measured as 10.44 μm by a Cilas™ 990 Laser Particulate Analyzer. The modulus of elasticity of the EPDM rubber was provided by the recycled rubber supplier as 6 MPa, elongation at break was 80–100%, its hardness was 37–40 Shore A, and the density was 1.4 g/cm^3.

2.2. Materials Processing and Experimental Characterization

The manufacturing process of the composites is illustrated in Figure 1, and more details can be found in our previous paper [4]. Sonication was performed to distribute GnPs and rubber particles more homogeneously in the epoxy matrix. Degassing of the molded final composite was required to eliminate the air bubbles that may have generated during the polymerization of epoxy.

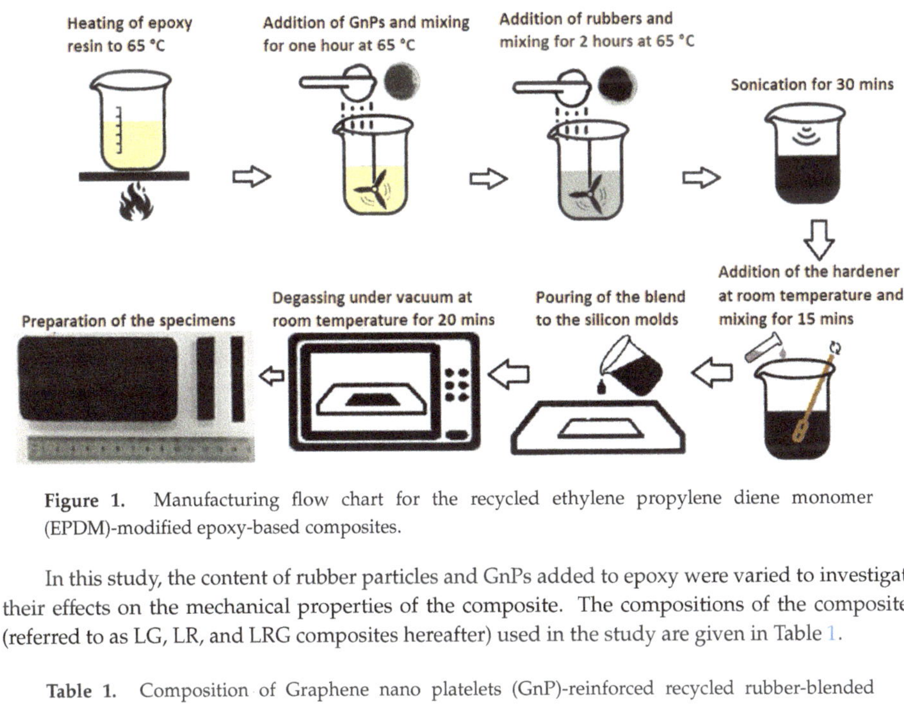

Figure 1. Manufacturing flow chart for the recycled ethylene propylene diene monomer (EPDM)-modified epoxy-based composites.

In this study, the content of rubber particles and GnPs added to epoxy were varied to investigate their effects on the mechanical properties of the composite. The compositions of the composites (referred to as LG, LR, and LRG composites hereafter) used in the study are given in Table 1.

Table 1. Composition of Graphene nano platelets (GnP)-reinforced recycled rubber-blended epoxy-based composites.

LRG Composites	GnP Content (wt. %)			
Rubber content (wt. %)	0%	0.5%	1.0%	1.5%
0%		LG0.5	LG1.0	LG1.5
10%	LR10	LR1G0.5	LR1G1.0	LR1G1.5
20%	LR20	LR2G0.5	LR2G1.0	LR2G1.5
30%	LR30	LR3G0.5	LR3G1.0	LR3G1.5

LRG: Epoxy + rubber + graphene nanoplatelets.

The densities of the composites were measured with a pycnometer, and Shore D hardness measurements were carried out according to the ASTM D 2240 standard. Quasi-static three-point bending tests (3PB-Instron 5569, Norwood, MA, USA) were performed in accordance to the ASTM D790 standard. Load on the specimen and midspan deflection were measured during each test. Midspan deflection of the specimen was measured by the crosshead position. In addition, fracture toughness parameters, such as critical stress intensity factor (K_{Ic}) and critical strain energy release rate (G_{Ic}), were investigated with single-edge notched-beam (SENB) specimens according to the ASTM D5045 standard. At least five specimens for each composition were used. The SEM fractography (Scope/JSM-6010LA Jeol®, Tokyo, Japan) was performed on the fracture surfaces to identify the toughening and damage mechanisms.

3. Results and Discussion

3.1. Physical Characteristics of the Manufactured Composites

The measured densities of the LRG composites are given in Table 2.

Table 2. Density of the manufactured composites.

Composition Name	Density (g/cm^3)	Composition Name	Density (g/cm^3)	Composition Name	Density (g/cm^3)
LR10	1.120	LR20	1.115	LR30	1.035
LG0.5	1.145	LG1.0	1.154	LG1.5	1.160
LR1G0.5	1.130	LR2G0.5	1.125	LR3G0.5	1.047
LR1G1.0	1.155	LR2G1.0	1.130	LR3G1.0	1.076
LR1G1.5	1.159	LR2G1.5	1.142	LR3G1.5	1.113

As expected, composite densities increased with the increasing GnP content. Density of the ternary composites (LRG) were bracketed by the densities of the binary (LG and LR) composites. Moreover, in binary group composites (LR groups), increasing rubber content decreased the density of the composites. When this trend is compared with the Shore D hardness, it is observed that the hardness of the composites was reduced. The density reduction also lowered the mechanical performance of the binary composites. Strain at break and the maximum flexural stress suffered from the decreasing density in the LR groups. However, in the ternary group composites (LRG), the composite performance cannot be simply associated to the density. More intricate mechanisms are involved in the composite performance together with density.

Another physical characteristic of the composites, surface hardness, is presented in Table 3.

Table 3. Surface hardness measurement of the manufactured composites in Shore D scale.

Composition Name	Shore D	Composition Name	Shore D	Composition Name	Shore D
LR10	74.8 ± 0.3	LR20	72.8 ± 0.3	LR30	64.4 ± 0.4
LG0.5	70.6 ± 0.5	LG1.0	76.2 ± 0.4	LG1.5	75.6 ± 0.2
LR1G0.5	74.2 ± 0.4	LR2G0.5	75.8 ± 0.1	LR3G0.5	72.4 ± 0.1
LR1G1.0	72.6 ± 0.2	LR2G1.0	75.0 ± 0.2	LR3G1.0	71.6 ± 0.5
LR1G1.5	74.0 ± 0.5	LR2G1.5	75.8 ± 0.2	LR3G1.5	70.2 ± 0.1

It can be seen that the addition of 0.5 wt. % GnP did generally enhance the composite hardness due to the fact of their hard nature compared to rubber and epoxy. However, there was no remarkable change in hardness when reinforcement contents were increased further to 1.0 wt. % and then to 1.5 wt. %. As GnPs have high affinity and strong van der Waals forces, they tend to agglomerate when their content is increased. For this reason, it is difficult to distribute GnPs homogeneously in the microstructure and have a proportional increase in the hardness.

3.2. Mechanical Characterization of the Manufactured Composites by Means of 3PB Tests

The results from the three-point bending tests are given in Figure 2 for one sample from each composite group, and all results are summarized in Table 4 with their standard deviations.

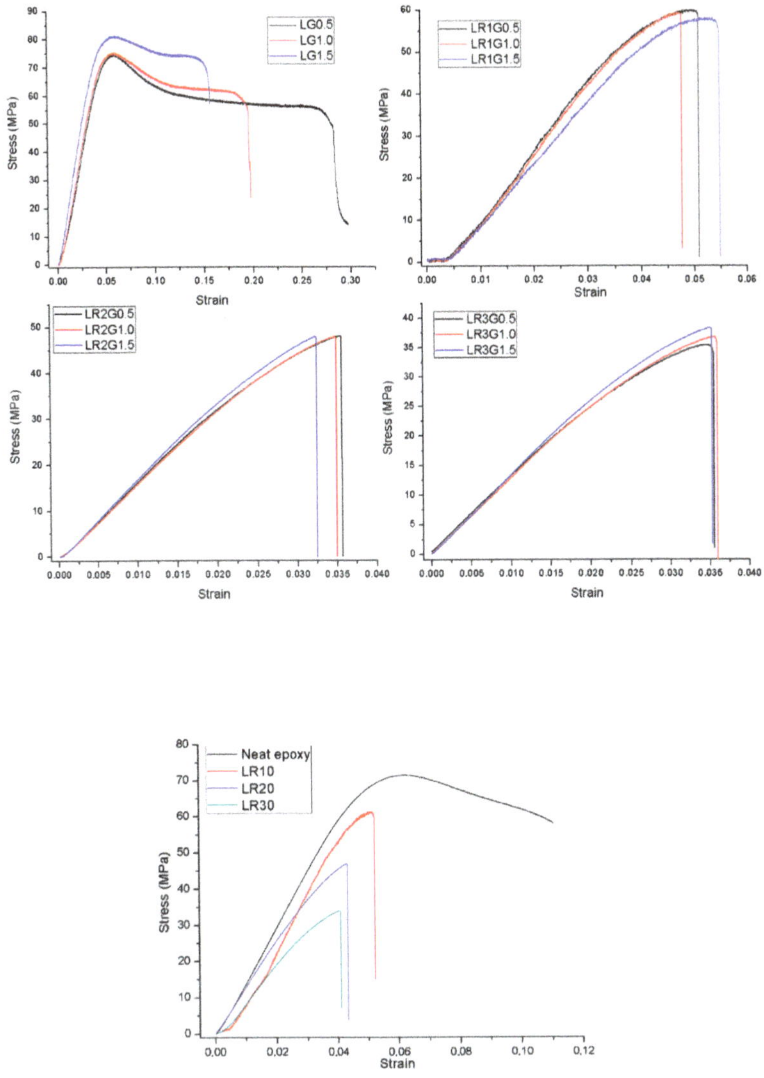

Figure 2. Engineering stress–strain curves of the manufactured composites.

Table 4. Three-point bending (3PB) test results.

Composition Name	Ultimate Flexural Stress (MPa)	Flexural Modulus (MPa)	Strain at Break
Neat epoxy	78.96 ± 1.22	1465.83 ± 145.05	0.13 ± 0.017
LG0.5	63.05 ± 11.58	1297.28 ± 193.76	0.064 ± 0.004
LG1.0	74.11 ± 1.78	1439.75 ± 101.60	0.061 ± 0.003
LG1.5	78.97 ± 2.49	1582.23 ± 111.73	0.057 ± 0.004
LR10	61.58 ± 1.64	1454.71 ± 16.28	0.049 ± 0.002
LR20	48.33 ± 1.02	1149.64 ± 20.74	0.045 ± 0.001
LR30	34.27 ± 3.77	478.25 ± 64.13	0.037 ± 0.001
LR1G0.5	59.65 ± 0.54	1478.84 ± 84.15	0.055 ± 0.004
LR1G1.0	59.18 ± 0.20	1475.10 ± 25.70	0.046 ± 0.001
LR1G1.5	59.08 ± 0.67	1294.78 ± 29.11	0.057 ± 0.002
LR2G0.5	47.94 ± 0.25	1417.84 ± 13.69	0.035 ± 0.001
LR2G1.0	48.49 ± 1.01	1474.58 ± 19.79	0.034 ± 0.001
LR2G1.5	48.61 ± 0.74	1537.54 ± 28.97	0.033 ± 0.001
LR3G0.5	35.53 ± 0.55	1064.35 ± 20.10	0.035 ± 0.001
LR3G1.0	36.71 ± 0.42	1079.57 ± 8.66	0.035 ± 0.001
LR3G1.5	38.68 ± 0.24	1135.30 ± 6.65	0.030 ± 0.001

Figure 2 shows that for LG composites (epoxy and GnPs), the strain at break increased with increasing GnP content, whereas for the LR composites (epoxy and rubber), the strain at break decreased with increasing rubber content. It is also observed in Table 4 that GnPs did not have a significant effect on the strength of the composites.

The increase in rubber content resulted in a drop in both the strength and strain at break of the composites. This tendency can be related to the poor interfacial adhesion of the recycled EPDM and epoxy blends. Because the EPDM rubber was vulcanized during the post-processing phase of its manufacturing cycle, it lacked free links on its surface. Thereby, the recycled EPDM lacked free links also, making it difficult to have a chemical bond with epoxy. Because of the incompatibility between recycled EPDM and epoxy, composite interfaces may contain some voids, and this gives rise to low stress transfer from the matrix to the rubber particles reducing the global rigidity of the compounds. Moreover, in the course of the solidification, different contractions of rubber and epoxy can bring some inequalities in the internal stress balance which leads to void formations at the composite interfaces. In addition, by the increasing content of the recycled rubber particles, the possibility to observe any agglomerations increases. Therefore, these agglomerations create the weak parts of the composite. As a consequence, when the composites are subjected to any loading, these abovementioned voids and agglomerations constitute the weak points of the composite where cracks can initiate. This state produces premature failure. Also, low rigidity of recycled EPDM has an effect in the drop of the mechanical properties of the composites.

The strain at break generally increased for 0.5–1.5 wt. % GnP loading for 10 wt. % recycled rubber content which indicates better chain mobility. However, with a further increase in recycled rubber loading as the number of particles increase, chain mobility as well as elongation at break decreased. The poor flexural strength of LR2G and LR3G group composites was mainly due to the inhomogeneous dispersion of recycled rubber filler in the epoxy matrix and the presence of GnPs agglomerates in the composite which inhibited stress transmission and reduced the flexural strength of the composite.

These arguments are supported by the fracture surface images in Figure 3a,b, taken after the 3PB tests. Figure 3a indicates the fracture surface of a composition with 10 wt. % of rubber, whereas Figure 3b shows a composition with 30 wt. % rubber. In Figure 3b, the composite with 30 wt. % rubber content had more discontinuities, shown with the red circles and arrows, at the epoxy and rubber interfaces in comparison to the composite with 10 wt. % rubber content. In Figure 3a,b, rough areas show recycled rubber particles, whereas smooth areas indicate the epoxy matrix. In addition, in Figure 3b, compared to composites with lower rubber content, more agglomerates of rubber particles are observed. Because of the above factors, the increase in rubber content enhanced the amount of discontinuities, resulting in a reduction in strength and strain at break of the composites manufactured. Therefore, GnPs were introduced into the composite structure to compensate for these adverse effects of recycled rubber. The GnPs' rigid nature enhanced the composites' modulus of elasticity. However, the GnPs were not able to efficiently compensate for the mentioned adverse effects of recycled rubber for stress and strain at break [26,27].

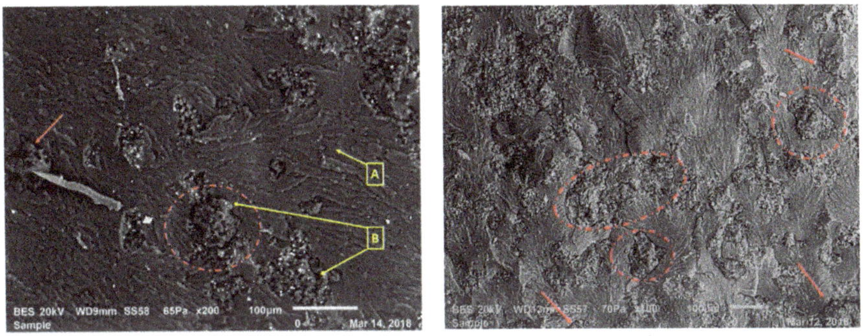

(a) (b)

Figure 3. Comparison of the fracture surfaces: (**a**) 10% rubber-consisting composition—LR1G0.5; (**b**) 30% rubber-consisting composition—LR3G0.5.

In addition, the flexural moduli of the LR1G0.5 and LR1G1.5 groups indicate some deviations from the expected values. This can stem from the issues related to composite manufacturing. For instance, during the mixture process, rubber particles and GnPs can aggregate in the microstructure, and these may create stress concentrations resulting in lower mechanical properties. Also, these composites are cured with a specific hardener at a predefined temperature. In this curing process, cross-links are created along the polymer chains. In general, if the density of the cross-links increases, the cured composite becomes more rigid. In this regard, during the curing procedure, some points, at the micro scale, can locally be cured either insufficiently or excessively. Therefore, these local differences may constitute the weak or strong points of the composites which influence the final mechanical properties. Apart from these anomalies, addition of GnPs improved the flexural modulus of the 20–30 wt. % recycled rubber containing epoxy blends. On the other hand, increasing rubber content degraded the flexural modulus of the binary epoxy recycled rubber composites ($E_{LR10} > E_{LR20} > E_{LR30}$).

3.3. Numerical Verification of the 3PB Tests

In this step of the study, experimental results were compared with numerical approaches. In this context, a modified Halpin–Tsai model considering the effects of nanoplatelets was used as the homogenization strategy. In the modified Halpin–Tsai model, shape factor and aspect ratio of the inclusions were taken into consideration. Otherwise, incorrect shape factor adoption may lead to erroneous results. The modified Halpin–Tsai model is given in Equations (1) and (2) [28,29]:

$$E_c = \left(\frac{3}{8} \frac{1 + (2L_G/3T_G)\,\eta_L V_G}{1 - \eta_L V_G} + \frac{5}{8} \frac{1 + 2\eta_T V_G}{1 - \eta_T V_G} \right) E_P \tag{1}$$

$$\eta_L = \frac{(E_G/E_P) - 1}{E_G/E_P + 2L_G/3T_G} \quad \eta_T = \frac{(E_G/E_P) - 1}{E_G/E_P + 2} \tag{2}$$

In these equations, E_C is the elasticity modulus of the final composite with randomly oriented graphene nanoplatelets, and E_G and E_P are the elasticity moduli of graphene and the matrix, respectively. The modulus of elasticity of the matrix was reckoned as the combination of the EPDM rubber particles with epoxy according to studies in the literature [2]. V_G, T_G, and L_G refer to the volume fraction, thickness, and length of graphene. The dimensions of the graphene sheets were provided by the supplier of the GnPs. On the other hand, to estimate the modulus of elasticity of the epoxy–recycled rubber blend, classical H–T equations were used, as below:

$$\frac{E_{m2}}{E_{matrix}} = \frac{1+\xi\eta\varphi_R}{1-\xi\eta\varphi_R},\ \eta = \frac{\frac{E_R}{E_{matrix}}-1}{\frac{E_R}{E_{matrix}}+\xi}, \tag{3}$$

Here, E_{m2}, E_R, and E_{matrix} are the modulus of elasticity of the epoxy–rubber blend, recycled rubbers, and the epoxy matrix, respectively, and E_R was taken as 6 MPa. φ_R is the volume fraction of the rubber particles, the shape factor ξ of the rubber particles was assumed to be 2 (spherical particles).

The moduli of the composites estimated according to Equations (1)–(3) are presented in Figure 4 along with the experimental results.

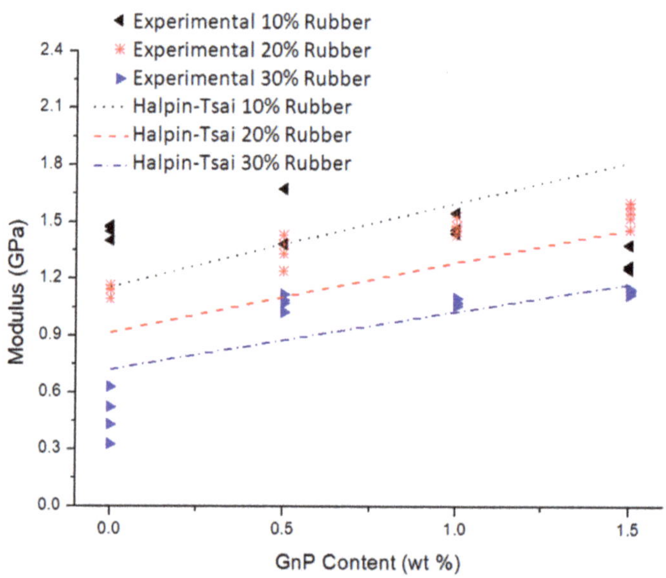

Figure 4. Experimental results and Halpin–Tsai model comparison of elasticity modulus of LRG composites by the increasing content of GnPs.

Figure 4 shows that the moduli of elasticity of the ternary composites (LRG groups) obtained by the Halpin–Tsai model had a reasonable agreement with the experimental results. In particular, the Halpin–Tsai model had better agreement with the experimental results for 20–30 wt. % rubber-containing composites. In general, for the compositions of LR2G and LR3G groups, the Halpin–Tsai model underestimated the moduli of elasticity. The synergistic effect of GnPs and rubber at the higher contents may be a reason for getting higher moduli experimentally than that predicted by the Halpin–Tsai model. The unanticipated decline in the experimental results is attributed to agglomeration of GnPs and recycled rubber particles at the higher contents.

After using the Halpin–Tsai for the homogenization, 3PB tests were simulated under the same test conditions by using the Abaqus™ FEM solver. These FEM calculations were performed mainly to observe the stress field of the composites subjected to bending moments as well as to plot the numerically obtained stress–strain curves to compare with the experimental ones. Numerical simulations were only implemented for the LR2G1.5 and LR3G1.5 group compositions, as the modified Halpin–Tsai modelling yielded estimates for the moduli that compared satisfactorily with the experimental results.

In these simulations, material properties such as density, obtained from different characterization methods were introduced to the FEM solver, and the tests were simulated by dimensions of the specimens used in the simulations which were the same as the actual specimens used in the 3PB experiments. Boundary conditions were also implemented as in the 3PB tests. The displacements leading to the failure of the specimens (displacement at break) were imposed on the specimens in Abaqus™ as in the 3PB tests. In Table 5, the fundamental characteristics of the numerical calculations in Abaqus are given. After running the calculations in Abaqus™, results were obtained and compared with the experimental results.

Table 5. Parameters used in FEM.

FEM Parameters	
Contact properties	Loading tip—specimen: Frictionless, Hard contact
Mesh properties	C3D8R: A 8-node linear brick
Stress dependence	Isotropic

The stress field obtained from the FEM analysis for the LR2G1.5 specimen under 3PB loading is given in Figure 5. As expected, at midspan, at the bottom of the specimen maximum tensile stress was observed. This stress leads to crack initiation followed by crack propagation and eventually to the failure of specimen.

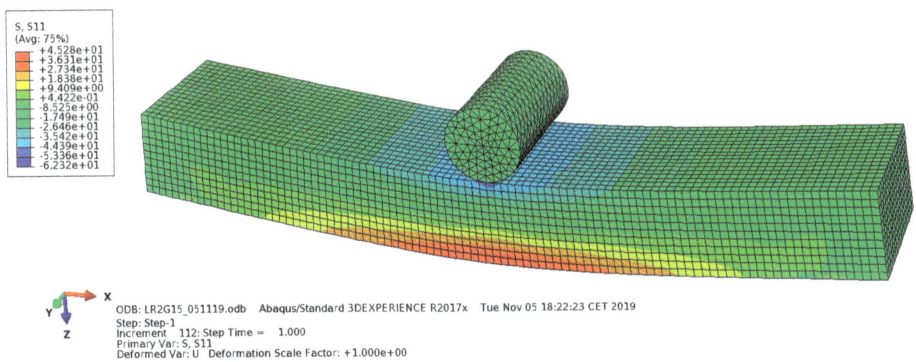

Figure 5. Stress field obtained from the finite element method (FEM) analysis for LR2G1.5.

Figure 6 depicts the experimental and numerical comparison of the stress–strain curves for the LR2G1.5 and LR3G1.5 group specimens. A reasonable estimate of the experimental results were observed by using Halpin–Tsai homogenization in this numerical modelling. The lack of yielding on the stress–strain curves indicates the brittle failure of the composites. In addition, the non-linearity in the curve can be associated to a combination of interfacial slippage and the expansion of plasticity in the matrix. By this way, the energy absorption capability can be increased while preserving the high stiffness of the composite [30].

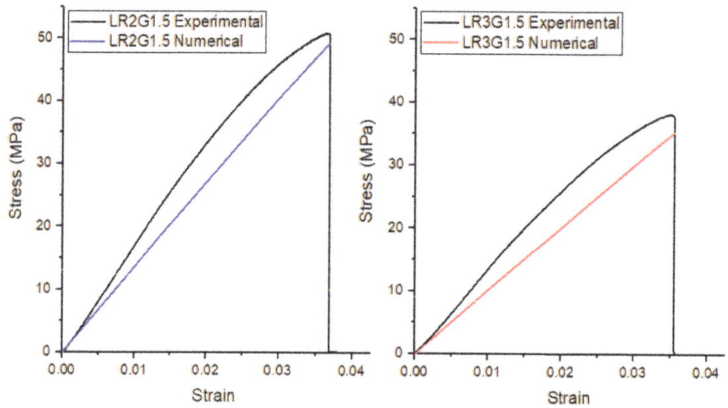

Figure 6. Comparison of the experimental results and FEM analyses.

3.4. Determination of the Fracture Toughness and Toughening Mechanisms Identification by Means of SEM

Due to the manufacturing and post-processing methods or during service, cracks or defects may occur in the materials. Also, some discontinuities or particles with undesired geometries inside the microstructure can behave as cracks, and they can be responsible for the fracture of materials. Therefore, it is not always the most reliable way to design critical aeronautical part, such as the brackets in the wings, by considering the yield strength of the manufactured composites, since fractures may occur in the presence of cracks at smaller loads. For this reason, using fracture mechanics is considered the best option to get through such problems [31]. Fracture toughness is a numerical designation of the resistance of material to crack propagation under load. After performing bending tests using SENB-type notched specimens, the fracture toughness (K_{Ic}) and fracture energy (G_{Ic}) of the manufactured composites are given in Figure 7.

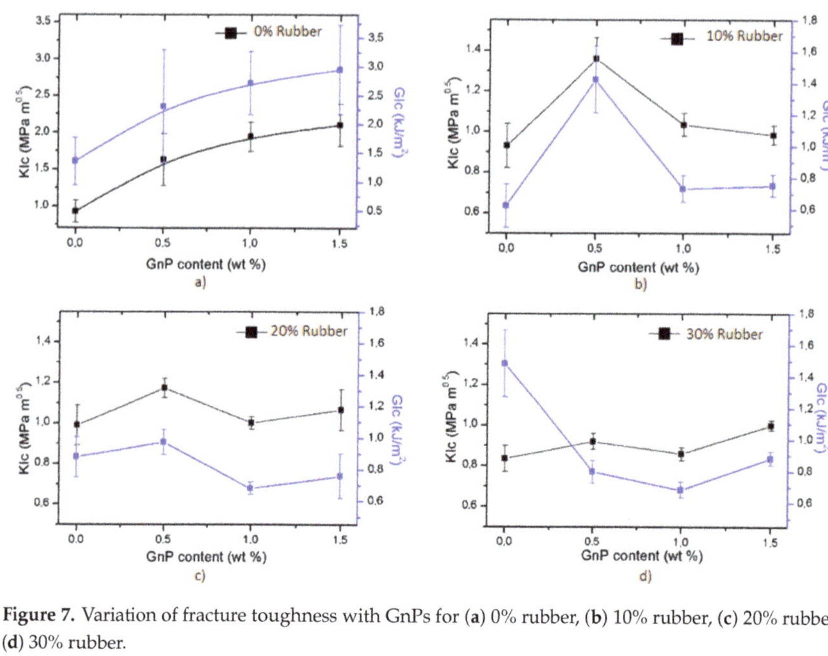

Figure 7. Variation of fracture toughness with GnPs for (**a**) 0% rubber, (**b**) 10% rubber, (**c**) 20% rubber, (**d**) 30% rubber.

In Figure 7a, the favorable impact of GnPs on fracture toughness of the epoxy resin is quite apparent. As explained in detail in the following sub-section, toughening mechanisms, including GnPs layer separation and crack deflection, can increase the fracture toughness of the epoxy resin. Moreover, increasing the content of GnPs improved the K_{Ic} of these binary composites. In addition, fracture energy (G_{Ic}) is a second order function of K_{Ic} (($G_{Ic} = \frac{{K_{Ic}}^2(1-v^2)}{E}$)) v: Poisson's ratio, E: Modulus of Elasticity), and this explains the similar trend of G_{Ic} to K_{Ic}. However, this quadratic form of G_{Ic} leads to larger error bars in Figure 7.

The variation of K_{Ic} and G_{Ic} was not as smooth for the epoxy–recycled rubber blends with GnPs (Figure 7b–d). As explained before, epoxy–recycled rubber blends may contain many discontinuities in the microstructure. Moreover, it is challenging to distribute GnPs uniformly in the epoxy matrix (high affinity of carbon atoms and van der Waals forces) which can influence the fracture toughness of the composites. Therefore, it is not surprising to observe fluctuations in the K_{Ic}-G_{Ic} of the epoxy–recycled rubber and GnP-containing ternary composites [32–34].

Lastly, in Figure 7d, for 30 wt. % rubber content, the relative magnitudes of K_{Ic} and G_{Ic} were different than the 10 and 20 wt. % rubber. This is due to the fact that the elastic modulus of LR30 was much lower than the elastic moduli of LR10 and LR20.

After mechanical tests, the fracture surfaces of the 3PB test specimens were observed via SEM. From these observations, various toughening mechanisms were determined.

One toughening mechanism was observed as crack deflection. If a crack comes across a hard particle or a different form of a reinforcement during its propagation, it finds alternative paths to maintain its propagation. As a consequence, a wavy crack propagation line is observed as seen in Figure 8c. The GnPs in the epoxy matrix may behave as stress concentrators, and they generate many micro-cracks. These micro-cracks increase the total fracture surface area because of crack deflection [35]. Also, a height difference between the deflected crack front and the original crack front is observed once the crack is deflected. This leads to a tortuous passage and explains the rough surfaces observed in Figure 8a [36,37].

Moreover, shear forces during the crack propagation facilitate the separation of the graphene sheets. To separate the GnP sheets, a certain amount of energy needs to be consumed and this energy is provided from the fracture energy. If a certain quantity of fracture energy is consumed, the energy needed to propagate the crack will not be enough for total rupture. Therefore, this phenomenon improves the fracture toughness. Also, the crack deflection and GnP layer separation create a combined toughening mechanism. This combined mechanism generates a characteristic "dimple-type" fracture surface which is shown in Figure 8a,b [38,39]. These dimple-type fracture surfaces are also accepted as crack initiation sites.

Another toughening mechanism is shown in Figure 9a as shear yielding (indicated with shear bands marked with the red arrow). Also, inside the red circle in Figure 9a, cavitated and torn rubber particles surrounded by epoxy matrix can be observed. This encircled zone is magnified in Figure 9b, and the local rough surfaces in this figure show an enhanced area of fracture in these composites. This circumstance is an indicator for mechanisms of crack deflection and cavitation.

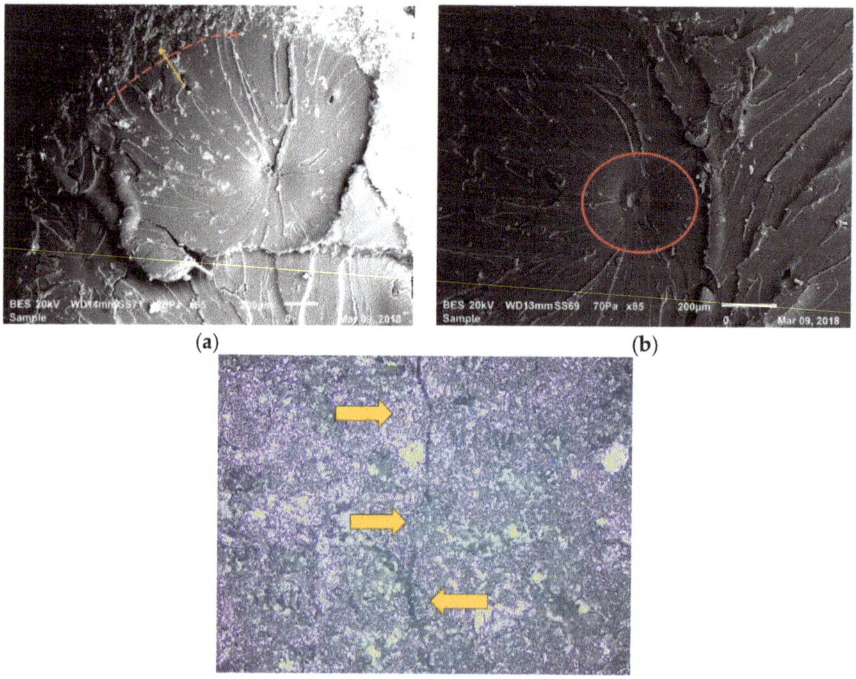

(a) (b) (c)

Figure 8. Fracture surface observation in LRG group composites: (**a**) tortuous passage in the fracture of GnP-reinforced epoxy-based composites; (**b**) dimple generation; (**c**) crack deflection.

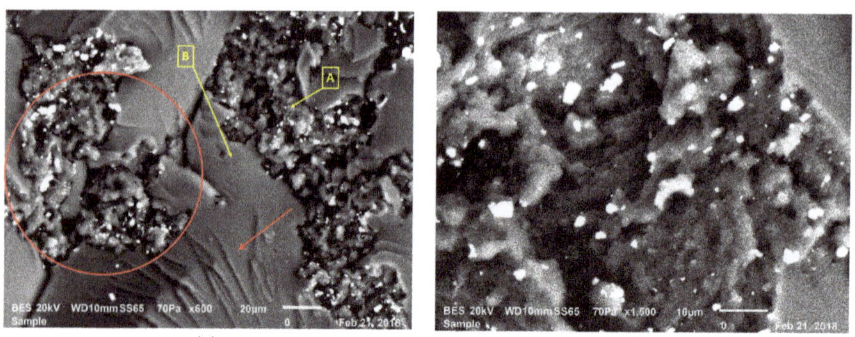

(a) (b)

Figure 9. SEM fractography on the LRG group's compositions: (**a**) shear band formation; (**b**) void observation.

Many factors can lead to cavitation such as the resin system, curing pressure, environmental conditions. Hence, it is challenging to attain a void-free composite section. In the epoxy matrix, plastic deformation of the polymer and debonding of the inclusions can remarkably alter the number and the size of the voids formed. In this research, to modify the epoxy matrix, GnPs and recycled EPDM rubber were used. The GnPs and recycled rubber have high bulk moduli ($K = E/3(1 - 2\nu)$). Because GnPs possess a high modulus of elasticity ($E \cong 1000$ GPa), recycled EPDM rubber also has a Poisson's ratio (v) of around 0.49. This states that GnPs and recycled EPDM particles are very rigid elastic bodies when they are subjected to triaxial stresses, and they are going to be highly resistant to any volumetric deformation. However, Poisson's ratio of the epoxy matrix is approximately 0.33 which is substantially lower than rubber. Moreover, the epoxy matrix strain softens after yielding which is seen typically in the glassy polymers, and accordingly, the yielded epoxy matrix is going to be relatively compliant and will plastically deform more smoothly. However, the "rigid" rubbery phase and GnPs will obstruct any significant plastic dilatation in the matrix, unless these GnPs and rubber particles are pulled-out from the epoxy matrix or if the rubber particles do not internally cavitate. Thereby, by the interfacial debonding of GnPs and rubbers, some of the stored strain energy is dissipated and it is followed by shear yielding and shear band formation of the epoxy matrix encircling the fillers.

Shear yielding identified by shear bands is an important mechanism that enhances the strength of the polymer in the case of ductile fracture. Even though the final composite indicates a brittle fracture, crack propagation includes a localized viscoelastic and plastic energy dissipating process around the crack tip because of the polymer matrix. As a result, energy absorption is promoted by this mechanism in the material, and it enhances the material's fracture toughness [40,41].

Following to debonding and shear yielding, cavitation arises in the epoxy matrix surrounding GnPs and rubber particles because of the localized plastic flow. Then, by the increasing the number of cavitations and plastic deformation of the epoxy matrix, some amount of the stored strain energy is dissipated which improves the composites' fracture toughness [42–45].

Lastly, when GnPs are found in front of the crack during crack propagation, twisted crack pathways are generated due to the GnPs as shown in Figure 10. In these regions, some of the fracture energy can be dissipated by the interaction of GnPs with the crack. In Figure 10, a lamellar structure (inside red ellipse) was observed as an indicator of the ductile yielding. This structure also indicates the transition between ductile and brittle states [46]. As a consequence, if there is ductile yielding, more energy can be dissipated during the fracture which increases the fracture toughness of the composites [47].

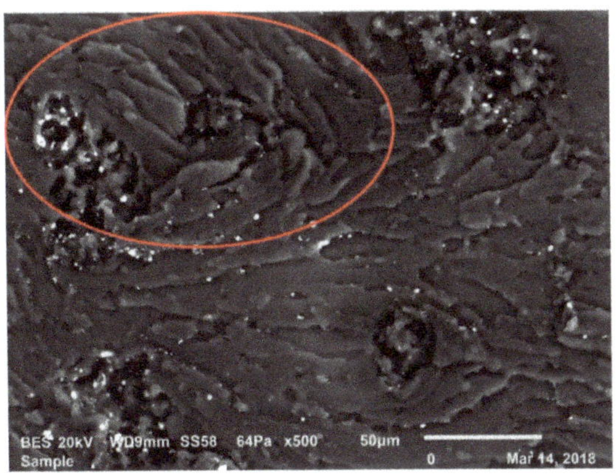

Figure 10. Ductile yielding in the LRG group's compositions.

Fracture surface observation on the ternary composites showed different toughening mechanisms provided by GnPs and rubbers. The GnP layer separation and crack deflection were seen as the main mechanisms that improved the fracture toughness of the composites. In addition, cavitation and shear yielding were promoted by the presence of the recycled rubbers and GnPs. In conclusion, a good combination of different toughening mechanisms improved the fracture toughness of the manufactured composites.

The results presented in this study constitute a permitted part of an ongoing project with the aeronautic industry. The complete design of the considered element could not be presented in this manuscript in detail. The idea to use recycled rubber with the combination of epoxy and graphene nanoplatelets was presented in this manuscript. Based on the comparison of numerical and experimental results, the mechanical strength of the manufactured composites can be improved further by using more advanced manufacturing facilities enabling them to resist the applied loads more effectively. Therefore, these composites can provide a cost-efficient solution for aircraft manufacturers.

4. Conclusions

A solution blending method was used to manufacture epoxy-based novel composites for aeronautical applications. This method is a practical solution for manufacturing such composites at laboratory scale, and it can be scaled up to an industrial level easily. Increasing the rubber content brings a reduction in the density of the epoxy–recycled rubber blends, and this can be considered a positive outcome for the sake of the lightweight property of the composites. However, due to the interface issues and agglomerations, higher rubber rates can be avoided for the composites targeting mechanical performance. Otherwise, surface modifiers can be used on recycled rubbers to increase their affinity with the epoxy resin. In addition, the positive effect of GnPs on the fracture toughness and the elastic modulus was more apparent by the increasing rubber content. Therefore, GnP reinforcement becomes more reasonable for the increased rubber content.

Halpin–Tsai homogenization proposes a rapid estimation of the elasticity modulus of the ternary group composites, and it can be utilized to simulate the mechanical behavior of the composites in the elastic region.

It was seen that GnPs were involved in many toughening mechanisms which brought about significant improvement in the fracture toughness of the epoxy. In particular, the improvement in the mechanical performance with a very low content of GnP indicates the potential of this kind of composite.

By considering all the results in this study, these novel composites can be used in auxiliary components in the wings of an aircraft as well as luggage weather strip and radiator lining manufacturing in the automotive industry after eliminating issues with homogeneity. Moreover, manufactured composites may offer potential applications not only in aeronautics but also in the building engineering industry as smart building materials. One interesting hint should be given for aeronautic manufacturers here: the very high electrical conductivity of GnPs brings a new and original idea for the multifunctionality of these novel composites, and it may reduce the damage risk from lightning strikes.

Author Contributions: Conceptualization, A.B.I., E.B.; methodology, A.B.I.; software, A.B.I.; investigation, A.B.I.; writing—original draft preparation, A.B.I.; writing—review and editing, A.B.I., E.B., I.M.; supervision, E.B., I.M. All authors have read and agreed to the published version of the manuscript.

Funding: This research was funded by SUPMECA Paris and SUPMECA-Paris Research Department.

Acknowledgments: We acknowledge G. Zambelis from Airbus-Helicopter/Paris and O. Friderikos from LMT Cachan who supported this project. We thank them for their technical help and valuable discussions.

Conflicts of Interest: The authors declare no conflict of interest.

References

1. May, C. *Epoxy Resins: Chemistry and Technology*, 2nd ed.; CRC Press: London, UK, 1987; pp. 46–121.
2. Johnsen, B.B.; Kinloch, A.J.; Mohammed, R.D.; Taylor, A.C.; Sprenger, S. Toughening mechanisms of nanoparticle-modified epoxy polymers. *Polymer* **2007**, *48*, 530–541. [CrossRef]
3. Liang, Y.L.; Pearson, R.A. The toughening mechanism in hybrid epoxy-silica-rubber nanocomposites (HESRNs). *Polymer* **2010**, *51*, 4880–4890. [CrossRef]
4. Irez, A.B.; Miskioglu, I.; Bayraktar, E. Toughening Mechanisms on Recycled Rubber Modified Epoxy Based Composites Reinforced with Graphene Nanoplatelets. In *Mechanics of Composite, Hybrid and Multifunctional Materials*, 1st ed.; Thakre, P.R., Singh, P.R., Slipher, G., Eds.; Springer International Publishing: Cham, Switzerland, 2019; Volume 5, pp. 283–290. [CrossRef]
5. Adhikari, B.; De, D.; Maiti, S. Reclamation and recycling of waste rubber. *Prog. Polym. Sci.* **2000**, *25*, 909–948. [CrossRef]

6. Fiksel, J.; Bakshi, B.R.; Baral, A.; Guerra, E.; De Quervain, B. Comparative life cycle assessment of beneficial applications for scrap tires. *Clean Technol. Environ. Policy* **2011**, *13*, 19–35. [CrossRef]
7. Fang, Y.; Zhan, M.; Wang, Y. The status of recycling of waste rubber. *Mater. Des.* **2001**, *22*, 123–128. [CrossRef]
8. Isayev, A.I. Recycling of Rubbers. In *Science and Technology of Rubber*, 3rd ed.; Mark, J., Erman, B., Eirich, F., Eds.; Academic Press: Cambridge, MA, USA, 2005; pp. 663–701.
9. Myhre, M.; MacKillop, D.A. Rubber Recycling. *Rubber Chem. Technol.* **2002**, *75*, 429–474. [CrossRef]
10. Caldona, E.B.; De Leon, A.C.C.; Pajarito, B.B.; Advincula, R.C. A Review on Rubber-Enhanced Polymeric Materials. *Polym. Rev.* **2017**, *57*, 311–338. [CrossRef]
11. Mangaraj, D. Rubber recycling by blending with plastics. In *Rubber Recycling*, 1st ed.; De, S.K., Isayev, A., Khait, K., Eds.; CRC Press: Boca Raton, FL, USA, 2005; pp. 272–324.
12. Valášek, P.; Müller, M. EPDM rubber material utilization in epoxy composite systems. *Agron. Res.* **2014**, *12*, 291–298.
13. Al-Aqeeli, N. Fabrication and Assessment of Crumb-Rubber-Modified Coatings with Anticorrosive Properties. *Materials* **2015**, *8*, 181–192. [CrossRef]
14. Roche, N.; Ichchou, M.N.; Salvia, M.; Chettah, A. Dynamic Damping Properties of Thermoplastic Elastomers Based on EVA and Recycled Ground Tire Rubber. *J. Elastomers Plast.* **2011**, *43*, 317–340. [CrossRef]
15. Zhu, S.-H.; Penlidis, A.; Tzoganakis, C.; Ginzel, E. Ultrasonic properties and morphology of devulcanized rubber blends. *J. Appl. Polym. Sci.* **2012**, *124*, 2062–2070. [CrossRef]
16. Li, B.; Zhou, H.; Huang, G. A novel impedance matching material derived from polymer micro-particles. *J. Mater. Sci.* **2007**, *42*, 199–206. [CrossRef]
17. Xue, Q.-J.; Wang, Q.-H. Wear mechanisms of polyetheretherketone composites filled with various kinds of SiC. *Wear* **1997**, *213*, 54–58. [CrossRef]
18. Bagheri, R.; Marouf, B.T.; Pearson, R.A. Rubber-Toughened Epoxies: A Critical Review. *Polym. Rev.* **2009**, *49*, 201–225. [CrossRef]
19. Gong, L.-X.; Zhao, L.; Tang, L.-C.; Liu, H.-Y.; Mai, Y.-W. Balanced electrical, thermal and mechanical properties of epoxy composites filled with chemically reduced graphene oxide and rubber nanoparticles. *Compos. Sci. Technol.* **2015**, *121*, 104–114. [CrossRef]
20. Geim, A.K.; Novoselov, K.S. The rise of graphene. *Nat. Mater.* **2007**, *6*, 183–191. [CrossRef] [PubMed]
21. Balandin, A.A.; Ghosh, S.; Bao, W.; Calizo, I.; Teweldebrhan, D.; Miao, F.; Lau, C.N. Superior thermal conductivity of single-layer graphene. *Nano Lett.* **2008**, *8*, 902–907. [CrossRef]
22. Scarpa, F.; Adhikari, S.; Srikantha Phani, A. Effective elastic mechanical properties of single layer graphene sheets. *Nanotechnology* **2009**, *20*, 065709. [CrossRef]
23. Wichmann, M.H.G.; Schulte, K.; Wagner, H.D. On nanocomposite toughness. *Compos. Sci. Technol.* **2008**, *68*, 329–331. [CrossRef]
24. Srivastava, I.; Koratkar, N. Fatigue and fracture toughness of epoxy nanocomposites. *JOM J. Miner. Met. Mater. Soc.* **2010**, *62*, 50–57. [CrossRef]
25. Liang, Y.L.; Pearson, R.A. Toughening mechanisms in epoxy–silica nanocomposites (ESNs). *Polymer* **2009**, *50*, 4895–4905. [CrossRef]
26. Moghaddamzadeh, S.; Rodrigue, D. The effect of polyester recycled tire fibers mixed with ground tire rubber on polyethylene composites. Part II: Physico-mechanical analysis. *Prog. Rubber Plast. Recycl. Technol.* **2018**, *34*, 128–142. [CrossRef]
27. Medina, N.F.; Garcia, R.; Hajirasouliha, I.; Pilakoutas, K.; Guadagnini, M.; Raffoul, S. Composites with recycled rubber aggregates: Properties and opportunities in construction. *Constr. Build. Mater.* **2018**, *188*, 884–897. [CrossRef]
28. Li, W.; Dichiara, A.; Bai, J. Carbon nanotube–graphene nanoplatelet hybrids as high-performance multifunctional reinforcements in epoxy composites. *Compos. Sci. Technol.* **2013**, *74*, 221–227. [CrossRef]
29. Zhao, X.; Zhang, Q.; Chen, D.; Lu, P. Enhanced Mechanical Properties of Graphene-Based Poly(vinyl alcohol) Composites. *Macromolecules* **2010**, *43*, 2357–2363. [CrossRef]
30. Dwaikat, M.M.S.; Spitas, C.; Spitas, V. Predicting nonlinear stress–strain curves of unidirectional fibrous composites in consideration of stick–slip. *Compos. Part B Eng.* **2013**, *44*, 501–507. [CrossRef]
31. Yalcin, D. Fracture Characterization of Knitting Fabric Reinforced Laminated Composites. Master's Thesis, Usak University, Usak, Turkey, 2016.

32. Quan, D.; Ivankovic, A. Effect of core–shell rubber (CSR) nano-particles on mechanical properties and fracture toughness of an epoxy polymer. *Polymer* **2015**, *66*, 16–28. [CrossRef]
33. Irez, A.B.; Bayraktar, E.; Miskioglu, I. Flexural fatigue damage analyses of recycled rubber—Modified epoxy-based composites reinforced with alumina fibres. *Fatigue Fract. Eng. Mater. Struct.* **2019**, *42*, 959–971. [CrossRef]
34. Irez, A.B. Conception, Élaboration et Caractérisation des Composites Modifiées par Incorporation de Particules de Caoutchouc Recyclées et Devulcanisées à Base D'époxy: Une Approche Expérimentale Pour des Mécanismes de Renforcement. Ph.D. Thesis, Université-Paris Saclay, Paris, France, 2018.
35. Chandrasekaran, S.; Seidel, C.; Schulte, K. Preparation and characterization of graphite nano-platelet (GNP)/epoxy nano-composite: Mechanical, electrical and thermal properties. *Eur. Polym. J.* **2013**, *49*, 3878–3888. [CrossRef]
36. Ladani, R.B.; Wu, S.; Kinloch, A.J.; Ghorbani, K.; Zhang, J.; Mouritz, A.P.; Wang, C.H. Multifunctional properties of epoxy nanocomposites reinforced by aligned nanoscale carbon. *Mater. Des.* **2016**, *94*, 554–564. [CrossRef]
37. Park, Y.T.; Qian, Y.; Chan, C.; Suh, T.; Nejhad, M.G.; Macosko, C.W.; Stein, A. Epoxy Toughening with Low Graphene Loading. *Adv. Funct. Mater.* **2015**, *25*, 575–585. [CrossRef]
38. Chandrasekaran, S.; Sato, N.; Tölle, F.; Mülhaupt, R.; Fiedler, B.; Schulte, K. Fracture toughness and failure mechanism of graphene based epoxy composites. *Compos. Sci. Technol.* **2014**, *97*, 90–99. [CrossRef]
39. Wang, F.; Drzal, L.T.; Qin, Y.; Huang, Z. Enhancement of fracture toughness, mechanical and thermal properties of rubber/epoxy composites by incorporation of graphene nanoplatelets. *Compos. Part A Appl. Sci. Manuf.* **2016**, *87*, 10–22. [CrossRef]
40. Kinloch, A.J.; Young, R.J. *Fracture Behaviour of Polymers*, 1st ed.; Springer: Dordrecht, The Netherlands, 1995; pp. 107–146.
41. Manjunatha, C.M.; Taylor, A.C.; Kinloch, A.J.; Sprenger, S. The cyclic-fatigue behaviour of an epoxy polymer modified with micron-rubber and nano-silica particles. *J. Mater. Sci.* **2009**, *44*, 4487–4490. [CrossRef]
42. Ravindran, A.R.; Ladani, R.B.; Wu, S.; Kinloch, A.J.; Wang, C.H.; Mouritz, A.P. Multi-scale toughening of epoxy composites via electric field alignment of carbon nanofibres and short carbon fibres. *Compos. Sci. Technol.* **2018**, *167*, 115–125. [CrossRef]
43. Huang, Y.; Kinloch, A.J. The role of plastic void growth in the fracture of rubber-toughened epoxy polymers. *J. Mater. Sci. Lett.* **1992**, *11*, 484–487. [CrossRef]
44. Williams, J.G. Particle toughening of polymers by plastic void growth. *Compos. Sci. Technol.* **2010**, *70*, 885–891. [CrossRef]
45. Singh, K.; Nanda, T.; Mehta, R. Addition of nanoclay and compatibilized EPDM rubber for improved impact strength of epoxy glass fiber composites. *Compos. Part A Appl. Sci. Manuf.* **2017**, *103*, 263–271. [CrossRef]
46. Krishnan, P. Rheology of Epoxy/Rubber Blends. In *Handbook of Epoxy Blends*, 1st ed.; Parameswaranpillai, J., Hameed, N., Pionteck, J., Woo, E., Eds.; Springer International Publishing: Cham, Switzerland, 2017; Volume 1, pp. 185–210. [CrossRef]
47. Irez, A.B.; Zambelis, G.; Bayraktar, E. A New Design of Recycled Ethylene Propylene Diene Monomer Rubber Modified Epoxy Based Composites Reinforced with Alumina Fiber: Fracture Behavior and Damage Analyses. *Materials* **2019**, *12*, 2729. [CrossRef]

polymers

MDPI

Article

Nano-Level Damage Characterization of Graphene/Polymer Cohesive Interface under Tensile Separation

S. S. R. Koloor [1,2,*], S. M. Rahimian-Koloor [3], A. Karimzadeh [1], M. Hamdi [4,5], Michal Petrů [2] and M. N. Tamin [1,*]

1 School of Mechanical Engineering, Universiti Teknologi Malaysia, Johor Bahru 81310, Malaysia
2 Institute for Nanomaterials, Advanced Technologies and Innovation, Technical University of Liberec, Studentska 2, 461 17 Liberec, Czech Republic
3 Molecular Simulation Research Laboratory, Department of Chemistry, Iran University of Science and Technology, Tehran 16846, Iran
4 Centre of Advanced Manufacturing and Material Processing, Department of Mechanical Engineering, Faculty of Engineering, University of Malaya, Kuala Lumpur 50603, Malaysia
5 Chancellory Office, National University of Malaysia, Bangi 43600, Selangor, Malaysia
* Correspondence: s.s.r.koloor@gmail.com (S.S.R.K.); nasirtamin@utm.my (M.N.T.); Tel.: +60-17-761-9129 (S.S.R.K.); +60-12-778-1410 (M.N.T.)

Received: 30 June 2019; Accepted: 24 July 2019; Published: 2 September 2019

Abstract: The mechanical behavior of graphene/polymer interfaces in the graphene-reinforced epoxy nanocomposite is one of the factors that dictates the deformation and damage response of the nanocomposites. In this study, hybrid molecular dynamic (MD) and finite element (FE) simulations of a graphene/polymer nanocomposite are developed to characterize the elastic-damage behavior of graphene/polymer interfaces under a tensile separation condition. The MD results show that the graphene/epoxy interface behaves in the form of elastic-softening exponential regressive law. The FE results verify the adequacy of the cohesive zone model in accurate prediction of the interface damage behavior. The graphene/epoxy cohesive interface is characterized by normal stiffness, tensile strength, and fracture energy of 5×10^{-8} (aPa·nm^{-1}), 9.75×10^{-10} (nm), 2.1×10^{-10} (N·nm^{-1}) respectively, that is followed by an exponential regressive law with the exponent, $\alpha = 7.74$. It is shown that the commonly assumed bilinear softening law of the cohesive interface could lead up to 55% error in the predicted separation of the interface.

Keywords: adhesives; cohesive zone model; finite element method; graphene-polymer nanocomposite; graphene/polymer interface; molecular dynamics; regressive softening law

1. Introduction

The ever-increasing demands for the usage of graphene-reinforced polymer composites in advanced instruments such as gas sensors, electrodes for batteries, solar cells, etc., require in-depth mechanical characterization of these nanomaterials [1–3]. Graphene nanostructure is a two-dimensional sheet of carbon atoms in a honeycomb hexagonal arrangement, which have shown exceptional potential for enhancing the thermal, mechanical, and electrical properties of polymers [4–6]. The excellent properties of the graphene sheets have motivated researchers to further explore the behavior of this nanostructure and its composition with polymers, which require new development of experimental methods, analytical approach, and numerical investigation [2,7–10]. The composition of graphene and polymer creates a nanocomposite with excellent mechanical properties, which is derived not only from the properties of the graphene and the polymer, but also from the interaction of these two constituent materials [2,8,11,12]. The load transfer between the reinforcement and the matrix in a composite is strongly affected by the interface between the two phases [13–15]. Therefore, the interface properties have a significant influence on the global behavior of nanocomposite materials, and should be considered in the design and simulation of the nanocomposite structures [14–17].

While numerous investigations have been performed on the mechanical characterization of the graphene, polymers, and graphene-polymer nanocomposites [4,14,18–22], a limited study was reported on the behavior of the graphene/polymer interface [10,16,23]. The behavior of the graphene/polymer interface is traceable to the atomic interactions between the two materials, thus could be quantified through simulations at the atomic level [10]. Among the various atomistic approaches, molecular dynamics (MD) and molecular structural mechanics (MSM) have been used to simulate the physical and mechanical behaviors of graphene-polymer nanocomposites [16,21,24–26]. In addition, the atomistic methods have been combined with the continuum-level approaches through multiscale modelling to quantify the mechanical response of nanocomposite materials [8,27,28]. In the continuum-based approach, the mechanical behavior of graphene-polymer nanocomposite is calculated using the multiscale models with random distribution of graphene structures in the polymer matrix [14,22,28,29]. In these models, the graphene/polymer interface was simulated either as perfectly bonded [23], using a spring model that behaves according to Lennard–Jones (L-J) potential [28], with the effective interface layer model [22,30], using a linear spring model for imperfect interface [14], or with the cohesive imperfect zone model [29]. The embedded graphene in the graphene-polymer nanocomposite exhibits different mechanical properties compared to the isolated one because the atomic interactions cause the formation of an imperfect bond between the polymer and the graphene [8]. Different studies on graphene- and nanocarbon fiber-reinforced polymer matrix composites reported a significant variation in the mechanical properties of the materials due to the Van der Waals interactions of the nanostructure with the surrounding polymer atoms during the curing process [8,19,31]. Such interaction creates a distinct interface between the nanostructured graphene and polymer phases, as identified by an irregular distribution of the polymer with an asymmetric density around the nanostructured graphene.

The cohesive zone model (CZM) is often used to simulate the elastic-damage behavior of the interface in composites at different length scales [13,16,29,32–34]. In the cohesive zone modelling, the interface behaves with linear elastic response to damage initiation at the maximum stress level, followed by damage evolution as represented by various forms of the softening laws [33,35–38]. In MD simulation, the cohesive law defines the atomic interaction at the micro-/nano-interface. It has been employed in the simulation of the deformation and fracture of graphene grain boundary [34], perfectly flat graphene sheet interacting with polyethylene [16], and embedded-atom method potentially for interface behavior of planar, tilt nano-grain boundary interface [39]. In addition, the hybrid MD-finite element (FE) simulation has been used to simulate the atomistic mechanism of the grain-boundary debonding of solid materials using the cohesive law [40]. The research has demonstrated the flexibility of the CZM to capture the linear-nonlinear behavior of the interface in single- and multi-phase materials, and at different length scales.

The objective of this study is to establish an accurate methodology for predicting the properties and mechanical response of the graphene/polymer interface in graphene-reinforced epoxy nanocomposites employing both the MD and FE simulation procedures. In the modeling process, the formation of chemical bonding at the graphene/epoxy interface during the curing of the nanocomposite is simulated using the MD approach. An in-house code is developed to simulate the polymer cross-linking process during the curing of the nanocomposite. Consequently, the characteristic of the graphene/epoxy cohesive interface under the tensile separation loading is established. The resulting properties of this nanoscale cohesive interface are then prescribed for the cohesive elements of the FE simulation approach. This enables the accurate FE prediction of the properties and the mechanics of deformation of the graphene-reinforced epoxy nanocomposites in the multiscale simulation.

2. Computational Methods

2.1. Molecular Dynamics Simulation

The elastic and damage behaviors of the graphene/polymer interface in the nanocomposite were determined through the MD simulation of the nanocomposite. For this purpose, a representative volume element (RVE) of the graphene-polymer nanocomposite with the specifications provided in Table 1, was created for modelling the curing process of the polymer matrix in the presence of the graphene. The simulation of the polymer curing process and the tensile loading of the nanocomposite using the MD modelling are described in the following subsections.

Table 1. The equilibrated representative volume element (RVE) specifications of the graphene-epoxy nanocomposite.

Configuration of RVE	Graphene Sheet Length (nm)	Box Volume (nm^3)	Number of Epoxy Molecules	Number of Hardener Molecules	Density after Curing Process and Final NPT Equilibration (g/cm^3)
Long	4.540 × 4.520	143.653	258	86	1.1865

2.1.1. Cross-Linking of the Polymer during the Curing Process

The initial molecular model of the polymer was created in PACKMOL software. The model was then submitted to the large scale atomic/molecular massively parallel simulator (LAMMPS) software to model the curing process of the polymer [41,42]. The polymer matrix consists of Diglycidyl Ether of Bisphenol F (EPON 862) and Triethylenetetramine (TETA) as the resin and hardener respectively, with a stoichiometric ratio of 3:1. In an epoxy reaction with the hardener group, the lone pair of nitrogen atoms is attached to the open epoxy rings. However, such an in-situ reaction could not be modelled in the classical MD simulation based on Newton's laws, which requires the chemical bonds and the subsequent atomic structures to be defined prior to the simulation and remain unchanged until the end of the process. Therefore, an algorithm was developed to simulate the reaction between hardener and EPON through the cross-linking process. In this algorithm, the epoxy rings of the EPON molecule were opened to form activated EPON 862, to simplify this reaction before packing, as illustrated in Figure 1a. Then, the EPON and TETA molecules with optimized geometries were loaded with the ratio of 390:130 inside the simulation box. This procedure was done at the density of 0.8 g/cm^3 in order to restrict the formation of unreal configuration and huge forces which occur due to the ring spearing and atoms overlapping [43,44]. The density of the molecular system was increased to the real density through the initial equilibration level by the isothermal-isobaric ensemble (NPT) [23,45,46]. Figure 1b shows the variation of the RVE density with respect to the simulation time after the initial equilibration. The curing process was performed subsequently on this equilibrated molecular system.

In the modelling of the nanocomposites RVE, the graphene sheet was inserted along the x-y plane such that its center-of-mass is placed on the center of the simulation box, while the polymer molecules were positioned around the graphene sheet. Subsequently, the curing process was applied through the static cross-linking procedure [43,47,48]. This was executed through a written FORTRAN programming code, and compiled in the Linux platform. The code consists of the following steps:

1. It was assumed that the primary and secondary amines have the same reactivity with the reaction cutoff distance of 5 Å.
2. The system was checked for the bond formation at every 50 ps after each equilibration.
3. The constants in the equation of the newly-formed bonds were decreased in the first equilibration level and progressively elevated to the real values [47].
4. The outputs were examined to update the topology parameters according to the new bonds.
5. An annealing process was implemented to release the residual stresses. In this process, the temperature was increased to 450 K and then cooled to 300 K gradually over 2 × 106 time-steps.

6. Steps 2 through 5 were iterated to obtain about 80% cross-linking.
7. Finally, the last NPT equilibration in an interval time of 3 ns was applied to the RVE.

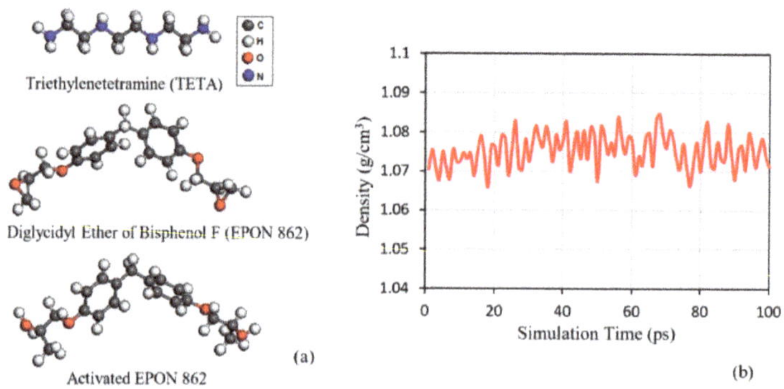

Figure 1. (**a**) Molecular structure of the polymer matrix used in the molecular dynamics (MD) simulation, (**b**) variation of RVE density against time after the initial isothermal-isobaric ensemble (NPT) equilibration.

The MD simulation procedure of the curing process is verified by comparing the predicted elastic properties of the epoxy polymer with those of measured values. In this process, an RVE of the epoxy polymer is created and cured to establish the intrinsic properties. The cured epoxy model is then loaded in tension to establish the force-displacement response in the elastic range, thus the elastic properties of the polymer could be determined. The details of the modelling steps are discussed elsewhere [31]. The close comparison of the predicted and measured properties [49,50], as listed in Table 2, serves to validate the MD modelling of the curing process.

Table 2. Properties of the epoxy polymer obtained from the MD simulation and experiments.

Parameter	MD Simulation	Experiments
Density, g/cm^3	1.14	1.16
Poisson's ratio	0.39	0.3–0.4
Young's modulus, GPa	2.77	2.4–3.4
Shear modulus, GPa	1.03	1.0–1.5

2.1.2. Tensile Separation Process

The MD simulation of the graphene-polymer nanocomposite under tensile loading was performed using LAMMPS software [42] with a pair cut-off distance of 12 Å and a time step size of 1 fs. The graphene was modelled based on the Adaptive Intermolecular Reactive Empirical Bond Order (AIREBO) Potential of Stuart as [51]:

$$V = \frac{1}{2}\sum_{i}\sum_{i\neq j}\left[V_{ij}^{REBO} + V_{ij}^{LJ} + \sum_{k\neq i,j}\sum_{l\neq i,j,k} V_{kijl}^{TORSION}\right] \quad (1)$$

where, V_{ij}^{REBO} is the Reactive Empirical Bond Order (REBO) potential by Brenner, V_{ij}^{LJ} defines longer-ranged interactions ($2 < r <$ cutoff) by an equation similar to standard L-J potential, and $V_{kijl}^{TORSION}$ describes different explicit 4-body dihedral angles [52].

The polymer matrix and the Van der Waals interactions between the graphene and polymer atoms were simulated based on the ab initio polymer consistent force field (PCFF) [23,53,54]. The PCFF force field is able to accurately represent the interactions between the sp^2 carbon atoms of graphene and all of the polymer atoms in the condensed graphene-polymer nanocomposites [23,54]. In the PCFF force field, the non-bonded Van der Waals interaction between the same type atoms (denoted by *ii*) is described using the L-J potential as:

$$V = \varepsilon\left[2\left(\frac{\sigma}{r}\right)^9 - 3\left(\frac{\sigma}{r}\right)^6\right] \qquad r < r_c \tag{2}$$

where, V is the interaction energy, σ represents the distance at which the interatomic interactions are zero, ε is the depth of the potential well, and r and r_c are inter atomic and cutoff distances, respectively. In computational chemistry, the interaction energy between two dissimilar non-bonded atoms is obtained using the mixing rule [55]. Therefore, the values of ε and σ for two atoms of different types of i and j were calculated based on the sixth-power mixing rule, as:

$$\varepsilon_{ij} = \frac{2\left(\sqrt{\varepsilon_i\varepsilon_j}\right)\left(\sigma_i^3\sigma_j^3\right)}{\sigma_i^6 + \sigma_j^6} \qquad \sigma_{ij} = \left(\frac{\left(\sigma_i^6 + \sigma_j^6\right)}{2}\right)^{1/6} \tag{3}$$

The RVE was modelled as a long structure of embedded graphene in the epoxy matrix, with a specified periodic boundary condition [31], as shown in Figure 2. In this model, the temperature was equilibrated at 300 K and the pressure was adjusted to one atmosphere in all directions, using the Nose–Hoover style of non-Hamiltonian equations of motion. Consequently, the pressure was maintained uniform and equal in the x- and y-directions (graphene in-plane directions), and independently in the z-direction. The MD simulation of the nanocomposite RVE is used to establish the evolution of stress with the tensile deformation of the graphene/polymer interface. The polymer atoms of the model, as shown in Figure 2, are displaced along the z-direction, such that the distance between any two layers of the atom in the x-y-plane remains fixed. The atoms in a plane are free to move in the x-y-plane. Under this periodic boundary condition, the RVE box dimension is displaced in the Z-direction with a rate of 10^{-7} ns^{-1}, the other faces were maintained under one atmospheric pressure [56,57].

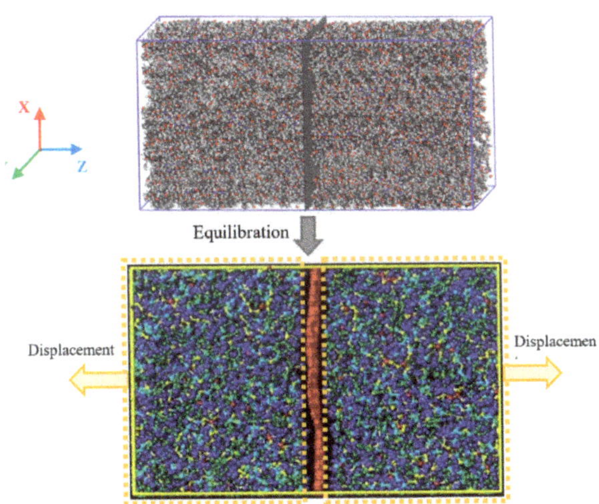

Figure 2. The simulation box containing long embedded graphene with the applied displacement of the nanocomposite.

Following each loading step, the stress components, τ_{pq}, were calculated using Virial expression as [58,59]:

$$\tau_{pq} = -\frac{1}{v}\left[\left(\sum_{i=1}^{N} m_i\left(\mathrm{v}_{ip}\ \mathrm{v}_{iq}\right)\right) + \left(\sum_{i<j} \mathrm{r}_{ijp}\mathrm{f}_{ijq}\right)\right] \quad (4)$$

where, m and v are the mass and velocity of atom i respectively, v is the total volume of the RVE, and f denotes the force between atoms with i and j indices. The vector τ_{pq} represents the stress components τ_{xx}, τ_{yy}, τ_{zz}, τ_{xy}, τ_{xz}, and τ_{yz}.

2.2. Continuum Mechanics Simulation

The MD model of the graphene-polymer nanocomposite, as shown in Figure 2, is employed in the continuum mechanics analysis through the FE simulation. The graphene/polymer interface was modelled with the cohesive behavior using the cohesive element.

2.2.1. Cohesive Zone Model

The mechanical behavior of the interface between two continuum bodies can be simulated using the CZM with a bilinear traction-separation law for single- (normal or shear) and mixed-mode loading conditions [33,35]. Under the tensile loading, the elastic-softening behavior of the interface is illustrated in Figure 3. The interface material point behaves linearly until damage initiation occurs at the maximum traction. This is followed by the softening process until the tensile separation of the material point. The softening process could be described by a gradual decreasing stress with the applied displacement through a linear, regressive, or progressive curve, as illustrated in Figure 3.

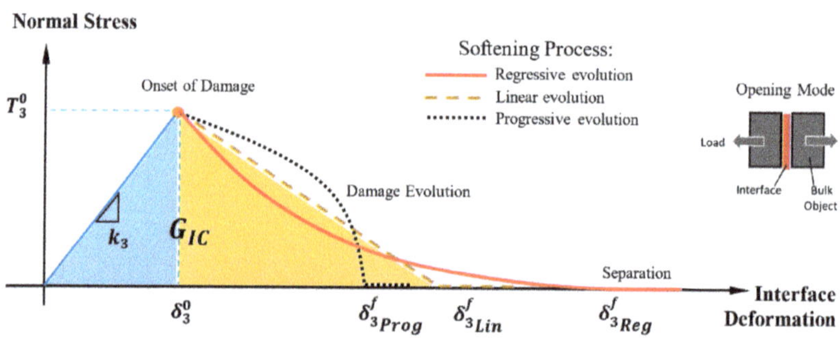

Figure 3. The cohesive softening law to describe interface behavior in the tensile separation mode.

The initial linear elastic behavior of the interface is described by:

$$\begin{Bmatrix} t_3 \\ t_1 \\ t_2 \end{Bmatrix} = \begin{bmatrix} k_3 & 0 & 0 \\ 0 & k_1 & 0 \\ 0 & 0 & k_2 \end{bmatrix} \begin{Bmatrix} \delta_3 \\ \delta_1 \\ \delta_2 \end{Bmatrix} \quad (5)$$

where, t_i and δ_i are the traction and separation at an interface point, and indices 3, 1, and 2 refer to mode I (normal) load, mode II, and mode III (shear) loadings, respectively. The parameter k_i represents the cohesive stiffness of the interface in the respective load direction. In this study, the pure tensile loading mode was considered, thus the traction-displacement relation could be simplified to $t_3 = k_3\delta_3$.

2.2.2. Damage Initiation and Propagation Criterion

A quadratic stress-based criterion was used to indicate the damage initiation in the mixed mode condition as:

$$\sqrt{\left(\frac{t_3}{T_3^0}\right)^2+\left(\frac{t_2}{T_1^0}\right)^2+\left(\frac{t_1}{T_2^0}\right)^2}=d \tag{6}$$

where, T_i^0 represents the cohesive strength, and d is the damage initiation variable. The variable d varies from "zero" for the pristine interface to "one" indicating the onset of interface of damage. In the pure tensile mode, Equation (6) was simplified to $\left(\frac{\langle t_3\rangle}{T_3^0}\right)^2=d$, The total energy dissipated during the elastic-softening deformation of the interface, as represented by the area under the traction-displacement curve, represents the critical strain energy release rate, (G_{IC}, Figure 3). This is also representing the fracture energy of the interface which is fixed for any softening law being used for the analysis. The strain energy release rate, G_i, in each loading mode could be calculated by the following equation:

$$G_i=\int_0^{\delta_{im}^f} t_i d\delta_i \qquad i=1,\ 2,\ 3 \tag{7}$$

where, δ_{im}^f is the relative displacement at failure. This displacement is a dependent parameter to the type of interface softening process, as shown in Figure 3. The graphene-polymer interface behavior in this study was computed using the regressive softening law in an exponential decay form. The damage propagation variable, d_{pi} was defined as [60]:

$$d_{pi}=1-\left(\frac{\delta_i^0}{\delta_i}\right)\left(1-\frac{1-e^{\left(-\alpha\left(\frac{\delta_i-\delta_i^0}{\delta_i^f-\delta_i^0}\right)\right)}}{1-e^{(-\alpha)}}\right) i=1, 2,\ 3 \tag{8}$$

where, α is a non-dimensional interface parameter that determined by the rate of damage evolution. This parameter was calculated from the exponential decay curve of the interface softening behavior through the MD simulation of the graphene-polymer interface.

2.2.3. Finite Element Simulation

The geometrical model of the RVE consisting of a graphene sheet bonded by two blocks of epoxy polymer is shown in Figure 4a. The geometrical properties of the nanocomposite model are as provided in Table 1. The interface thickness was determined from the MD simulation of the curing process (see Section 3.1). Due to the symmetry of the RVE model, only one-half of the model with one-half thickness of the graphene layer was represented in the FE simulation. The graphene layer and the polymer block were discretized using continuum elements (8-node, linear, reduced integration elements (C3D8R)). A layer of 8-node cohesive elements (COH3D8) were employed to represent the graphene/epoxy interface, as illustrated in Figure 4b. A mesh convergence study was performed to ensure that the FE-calculated variables are independent of the size of the elements. This resulted in small-size elements in the region next to the interface and gradually larger elements away from the interface. In addition, the FE model was partitioned into different volumes, each discretized with a slightly different mesh density to avoid the effects of symmetrical mesh distribution on the computed results [61].

The hyper-elastic behavior of the epoxy and the properties of the graphene used in the FE simulation were reported elsewhere [8]. The properties of the CZM were obtained through the MD simulation (see Section 3.2), performed prior to the FE analysis. A symmetric boundary condition was assigned to the outer surface of the graphene later, as illustrated in Figure 4b. A reference point was identified at the centroid of the polymer block to "tie" all the elements of the polymer. A global displacement in the z-direction is prescribed for the reference point so as to reproduce identical loading conditions that were used in the MD simulation. The traction-displacement curve of the graphene/epoxy interface could then be extracted.

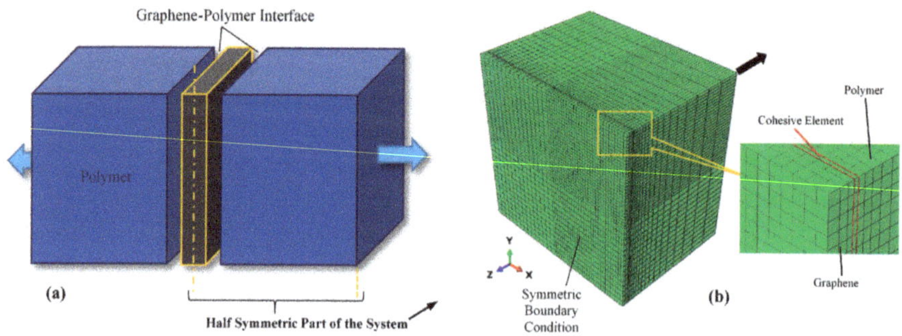

Figure 4. (**a**) Schematic view of the graphene-polymer system, and (**b**) finite element (FE) model of the half symmetric part of the nanocomposite system.

3. Results and Discussion

The computational results of the graphene-polymer nanocomposite model, performed in MD and FE simulation environments, are presented and discussed in the following subsections. The mechanics of deformation of the graphene/polymer interface under the tensile loading is described. The validation aspect of the FE simulation approach in capturing the physics of the interface failure process is deliberated.

3.1. Thickness of the Graphene and Graphene/Polymer Interface

The density profile of the epoxy polymer across the cured polymer/graphene/polymer phases of the RVE model is shown in Figure 5. The average density of the epoxy is estimated at 1.17×10^{12} ng/nm^3, as illustrated by the dashed line and without considering the density readings below 0.65×10^{12} ng/nm^3. It is noted that a 0.3% variation of the average density of the epoxy is calculated over the sampling distance of 7 nm, particularly contributed by the higher density gradient next to the graphene/polymer interface. The polymer density diminishes to zero across a small gap to the location of the graphene. This density gradient indicates the non-uniform distribution of the polymer atoms in the nearby graphene/polymer interface.

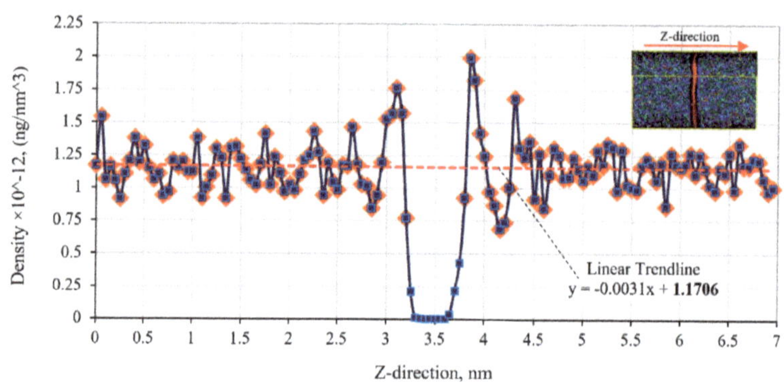

Figure 5. Variation of the epoxy polymer density along the RVE length of the graphene-polymer nanocomposite.

The profile was used to establish the boundary of the graphene and polymer constituents. The RVE was segmented into equal spacing of 0.05 nm along the z-direction (across the graphene/polymer interface). A zero-density gap with a thickness of 0.40 nm was determined at the central location of the nanocomposite RVE. The graphene thickness was assumed to be similar to the distance between the graphene sheets in the graphite structure (0.34 nm) and located in the middle of the nanocomposite RVE. Thus, the graphene/epoxy interface thickness on each side of the graphene is determined as having the thickness of 0.03 nm. The epoxy matrix on each side of the graphene was considered as 3.3 nm-thick with the cross-sectional area of 4.52×4.54 nm^2 (see Table 1) for the RVE. The calculated thickness of the graphene/epoxy interface will be used in the FE model of the RVE with finite-thickness cohesive elements.

3.2. Nanomechanical Behaviors of Graphene/Polymer Interface

The nanomechanical behavior of the graphene/epoxy interface, predicted using the MD simulation, is shown by the stress-displacement curve in Figure 6a. The results showed that, throughout the applied tensile loading (displacement-controlled), a non-linear increase in the interface stress to the peak level of 9.75×10^{-10} aPa is achieved. The stress decreases exponentially with the continuous displacement. A similar cohesive stress-displacement response with a nonlinear spring model was predicted for the interface of a pristine graphene/one-layer polymer using the MD approach [10,16]. At the atomic level, the interaction between the neighboring atoms could be described in terms of the interaction energy, as shown in Figure 6b, for two atoms with $\varepsilon = 3.4$ Å and $\sigma = 10$ meV. The interaction energy level is at a minimum (also known as the potential well) when the atoms are at their equilibrium distance apart, such that the cohesive (attractive) and repulsive forces between the atoms are balanced. Under the applied separation forces, the interaction energy decreases in an exponential manner, but the energy is restored upon unloading. If the loading continued, the separation energy would diminish to zero, denoting the atomic separation. The inflection point denotes the minimum interaction force as represented by the slope of the energy-distance curve (Figure 6b). Beyond this distance, the required interaction force of atomic separation that diminish exponentially could be interpreted as representing the degradation of the bond strength.

Analogously, while the initial deformation of the graphene/epoxy interface is elastic, beyond the displacement corresponding to the peak stress the decreasing stress-displacement curve also exhibits an exponential decay form, and constitutes of both recoverable and irrecoverable components (Figure 6a). Only when the tensile separation is large enough, the stress completely diminishes to cause the separation of the interface material point. The exponential decrease portion of the stress-displacement curve, as predicted by the MD simulation, is postulated to adequately represent the softening behavior of the cohesive graphene/epoxy interface at the continuum scale. Consequently, the softening characteristic of the CZM (see Figure 3) for the interface could be established based on the exponential decay curve, as illustrated in Figure 6c. The corresponding rate of the decay, as described in Equation (8), is characterized by the exponent, $\alpha = 7.74$.

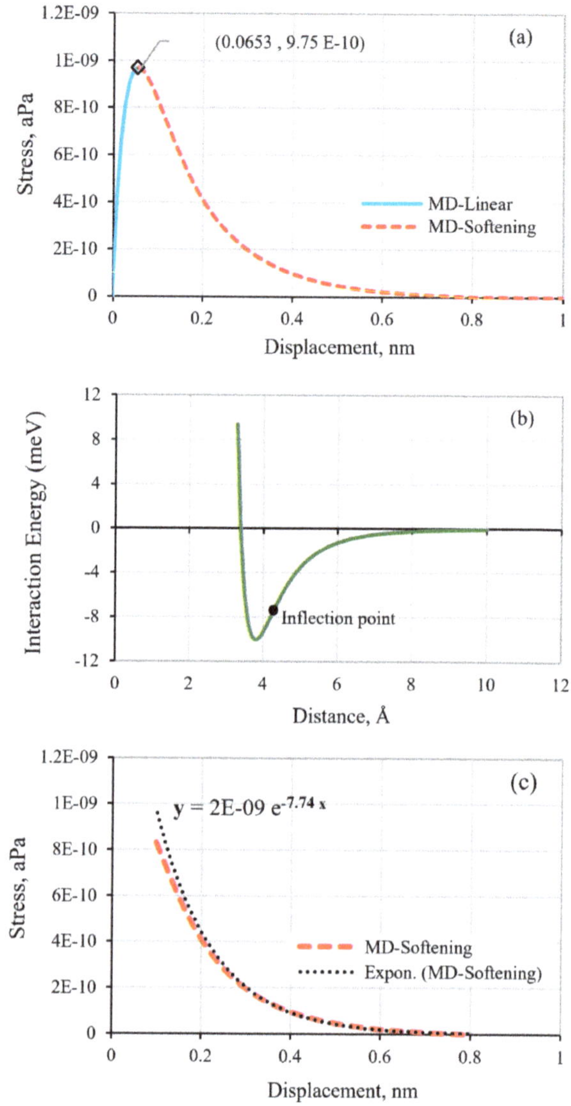

Figure 6. (**a**) The stress-displacement response, as calculated from the MD simulation, (**b**) interatomic energy-distance curve, and (**c**) the characteristic exponential decay softening law of the CZM.

Graphene/epoxy interface properties: The elastic properties of the interface were also determined from the MD-calculated stress-displacement curve of Figure 6a. The interface tensile strength, T_3^o, is defined by the peak stress level while the corresponding displacement represents the value at the onset of interface damage, δ_{33}^o. The initial tangential slope of the stress-displacement curve defines the penalty stiffness, k_3, of the interface. The area bounded by the curve represents the critical strain energy release rate, G_{IC}, for the tensile loading mode, which could be calculated by Equation (7). These properties along with the parameters of the exponential softening law are used to define the graphene/epoxy cohesive interface behavior for the FE simulation. The property values are listed in Table 3.

Table 3. The elastic and damage properties of graphene/polymer cohesive interface.

Parameter	Symbol (Unit)	Value
Tensile stiffness	k_3, (aPa·nm^{-1})	5×10^{-8}
Tensile strength	T_3^0, (aPa)	9.75×10^{-10}
Displacement at damage initiation	δ_{33}^o, (nm)	0.0653
Displacement at separation	δ_{33}^f, (nm)	0.8
Exponent for the regressive softening law, (Equation (8))	α	7.74
Critical Mode I strain energy release rate	G_{IC}, (N·nm^{-1})	2.1×10^{-10}

3.3. Response of the Graphene/Epoxy Interface Model

The FE-calculated stress-displacement response of the graphene/epoxy cohesive interface using the properties listed in Table 3 is shown in Figure 7. The close correlation of the CZM response with the input response from the MD simulation, with less than 3% difference, served to verify that appropriate procedures were employed in developing the FE model. The deviation is likely due to the different types of boundary condition specified for the MD and FE model resulting in slightly different stress levels for the onset of damage in each case. The intrinsic exponential-decay cohesive response of the graphene/epoxy interface, as established by the MD simulation, is faithfully reproduced by the FE model.

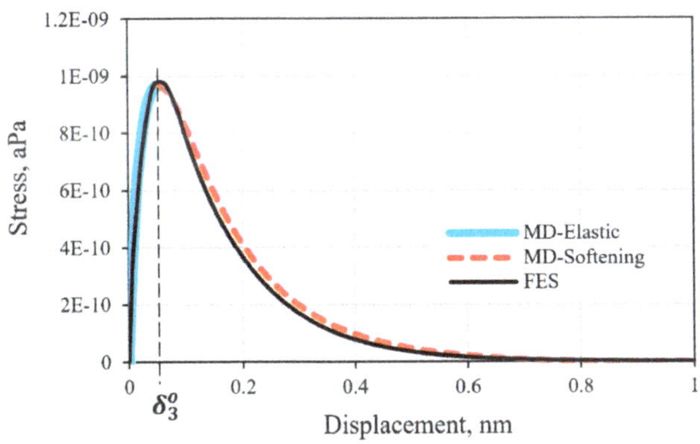

Figure 7. Comparison of the FE and MD stress-displacement response of the graphene/polymer interface.

In the absence of the MD-predicted exponential regressive behavior, a bilinear softening law is often employed as an estimate of the cohesive interface behavior. The response of the FE model with a bilinear softening law specified for the cohesive interface was examined. The CZM properties of tensile strength, penalty stiffness, and the critical Mode I strain energy release rate is maintained in the simulation to acknowledge the same physical interface. The resulting stress-displacement curves are compared in Figure 8. Results show that while insignificant differences of the elastic load-displacement response were displayed, the stress evolution during the damage process significantly differs. The bilinear softening law predicted a higher interface stress than the exponential regressive law by 25%, corresponding to the tensile displacement of 0.2 nm. In addition, the bilinear softening law would also predict an earlier separation of the cohesive interface. The interface separation is calculated at the displacement of 0.36 nm using the linear softening law, which corresponds to a 55% difference when compared to the prediction by the regressive softening rule of the interface at the displacement of 0.80 nm. Thus, the use of an inaccurate softening law and properties of the cohesive interface could lead to erroneous predicted properties, and the corresponding deformation mechanics and the failure process of the graphene-epoxy polymer nanocomposite.

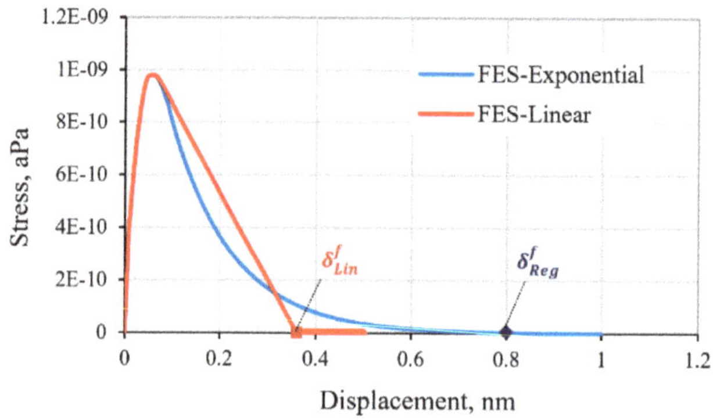

Figure 8. Comparison of the FE-predicted response of the cohesive interface with different softening laws.

The energy released by the graphene/epoxy interface during the interface failure process under the applied tensile loading is termed the damage dissipation energy (DDE). The evolution of the total DDE, for all the cohesive interface elements of the model with the two different softening laws of the cohesive interface, is compared in Figure 9. The DDE started to accumulate when the damage initiated at the attainment of the maximum load, corresponding to the damage initiation displacement, δ_0. The calculated evolution of the DDE follows the prescribed form of the regressive softening process, as illustrated in the figure. The initial rate of the energy dissipation is faster for the cohesive interface with exponential regressive than that of the linear softening law, as reflected by the initial slope of the curve. The evolving DDE saturates to a constant level when reaching the separation of the graphene/epoxy cohesive interface.

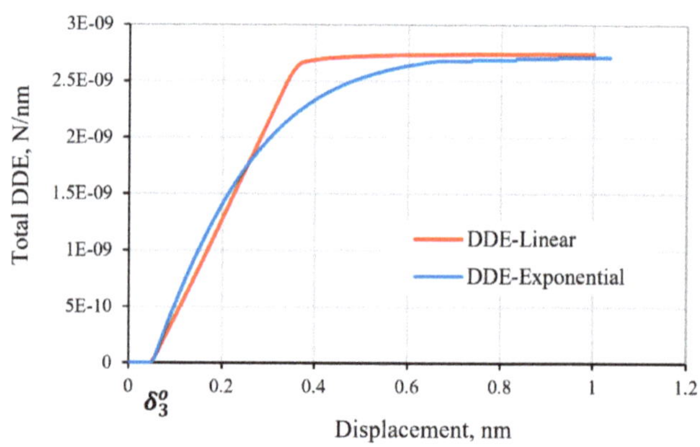

Figure 9. Evolution of the total damage dissipation energy (DDE) of the graphene/polymer interface with increasing tensile displacement.

4. Conclusions

The deformation and failure process of the graphene-polymer nanocomposites could be quantified through the FE simulation with the cohesive interface elements. The missing intrinsic properties and behavior of the cohesive interface could be determined from the MD simulation of the curing and subsequent loading of the nanocomposite RVE. This paper has demonstrated the procedures in establishing the intrinsic regressive softening law for the cohesive interface of the graphene/epoxy nanocomposites. Results show that the graphene/epoxy interface has a thickness of 0.03 nm, based on the density variation of the polymer across the graphene/epoxy interphase. Upon tensile loading of the RVE, an exponential-decay form of the cohesive interatomic forces was established. Analogously, the cohesive interface was described by the exponential regression law with the exponent, $\alpha = 7.74$. The assumed bilinear softening law of the cohesive interface, commonly employed in the microscale FE simulation of the nanocomposite behavior, could lead to erroneous results when compared to the MD-calculated exponential regression law. This was demonstrated by the different traction-displacement curves and the evolution of DDE to tensile separation of the graphene/epoxy interface of the nanocomposite RVE. A 55% difference in the predicted displacement at the separation of the interface was calculated when the bilinear softening law was assumed, compared to that using the exponential regressive damage law of the cohesive interface.

Author Contributions: Conceptualization, S.S.R.K., S.M.R.-K., A.K. and M.N.T.; methodology, S.S.R.K., S.M.R.-K. and A.K.; software, S.S.R.K. and S.M.R.-K.; validation, S.S.R.K., S.M.R.-K. and M.N.T.; formal analysis, S.S.R.K., S.M.R.-K., A.K. and M.N.T.; investigation, S.S.R.K., S.M.R.-K., A.K. and M.N.T.; writing—original draft preparation, S.S.R.K., S.M.R.-K., A.K.; writing—review and editing, S.S.R.K., A.K., M.N.T. and M.H.; supervision, M.N.T., M.H. and M.P.; project administration, S.S.R.K., M.N.T. and M.P.; funding acquisition, M.N.T., M.H. and M.P.

Funding: This work is funded by the Aerospace Malaysia Innovation Centre (AMIC) and the Universiti Teknologi Malaysia (UTM) under the Project No. AMIC/AM/P02-01 (UTM Grant No. 4C089 and 01M01, respectively), for the projects: "Advanced Testing of Laminated Composites for Aerospace Parts and Assemblies" and "Damage Assessment of Laminated Composites by Digital Image Correlation Technique". It is also partially supported by the Ministry of Education, Youth and Sports of the Czech Republic and the European Union (European Structural and Investment Funds–Operational Programme Research, Development and Education) in the frames of the project "Modular platform for autonomous chassis of specialized electric vehicles for freight and equipment transportation", Reg. No. CZ.02.1.01/0.0/0.0/16_025/0007293.

Acknowledgments: A. Karimzadeh is a researcher of UTM under the Post-Doctoral Fellowship Scheme of Posdoc-Q.J130000.21A2.04E49.

Conflicts of Interest: The authors declare no conflict of interest.

References

1. Du, Y.; Li, D.; Liu, L.; Gai, G. Recent achievements of self-healing graphene/polymer composites. *Polymers* **2018**, *10*, 114. [CrossRef] [PubMed]
2. Meng, F.; Huang, F.; Guo, Y.; Chen, J.; Chen, X.; Hui, D.; He, P.; Zhou, X.; Zhou, Z. In situ intercalation polymerization approach to polyamide-6/graphite nanoflakes for enhanced thermal conductivity. *Compos. Part B Eng.* **2017**, *117*, 165–173. [CrossRef]
3. Meng, F.; Zhong, J.; Chen, Y.; Liu, X. The influence of cross-linking reaction on the mechanical and thermal properties of polyarylene ether nitrile. *J. Appl. Polym. Sci.* **2011**, *120*, 1822–1828. [CrossRef]
4. Kim, H.; Abdala, A.A.; Macosko, C.W. Graphene/polymer nanocomposites. *Macromolecules* **2010**, *43*, 6515–6530. [CrossRef]
5. Stankovich, S.; Dikin, D.A.; Dommett, G.H.B.; Kohlhaas, K.M.; Zimney, E.J.; Stach, E.A.; Piner, R.D.; Nguyen, S.T.; Ruoff, R.S. Graphene-based composite materials. *Nature* **2006**, *442*, 282. [CrossRef] [PubMed]
6. Huang, C.-L.; Wu, H.-H.; Jeng, Y.-C.; Liang, W.-Z. Electrospun graphene nanosheet-filled poly(trimethylene terephthalate) composite fibers: Effects of the graphene nanosheet content on morphologies, electrical conductivity, crystallization behavior, and mechanical properties. *Polymers* **2019**, *11*, 164. [CrossRef]
7. Kim, S.; Do, I.; Drzal, L.T. Multifunctional xgnp/lldpe nanocomposites prepared by solution compounding using various screw rotating systems. *Macromol. Mater. Eng.* **2009**, *294*, 196–205. [CrossRef]
8. Rahimian-Koloor, S.M.; Moshrefzadeh-Sani, H.; Hashemianzadeh, S.M.; Shokrieh, M.M. The effective stiffness of an embedded graphene in a polymeric matrix. *Curr. Appl. Phys.* **2018**, *18*, 559–566. [CrossRef]
9. Díez-Pascual, M.A.; Luceño Sánchez, A.J.; Peña Capilla, R.; García Díaz, P. Recent developments in graphene/polymer nanocomposites for application in polymer solar cells. *Polymers* **2018**, *10*, 217. [CrossRef]
10. Dewapriya, M.A.N.; Rajapakse, R.K.N.D. Development of a homogenous nonlinear spring model characterizing the interfacial adhesion properties of graphene with surface defects. *Compos. Part B Eng.* **2016**, *98*, 339–349. [CrossRef]
11. Ahmad, A.I.; Koziol, K.K.; Deveci, S.; Kim, H.-K.; Kumar, V.R. Advancing the use of high-performance graphene-based multimodal polymer nanocomposite at scale. *Nanomaterials* **2018**, *8*, 947. [CrossRef] [PubMed]

12. Meng, F.; Ishida, H.; Liu, X. Introduction of benzoxazine onto the graphene oxide surface by click chemistry and the properties of graphene oxide reinforced polybenzoxazine nanohybrids. *RSC Adv.* **2014**, *4*, 9471–9475. [CrossRef]
13. Agnihotri, P.K.; Kar, K.K.; Basu, S. Cohesive zone model of carbon nanotube-coated carbon fiber/polyester composites. *Model. Simul. Mater. Sci. Eng.* **2012**, *20*, 035014. [CrossRef]
14. Azoti, W.L.; Elmarakbi, A. Multiscale modelling of graphene platelets-based nanocomposite materials. *Compos. Struct.* **2017**, *168*, 313–321. [CrossRef]
15. Dandekar, C.R.; Shin, Y.C. Molecular dynamics based cohesive zone law for describing al–sic interface mechanics. *Compos. Part A Appl. Sci. Manuf.* **2011**, *42*, 355–363. [CrossRef]
16. Awasthi, A.P.; Lagoudas, D.C.; Hammerand, D.C. Modeling of graphene–polymer interfacial mechanical behavior using molecular dynamics. *Model. Simul. Mater. Sci. Eng.* **2008**, *17*, 015002. [CrossRef]
17. Kolanthai, E.; Bose, S.; Bhagyashree, K.S.; Bhat, S.V.; Asokan, K.; Kanjilal, D.; Chatterjee, K. Graphene scavenges free radicals to synergistically enhance structural properties in a gamma-irradiated polyethylene composite through enhanced interfacial interactions. *Phys. Chem. Chem. Phys.* **2015**, *17*, 22900–22910. [CrossRef]
18. Ji, X.-Y.; Cao, Y.-P.; Feng, X.-Q. Micromechanics prediction of the effective elastic moduli of graphene sheet-reinforced polymer nanocomposites. *Model. Simul. Mater. Sci. Eng.* **2010**, *18*, 045005. [CrossRef]
19. Shiu, S.-C.; Tsai, J.-L. Characterizing thermal and mechanical properties of graphene/epoxy nanocomposites. *Compos. Part B Eng.* **2014**, *56*, 691–697. [CrossRef]
20. Tsai, J.-L.; Tu, J.-F. Characterizing mechanical properties of graphite using molecular dynamics simulation. *Mater. Des.* **2010**, *31*, 194–199. [CrossRef]
21. Firouz-Abadi, R.D.; Moshrefzadeh-Sany, H.; Mohammadkhani, H.; Sarmadi, M. A modified molecular structural mechanics model for the buckling analysis of single layer graphene sheet. *Solid State Commun.* **2016**, *225*, 12–16. [CrossRef]
22. Pontefisso, A.; Mishnaevsky, L. Nanomorphology of graphene and cnt reinforced polymer and its effect on damage: Micromechanical numerical study. *Compos. Part B Eng.* **2016**, *96*, 338–349. [CrossRef]
23. Liu, F.; Hu, N.; Han, M.; Atobe, S.; Ning, H.; Liu, Y.; Wu, L. Investigation of interfacial mechanical properties of graphene-polymer nanocomposites. *Mol. Simul.* **2016**, *42*, 1165–1170. [CrossRef]
24. Lin, F.; Xiang, Y.; Shen, H.-S. Temperature dependent mechanical properties of graphene reinforced polymer nanocomposites – a molecular dynamics simulation. *Compos. Part B Eng.* **2017**, *111*, 261–269. [CrossRef]
25. Hosseini Kordkheili, S.A.; Moshrefzadeh-Sani, H. Mechanical properties of double-layered graphene sheets. *Comput. Mater. Sci.* **2013**, *69*, 335–343. [CrossRef]
26. Rissanou, N.A.; Power, J.A.; Harmandaris, V. Structural and dynamical properties of polyethylene/graphene nanocomposites through molecular dynamics simulations. *Polymers* **2015**, *7*, 390–417. [CrossRef]
27. Tsai, J.-L.; Tzeng, S.-H.; Chiu, Y.-T. Characterizing elastic properties of carbon nanotubes/polyimide nanocomposites using multi-scale simulation. *Compos. Part B Eng.* **2010**, *41*, 106–115. [CrossRef]
28. Montazeri, A.; Rafii-Tabar, H. Multiscale modeling of graphene- and nanotube-based reinforced polymer nanocomposites. *Phys. Lett. A* **2011**, *375*, 4034–4040. [CrossRef]
29. Taherzadeh, M.; Baghani, M.; Baniassadi, M.; Abrinia, K.; Safdari, M. Modeling and homogenization of shape memory polymer nanocomposites. *Compos. Part B Eng.* **2016**, *91*, 36–43. [CrossRef]
30. Dai, G.; Mishnaevsky, L. Graphene reinforced nanocomposites: 3d simulation of damage and fracture. *Comput. Mater. Sci.* **2014**, *95*, 684–692. [CrossRef]
31. Rahimian-Koloor, S.M.; Moshrefzadeh-Sani, H.; Shokrieh, M.M.; Hashemianzadeh, S.M. On the behavior of isolated and embedded carbon nano-tubes in a polymeric matrix. *Mater. Res. Express* **2018**, *5*, 025019. [CrossRef]
32. Elices, M.; Guinea, G.V.; Gómez, J.; Planas, J. The cohesive zone model: Advantages, limitations and challenges. *Eng. Fract. Mech.* **2002**, *69*, 137–163. [CrossRef]
33. Koloor, S.S.R.; Tamin, M.N. Mode-ii interlaminar fracture and crack-jump phenomenon in cfrp composite laminate materials. *Compos. Struct.* **2018**, *204*, 594–606. [CrossRef]
34. Guin, L.; Raphanel, J.L.; Kysar, J.W. Atomistically derived cohesive zone model of intergranular fracture in polycrystalline graphene. *J. Appl. Phys.* **2016**, *119*, 245107. [CrossRef]
35. Dávila, C.G.; Camanho, P.P. *Analysis of the Effects of Residual Strains and Defects on Skin/Stiffener Debonding Using Decohesion Elements*; NASA Langley Research Center: Hampton, VA, USA; Norfolk, VA, USA, 2003.

36. Campilho, R.D.S.G.; Banea, M.D.; Neto, J.A.B.P.; da Silva, L.F.M. Modelling adhesive joints with cohesive zone models: Effect of the cohesive law shape of the adhesive layer. *Int. J. Adhes. Adhes.* **2013**, *44*, 48–56. [CrossRef]
37. ASTM. *Astm d5528–13, Standard Test Method for Mode i Interlaminar Fracture Toughness of Unidirectional Fiber-Reinforced Polymer Matrix Composites*; ASTM International: West Conshohocken, PA, USA, 2013.
38. ASTM. *Astm d7905/d7905m-14, Standard Test Method for Determination of the Mode ii Interlaminar Fracture Toughness of Unidirectional Fiber-Reinforced Polymer Matrix Composites*; ASTM International: West Conshohocken, PA, USA, 2014.
39. Spearot, D.E.; Jacob, K.I.; McDowell, D.L. Non-local separation constitutive laws for interfaces and their relation to nanoscale simulations. *Mech. Mater.* **2004**, *36*, 825–847. [CrossRef]
40. Yamakov, V.; Saether, E.; Glaessgen, E.H. Multiscale modeling of intergranular fracture in aluminum: Constitutive relation for interface debonding. *J. Mater. Sci.* **2008**, *43*, 7488–7494. [CrossRef]
41. Martínez, L.; Andrade, R.; Birgin, E.G.; Martínez, J.M. Packmol: A package for building initial configurations for molecular dynamics simulations. *J. Comput. Chem.* **2009**, *30*, 2157–2164. [CrossRef] [PubMed]
42. Plimpton, S. Fast parallel algorithms for short-range molecular dynamics. *J. Comput. Phys.* **1995**, *117*, 1–19. [CrossRef]
43. Hadden, C.M.; Jensen, B.D.; Bandyopadhyay, A.; Odegard, G.M.; Koo, A.; Liang, R. Molecular modeling of epon-862/graphite composites: Interfacial characteristics for multiple crosslink densities. *Compos. Sci. Technol.* **2013**, *76*, 92–99. [CrossRef]
44. Aghadavoudi, F.; Golestanian, H.; Tadi Beni, Y. Investigating the effects of resin crosslinking ratio on mechanical properties of epoxy-based nanocomposites using molecular dynamics. *Polym. Compos.* **2017**, *38*, E433–E442. [CrossRef]
45. Alian, A.R.; Meguid, S.A. Molecular dynamics simulations of the effect of waviness and agglomeration of cnts on interface strength of thermoset nanocomposites. *Phys. Chem. Chem. Phys.* **2017**, *19*, 4426–4434. [CrossRef] [PubMed]
46. Alian, A.R.; Kundalwal, S.I.; Meguid, S.A. Multiscale modeling of carbon nanotube epoxy composites. *Polymer* **2015**, *70*, 149–160. [CrossRef]
47. Varshney, V.; Patnaik, S.S.; Roy, A.K.; Farmer, B.L. A molecular dynamics study of epoxy-based networks: Cross-linking procedure and prediction of molecular and material properties. *Macromolecules* **2008**, *41*, 6837–6842. [CrossRef]
48. Yarovsky, I.; Evans, E. Computer simulation of structure and properties of crosslinked polymers: Application to epoxy resins. *Polymer* **2002**, *43*, 963–969. [CrossRef]
49. Brown, E.N.; White, S.R.; Sottos, N.R. Microcapsule induced toughening in a self-healing polymer composite. *J. Mater. Sci.* **2004**, *39*, 1703–1710. [CrossRef]
50. Garcia, F.G.; Soares, B.G.; Pita, V.J.R.R.; Sánchez, R.; Rieumont, J. Mechanical properties of epoxy networks based on dgeba and aliphatic amines. *J. Appl. Polym. Sci.* **2007**, *106*, 2047–2055. [CrossRef]
51. Vijayaraghavan, V.; Wong, C.H. Temperature, defect and size effect on the elastic properties of imperfectly straight carbon nanotubes by using molecular dynamics simulation. *Comput. Mater. Sci.* **2013**, *71*, 184–191. [CrossRef]
52. Stuart, S.J.; Tutein, A.B.; Harrison, J.A. A reactive potential for hydrocarbons with intermolecular interactions. *J. Chem. Phys.* **2000**, *112*, 6472–6486. [CrossRef]
53. Sun, H.; Mumby, S.J.; Maple, J.R.; Hagler, A.T. An ab initio cff93 all-atom force field for polycarbonates. *J. Am. Chem. Soc.* **1994**, *116*, 2978–2987. [CrossRef]
54. Lordi, V.; Yao, N. Molecular mechanics of binding in carbon-nanotube–polymer composites. *J. Mater. Res.* **2000**, *15*, 2770–2779. [CrossRef]
55. Desgranges, C.; Delhommelle, J. Evaluation of the grand-canonical partition function using expanded wang-landau simulations. Iii. Impact of combining rules on mixtures properties. *J. Chem. Phys.* **2014**, *140*, 104109. [CrossRef] [PubMed]
56. Shenogina, N.B.; Tsige, M.; Patnaik, S.S.; Mukhopadhyay, S.M. Molecular modeling of elastic properties of thermosetting polymers using a dynamic deformation approach. *Polymer* **2013**, *54*, 3370–3376. [CrossRef]
57. Yang, S.; Yu, S.; Cho, M. Influence of thrower–stone–wales defects on the interfacial properties of carbon nanotube/polypropylene composites by a molecular dynamics approach. *Carbon* **2013**, *55*, 133–143. [CrossRef]

58. Lu, Q.; Bhattacharya, B. The role of atomistic simulations in probing the small-scale aspects of fracture—A case study on a single-walled carbon nanotube. *Eng. Fract. Mech.* **2005**, *72*, 2037–2071. [CrossRef]
59. Martyna, G.J.; Tobias, D.J.; Klein, M.L. Constant pressure molecular dynamics algorithms. *J. Chem. Phys.* **1994**, *101*, 4177–4189. [CrossRef]
60. Dassault Systemes. *Abaqus Documentation*; Dassault Systemes: Vélizy-Villacoublay, France, 2009.
61. Koloor, S.S.R.; Ayatollahi, M.; Tamin, M. Elastic-damage deformation response of fiber-reinforced polymer composite laminates with lamina interfaces. *J. Reinf. Plast. Compos.* **2017**, *36*, 832–849. [CrossRef]

polymers

MDPI

Article

High-Performance Biobased Unsaturated Polyester Nanocomposites with Very Low Loadings of Graphene

Chengguo Liu [1], Cuina Wang [2], Jijun Tang [3], Jing Zhang [3], Qianqian Shang [1], Yun Hu [1], Hongxiao Wang [1], Qiong Wu [2], Yonghong Zhou [1,*], Wen Lei [2,*] and Zengshe Liu [4,*]

1 Institute of Chemical Industry of Forest Products, Chinese Academy of Forestry, Key Lab of Biomass Energy and Material, Jiangsu Province, National Engineering Lab for Biomass Chemical Utilization, Key Lab on Forest Chemical Engineering, State Forestry Administration, Nanjing 210042, China; liuchengguo@icifp.cn (C.L.); Shang_qianqian@126.com (Q.S.); 15150509893@139.com (Y.H.); wanghongxiao_1@163.com (H.W.)

2 College of Science, Nanjing Forestry University, Nanjing 210037, China; wangcuina66@163.com (C.W.); wqzjj9394@163.com (Q.W.)

3 Department of Corrosion Prevention and Polymer Materials, College of Materials Science and Engineering, Jiangsu University of Science and Technology, Zhenjiang 212003, China; tangjijunnju@126.com (J.T.); badbush2003@163.com (J.Z.)

4 Bio-Oils Research, National Center for Agricultural Utilization Research, Agricultural Research Service, United States Department of Agriculture, 1815 N. University St., Peoria, IL 61604, USA

* Correspondence: zyh@icifp.cn (Y.Z.); njfuleiwen@163.com (W.L.); Kevin.Liu@ARS.USDA.GOV (Z.L.); Tel.: +86-25-85482777 (Y.Z.); +86-25-85427206 (W.L.); +1-309-6816104 (Z.L.)

† Mention of trade names or commercial products in this publication is solely for the purpose of providing specific information and does not imply recommendation or endorsement by the U.S. Department of Agriculture. USDA is an equal opportunity provider and employer.

Received: 30 October 2018; Accepted: 17 November 2018; Published: 20 November 2018

Abstract: Graphene-reinforced tung oil (TO)-based unsaturated polyester nanocomposites were prepared via in situ melt polycondensation intergrated with Diels–Alder addition. Functionalized graphene sheets derived from graphene oxide (GO) were then extracted from the obtained nanocomposites and carefully characterized. Furthermore, dispersion state of the graphene nanosheets in the cured polymer composites and ultimate properties of the resultant biobased nanocomposites were investigated. Mechanical and thermal properties of the TO-based unsaturated polyester resin (UPR) were greatly improved by the incorporation of GO. For example, at the optimal GO content (only 0.10 wt %), the obtained biobased nanocomposite showed tensile strength and modulus of 43.2 MPa and 2.62 GPa, and T_g of 105.2 °C, which were 159%, 191%, and 49.4% higher than those of the unreinforced UPR/TO resin, respectively. Compared to neat UPR, the biobased UPR nanocomposite with 0.1 wt % of GO even demonstrated superior comprehensive properties (comparable stiffness and T_g, while better toughness and thermal stability). Therefore, the developed biobased UPR nanocomposites are very promising to be applied in structural plastics.

Keywords: graphene; unsaturated polyester resins; tung oil; biobased polymer nanocomposites; in situ melt polycondensation

1. Introduction

Unsaturated polyester resins (UPRs) are widely utilized in industrial and domestic areas due to their low cost, ease of handling, and good balance of mechanical, thermal, electrical, and chemical resistant properties [1,2]. However, with the current concerns on exploring alternatives to petroleum and reducing environmental pollution, research is being increasingly directed to develop polymeric materials from renewable resources such as proteins, oils, and carbohydrates [3–5]. Among all the biomass-derived feedstocks, plant oils are the primary choice to prepare UPRs because of their abundance, low toxicity, biodegradability, and triglyceride structures suitable for further chemical modification [6,7]. As a result, plant oil-based UPRs have attracted considerable attention since 2000.

Blending plant oils or their derivatives with petroleum-based UPRs is an efficient strategy to to prepare oil-based UPRs [8–14]. The addition of flexible oil-based modifiers usually leads to an apparent improvement of toughness compared to petroleum-based UPRs, however, a large loss of stiffness is commonly observed in the resulting biobased UPRs when the content of oil-based modifiers are not so high (10–20 wt %), thus leading to the unbalance of stiffness and toughness for the biobased UPRs. A good solution to address the obstacle is to reinforce the bioresins with nanofillers [11–13,15,16]. Biobased UPRs enhanced with a small amount of nanoclays were shown to exhibit good enhancements in mechanical, thermal, and barrier properties, whereas stiffness and other properties were only partially recovered [12,13]. Hence, exploring other nanofillers which are more efficient in the improvement of stiffness is an important task for the real application of such biobased UPRs.

Graphene, a single-atom-thick 2D sheet of sp^2-hybridized carbon atoms, has been extensively utilized in polymer nanocomposites owing to its superior properties like high mechanical stiffness (~ 1.0 TPa) and large surface area (~ 2630 $m^2 \cdot g^{-1}$) [17–19]. Graphene has shown dramictical enhancement in electrical, mechanical, thermal, and barrier properties of polymer composites at low concentrations [18–21]. Usually, there are three methods for fabriction of graphene/polymer composites: solution mixing, melt compounding, and in situ polymerization [19,22]. Compared to the former two methods, the in situ polymerization not only can make graphene sheets well-dispersed in polymer matrix, but also enable chemical bonds formed between graphene and polymer readily. Therefore, a variety of graphene-based polymeric composites, such as graphene/polystyrene nanocomposites [23,24], graphene/epoxy nanocomposites [25–27] and graphene/UPR nanocomposites [28–30] have been prepared through this method.

In our previous work, we prepared tung oil (TO)-modified UPRs via intermolecular Diels–Alder (D–A) addition between unsaturated polyesters and TO triglycerides [31]. The obtained biobased UPRs also demonsatrated a large drop in stiffness when the TO content was larger than 7.4 wt % of UPR (with styrene). Thus, in this study, graphene was empolyed to reinforce a TO-based UPR. To the best of our konwledge, graphene has never been used to reinforce biobased UPRs. Using graphene oxide (GO) as starting material, we prepared the graphene-reinforced TO-based UPRs by in situ polymerization combined with Diels–Alder (D–A) addition. Our goal is to recovery the loss of stiffness for tung oil-based UPRs, and to see whether balanced stiffness–toughness can be achieved.

2. Experimentals

2.1. Materials

Tung oil was purchased from Jiangsu Donghu oil Co., Ltd. (Yancheng, China), which has a specific gravity of 0.935−0.940 at 25 °C. Graphene Oxide was obtained from Nanjing XFNANO Materials Tech Co., Ltd. (Nanjing, China), which has a lateral size of 0.5−5 μm and thickness of 0.8−1.2 nm. Maleic anhydride (MA), phthalic anhydride (PA), styrene (≥99%), and hydroquinone were obtained from Sinopharm Chemical Reagent Co., Ltd. (Shanghai, China) Ethanol (≥99%), propylene glycol (PG) (99%), and toluene (≥99.5%) were obtained from Nanjing Chemical Reagent Co., Ltd. (China) (≥99%) Dibutyl phthalate (≥99.5%) and *N,N*-dimethylaniline (≥99%) were obtained from Shanghai Lingfeng Chemical Reagent Co., Ltd. (Shanghai, China) Benzoyl peroxide (≥99%) was obtained from Shanghai Macklin Biochemical Co., Ltd. (Shanghai, China) The PG and styrene were dried by molecular sieves for at least one week before use.

2.2. Synthesis of Neat UPR

50.22 g of PG, 39.22 g of MA, 29.62 g of PA, and 0.119 g of hydroquinone were added into a 250 ml four-necked flask equipped with a mechanical stirrer, a thermometer, a N_2 gas inlet, and a fractionating device. The mixture was then heated to 60 °C, agitated at 60 °C for 0.5 h, heated to 160 °C under N_2 protection, and maintained at 160 °C for 1.5 h. Subsequently, the mixture was heated to 200 °C and reacted at 200 °C until the acid value of the system decreased to a set value (around 33 mgKOH/g). After that the reaction temperature was reduced to 120 °C and 0.1 g of hydroquinone was added and mixed for 0.5 h. Finally, the temperature was lowered to 90 °C and 58.30 g of styrene (about 35% of the total resin weight) was added and blended with the resultant mixture for 1 h. A colorless and transparent liquid resin was produced.

2.3. Synthesis of UPR/TO

The UPR/TO resin was synthesized via melt polycondensation incorporated with D–A addition. The procedure can be divided into two stages. The first stage involved the synthesis of unsaturated polyester, as described in the above section. The acid value also reached the set value. At the second stage, the reaction temperature was lowered to 120 °C, 0.1 g of hydroquinone was added, and 16.65 g of TO was added dropwise into the flask within 0.5 h. At last, 58.30 g of styrene was added into the mixture and mixed at 90 °C for 1 h. A light yellow and translucent liquid resin was obtained.

2.4. Preparation of UPR/TO/GO Composites

The synthesis of UPR/TO/GO nanocomposites was carried out in three basic stages, as shown in Figure 1. In the first stage, 50.22 g of PG and an appropriate amount of GO powder were put into a 250 mL four-neck flask and ultrasonicated for 2 h to achieve a homogeneous GO/PG dispersion. In the second stage, a reaction mixture of MA (39.22 g), PA (29.62 g), and hydroquinone (0.119 g) was added into the 250 mL flask and the reaction was conducted identically to the procedures indicated in the first stage of synthesizing UPR/TO. The third stage was exactly the same as the second stage of UPR/TO. At the end, a black and opaque liquid resin was attained. In our experiments, the content of TO in the obtained UPR/TO and UPR/TO/GO composites was always 10 wt % of the UPR resin (including styrene). The content of GO in the UPR/TO/GO composites was 0.05, 0.10, 0.15, 0.20 and 0.30 wt % of the UPR resin (including styrene), thus for simplicity the corresponding nanocomposites were denoted as UPR/TO/G0.05, UPR/TO/G0.10, UPR/TO/G0.15, UPR/TO/G0.20, and UPR/TO/G0.30, respectively.

During the manufacturing process, GO could be simultaneously grafted by unsaturated polyesters and thermally reduced due to the high temperature involved in the melt polycondeansation, and the grafted unsaturated polyesters could further graft with TO via D–A addition, thus new functionalized graphene sheets (FGS) would be produced (Figure 1b). Successive centrifugation/redissolution cycles were performed to separate FGS from the obtained polymer composites. Typically, 30 g of the UPR/TO/G0.15 polymer composite (without styrene) was dissolved in 150 mL of toluene/ethanol mixed solvent (50/50, *v*/*v*), stirred for 2 h to remove the absorbed polymers from the graphene surface, and then centrifuged at 11,000 rpm for 30 min to precipitate the graphene completely. The attained centrifugate was dissolved by toluene/ethanol solvent and separated by centrifugation repeatedly for 5 times. At last, the obtained centrifugate was washed with ethanol twice, then dried at 50 °C for 4 h and under vacuum for another 48 h. The resulting black solid material was labeled as FGS. In an effort to further determine the contents of graphene, UPR, and TO in the FGS, a referenced polymeric nanocomposite were prepared according to the fabricating procedure of UPR/TO/G0.15 but without adding TO. The obtained composite was named as UPR/G0.15. By the same repeating centrifugal-washing procedure, a similar black solid material was obtained, which was labeled as FGS-i since it was an intermediate compared to the FGS.

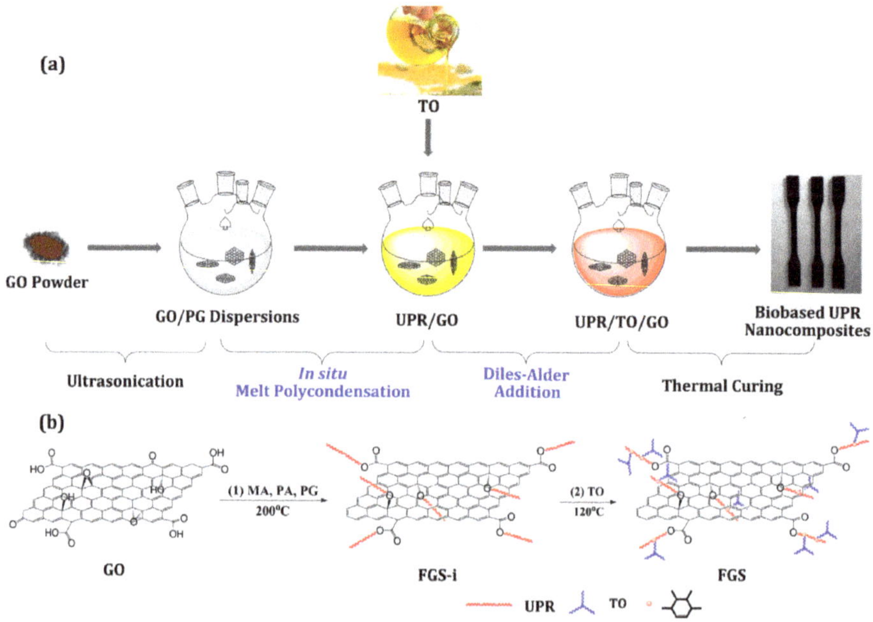

Figure 1. (**a**) The prepatration of graphene-reinforeced biobased unsaturated polyester nanocomposites via a combination of in situ melt polycondensation and Diels–Alder addition; (**b**) possible chemical changes of graphene oxide (GO) during the in situ preparation of biobased nanocomposites.

2.5. Curing of UPR, UPR/TO, and UPR/TO/GO Composites

The as-fabricated UPR, UPR/TO, and UPR/TO/GO composites were all cured in a same procedure. Typically, the resin samples were blended with the initiator (2 wt % of the total resin) for 30 min and with the promoter (0.2 wt % of the total resin) for 3 min, degassed, poured rapidly into homemade polytetrafluoroethylene molds, cured at room temperature for 3 h, and postcured at 80 °C for 12 h.

2.6. Characterization

Acid values were determined based on the procedures presented in GB/T 2895–1982. Atomic force microscopy (AFM) images were obtained using a SPM-9600 atomic force microscope (Shimadzu, Kyoto, Japan) in taping mode. FT-IR spectra were obtained using a Nicolet iS10 IR spectrometer (Thermo Fisher, Waltham, MA, USA). Thermogravimetric analysis (TGA) was conducted on a STA 409PC thermogravimetry instrument (Netzsch, Selb, Germany). X-ray photoelectron spectroscopy (XPS) was performed on a AXIS Ultra^{DLD} photoelectron spectrometer (Shimadzu, Kyoto, Japan). Raman spectra were collected on a DXR532 Raman spectrometer with a wavelength of 532 nm (Thermo Fisher, Waltham, MA, USA). Scanning electron microscopy (SEM) examinations were studied on an S-3400N scanning electron microscope (Hitachi, Tokyo, Japan). Transmission electron microscopy (TEM) examinations were carried out by a Tecnai G^{2}20 transmission electron microscope (FEI, Hillsboro, OR, USA). Tensile and flexural tests were conducted on a SANS7 CMT-4304 universal tester (Xinsansi, Shenzhen, China). Impact tests were conducted on a CEAST 9050 impact tester (Instron, Norwood, MA, USA). All the mechanical tests followed the procedures presented in the GB/T 2567-2008. Dynamic mechanical analysis (DMA) was performed on a Q800 solids analyzer (TA, New Castle, DE, USA) in three-point bending mode.

3. Results and Discussion

3.1. Structural Characterization of FGS

For polymer nanocomposites, it is well known that if the nanofiller and polymer matrix form chemical bonds, the interfacial interactions between the polymer and nanofiller will be improved greatly. To confirm whether chemical bonds are generated between the nanofiller and polymer, the chemical state of graphene nanosheets should be determined first. Thus FGS was extracted from the biobased UPR/TO/G0.15 composite by the repeating centrifugal-washing method and characterized by sedimentation experiments, AFM, FT-IR, TGA, XPS, and Raman spectroscopy, as discussed below. For comparison, GO and thermally reduced GO (RGO) (prepared by treating neat GO powder at 200 °C for 10 h, similar condition to the in situ polycondensation) were also characterized.

Firstly, sedimental experiments were taken to assess their solubility in organic solvent. GO, RGO, and FGS were dispersed in ethanol by stirring for 5 min and probe-sonicating for 5 min, and then precipitated for 30 min.

The resulting photographs are displayed in Figure 2a. Both FGS and RGO were black from appearance, while GO was golden yellow, indicating GO was reduced during the polycondensation. In addition, GO and RGO precipitated apparently, whereas FGS had no obvious preciptation. This difference indicated that FGS possessed better solubility than both GO and RGO, which may be attributed to that FGS were grafted by organic polymers. AFM was employed to further characterize the graphene samples. The typical tapping-mode AFM images of GO and FGS are presented in Figure 2b,c, respectively. The thickness of GO nanosheet was about 0.89 nm, which was in good accordance with the results from literature [28,32,33]. In contrast, the FGS possessed dinstinct thicknesses: at the thinest point it was only 6.18 nm, while at some other places it could reach dozens of nanometers (eg. 24.2 nm). The results suggested that graphene sheets were not only grafted by linear polymers like unsautrated polyesters, but could also be partially grafted by crosslinked polymers, since the grafted unsaturated polyesters could form crosslinking polymers with TO triglycerides [31]. The crosslinked polymers could be directly observed in the lower half of FGS's image (Figure 2c), where the nanosheet were rolled up in some extent due to the crosslinking effect of TO with the grafted polyesters. Figure 2d provides the FT-IR spectra of GO, RGO, and FGS. It can be seen that the spectrum of FGS was analegous to RGO but quite different from that of GO. The original bands in GO, such as carboxyls (3400–2500 cm^{-1}) and carbonyl (~ 1730 cm^{-1}), decreased obviously in intensity for FGS sample, indicating the reduction of GO to FGS. At last, samples of GO, RGO, FGS, and FGS-i were analyzed by TGA (Figure 2e). In the curve of GO there was two main weigh loss stages at around 100 and 250 °C, which are attributed to the removal of absorbed water and the pyrolysis of oxygen-containing functional groups, respectively [32,33]. The total weight loss below 250 °C was approximate 40%. In the curve of RGO, the degradation rate was slow, suggesting RGO was more stable than GO. In contrast, both the TGA curves of FGS-i and FGS exhibited only one main stage of weight loss at 270–450 °C, which is ascribed to the degradation of the grafted polyesters or biopolyesters. No apparent weight loss stages were found below 250 °C, suggesting that the original oxygenated groups on the surface were removed or grafted by polymers. From the main weight loss stage of FGS-i, the weight fraction of grafted unsaturated polyesters on the graphene surface could be estimated and was about 25%; while from the same stage of FGS, the total weight fraction of grafted polyesters and TO was about 35%. Thus it can be deduced that the FGS contained around 65% of graphene, 25% of unsaturated polyesters, and 10% of TO molecules.

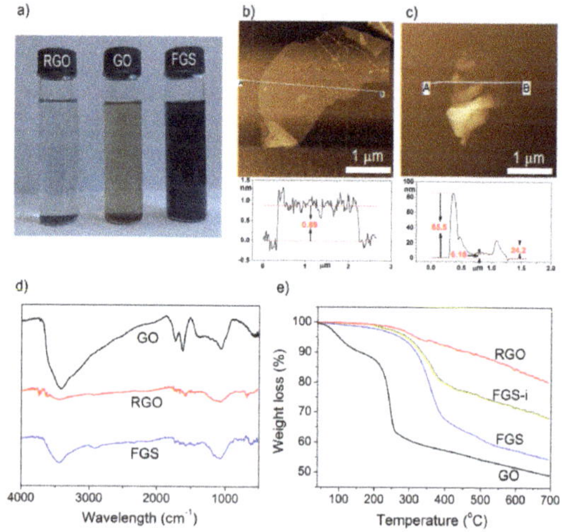

Figure 2. (**a**) Sedimentation experiment; (**b**,**c**) atomic force microscopy (AFM) images of GO and FGS; (**d**) FT-IR spectra of GO, thermally reduced GO (RGO), and FGS; (**e**) TGA curves of GO, RGO, FGS-i, and FGS.

XPS was conducted to further analyze the graphene-based compounds, as shown in Figure 3a–d. In the spectra of GO, only three peaks were depicted: sp^2 C=C (~285.0 eV), C–O (~287.0 eV), and O–C=O (~288.7 eV). The RGO spectra also comprised the three peaks, but the intensities of oxygen-containing peaks at 286.1 and 288.9 eV (relative to the C=C peak) decreased obviously. In both the spectra of FGS-i and FGS, the three peaks had a similar trend as those in RGO, indicating the reduction of GO. Differently, new peaks occurred at 285.4 or 285.6 eV, which means the ocurrence of sp^3 C–C bonds [34–37]. The bonds probably came from the grafted polyesters and TO molecules which involved many sp^3 C–C bonds. Compared to FGS-i, the intensity of new C–C peak in FGS spectra increased (relative to that of C=C peak) due to the incorporation of TO. The C/O atomic ratios for all the graphene nanomaterials, neat UPR, and TO were caculated and the related data are listed in Table 1. With the C/O ratios of UPR, RGO, and FGS-i, the weight content of grafted unsaturated polyesters onto graphene can be estimated and was about 28%; with the ratios of TO, FGS-i, and FGS, the grafted TO can also be determined and was around 5%. These values are close to the calculated values from the TGA results.

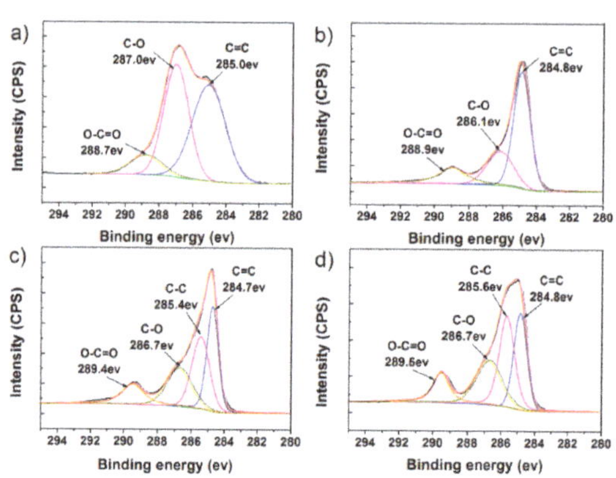

Figure 3. The C 1s peaks in XPS spectra of (**a**) GO, (**b**) RGO, (**c**) FGS-i, and (**d**) FGS.

Table 1. XPS data for all the graphene nanomaterials, neat UPR, and TO.

Sample	Relative Atomic Percentage (%)		C/O Atomic Ratios
	C	O	
GO	69.0	31.0	2.23
RGO	81.9	18.1	4.53
FGS-i	78.5	21.5	3.72
FGS	80.1	19.9	4.02
UPR	62.5	37.5	1.67
TO	90.5	9.5	9.5

Raman spectroscopy has been reported as a powerful probe for the structures of carbonaceous materials. Figure 4 gives the Raman spectra of GO, RGO, and FGS. The GO spectrum displayed a D band at 1344.1 cm^{-1} and G band at 1586.1 cm^{-1} which are ascribed to the breathing mode of κ-point phonons of A_{1g} symmetry and the first-order scattering of the the E_{2g} phonons, respectively [33,34]. In the spectrum of RGO, the G band blue-shifted to 1590.9 cm^{-1}, which was close to the value of raw graphite. In the spectrum of FGS, the D and G bands shifted to 1338.3 and 1575.5 cm^{-1}. The red-shift of G band can be attributed to that the defects of FGS were more apparent than those of GO. The intensity ratio of D and G bands (I(D/G)), corresponding to the disordered and ordered crystal structures of carbon, is inverse to the average size of sp^2 domains [33,34,36]. The I(D/G) value decreased from 1.28 for GO to 1.12 for RGO, which was in good agreement with the fact that RGO was reduced GO. In contrast, the value of FGS (1.33) was a little higher than that of GO. The reason for the slight increase may lie in that the grafted polymers make the defects of graphene more distinct than the orginal ones.

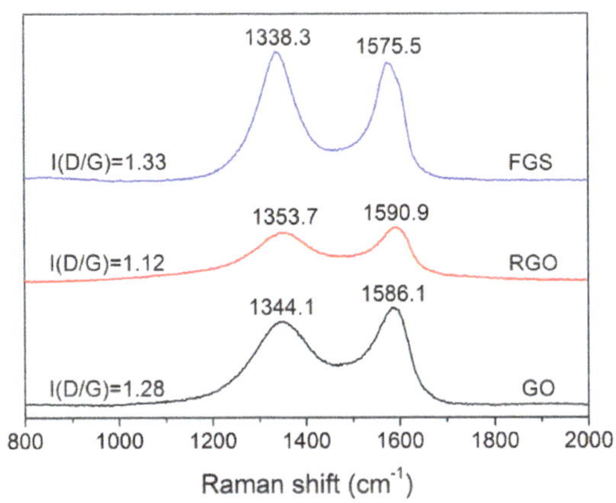

Figure 4. Raman spectra of GO, RGO, and FGS.

In conclusion, the original GO nanomaterial was grafted by unsaturated polyesters and simultanesously reduced during in situ polymerization, and followed by grafting TO triglycerides onto polyesters via D–A addition. Therefore, it can be inferred that the interfacial interactions between the polymer matrix and graphene nanosheets were strong.

3.2. Dispersions of Graphene in Polymer Matrix

Apart from the interaction between the nanofiller and polymer, disperison state of the nanofiller in the polymer matrix is another key factor that influences the ultimate properties of fabricated composites. SEM and TEM techniques can provide the direct observation of dispersion behaviours of nanofillers. Figure 5a,b shows the SEM images of tensile fracture surfaces of the biobased nanocomposites with GO contents of 0.10% and 0.30%, respectively. Before the SEM observation the fracture surface of tensile-testing specimen were etched in a 10% NaOH/ethanol solution at 40 °C for 2 h. The FGS was evenly dispersed in the UPR/TO/G0.10 composite, whereas the graphene sheets were much closer to each other in the UPR/TO/G0.30 composite. Moreover, the protruding graphene sheets in both the nanocomposites were still coated with adsobed resins after etching, reflecting the strong filler–polymer interfacial interaction [28,38]. It should be mentioned that phase separation like craters occurred in both the polymer matrices, which is attibuted to the excess of TO [31,39]. Figure 5c,d presents the TEM images of the UPR/TO/G0.10 and UPR/TO/G0.30 samples. The graphene nanosheets were exfoliated much better and dispersed more homogeneously in the UPR/TO/G0.10 composite (as indicated by the red arrows) than in the UPR/TO/G0.30 composite, which was consistent with the SEM results. The nanosheets in UPR/TO/G0.30 demonstrated some extent of aggregation. All these morphologies would affect ulitmate properties of the cured composites.

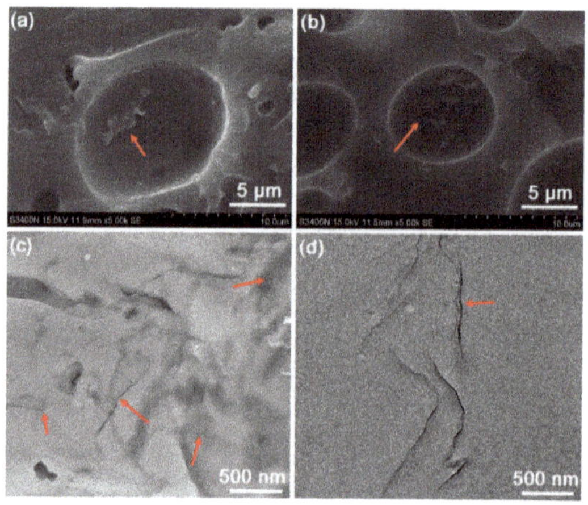

Figure 5. SEM images of the etched UPR/TO/GO composites with GO contents of (**a**) 0.10% and (**b**) 0.30%; TEM images of the UPR/TO/GO composites with GO contents of (**c**) 0.10% and (**d**) 0.30%.

3.3. Properties of the UPR/TO/GO Biobased Nanocomposites

In this work, the most noteworthy questions are whether the stiffness of the biobased UPRs can be completely recovered by the incorporation of graphene and whether the comprehensive properties of the resulting composites can compare to those of petroleum-based UPR. Thus mechanical and thermal properties of the obtained biobased UPR nanocomposites were investigated carefully.

Figure 6a illustrates the typical stress–strain curves of neat UPR, UPR/TO, and UPR/TO/G0.10. The UPR/TO/G0.10 composite showed a comparable stress to that of neat UPR, indicating that the prepared composite could be as good as the neat UPR in stiffness. Meanwhile the strain of UPR/TO/G0.10 was larger than that of UPR, suggesting the toughness of UPR/TO/G0.10 was even better. In addition, compared to the UPR/TO, although the incorporation of GO resulted in the decrease of strain, the stress of UPR/TO was improved remarkably. The improvement can be attributed to the good exfoliation of FGS and the strong interaction between FGS and polymer matrix, which can efficiently transfer the load from the matrix to the graphene nanofiller [19,22,28,32,33,40].

Figure 6b presents tensile and flexural strengths of neat UPR, UPR/TO, and UPR/TO/GO composites. The tensile and flexural strengths of neat UPR were 43.5 and 69.7 MPa, respectively. By the addition of 10% TO, the values decreased to 16.7 and 25.3 MPa, respectively, which results from a second phase generated by unreacted tung oil [31,39], as shown in Figure 5a,b. As the GO content increased, both the strengths increased firstly and decreased later. The increase is caused by the enhancement effect of the FGS filler, while the decrease probably results from the poor dispersion of graphene like aggregation. The composites possessed an optimal loading level at 0.10% of GO.

At this concentration the tensile and flexural strengths reached 43.2 MPa and 66.6 MPa, which was only 0.59% and 4.45% inferior to those of neat UPR, respectively. Nevertheless, they increased by 159% and 163% compared to UPR/TO, respectively, which can be ascribed to the sufficient exfoliation of graphene nanosheets and the strong interaction of FGS–polymer too.

Figure 6c demonstrates tensile and flexural moduli of neat UPR, UPR/TO, and UPR/TO/GO composites. These properties can represent the stiffness of polymer materials. The tensile and flexural moduli of the neat UPR were 2.65 and 2.03 GPa, respectively. By the incorporation of 10% TO, these values decreased to 0.90 and 0.63 GPa, respectively, which is due to the phase separation too. For the UPR/TO/GO composites, the moduli showed a similar trend as the strengths when the GO concentration grew. The composite at the GO content of 0.10% showed a tensile modulus of 2.62 GPa, which was 1.13% lower than that of UPR. Yet the flexural modulus of the composite reached 2.20 GPa, which was 8.37% higher than that of neat UPR. Therefore, the stiffness of the UPR/TO/G0.10 composite could be regarded as equivalent to that of neat UPR. Notably, the two values increased by 191% and 250% compared to those of UPR/TO, respectively, which can be also attributed to the complete exfoliation of FGS and the strong interfacial interaction between graphene and polymer matrix.

Figure 6d provides impact strength and tensile breaking strain of neat UPR, UPR/TO, and UPR/TO/GO composites. The two values of neat UPR were 2.58 kJ/m^2 and 2.03%, respectively. After the addition of 10% TO, they increased rapidly to 9.09 kJ/m^2 and 5.82%, respectively, which is caused by the effect of phase separation [8,9,31]. As the content of GO increased, both the data decreased gradually. As is known to us, the addition of nanofillers leads to a decrease of toughness [16,25,41]. The two values for the UPR/TO/G0.10 composite were 4.40 kJ/m^2 and 3.24%, which were 70.5% and 50.2% higher than those of pure UPR, respectively. These results suggested that the toughness and flexibility of UPR/TO/G0.10 were obviously better than those of neat UPR.

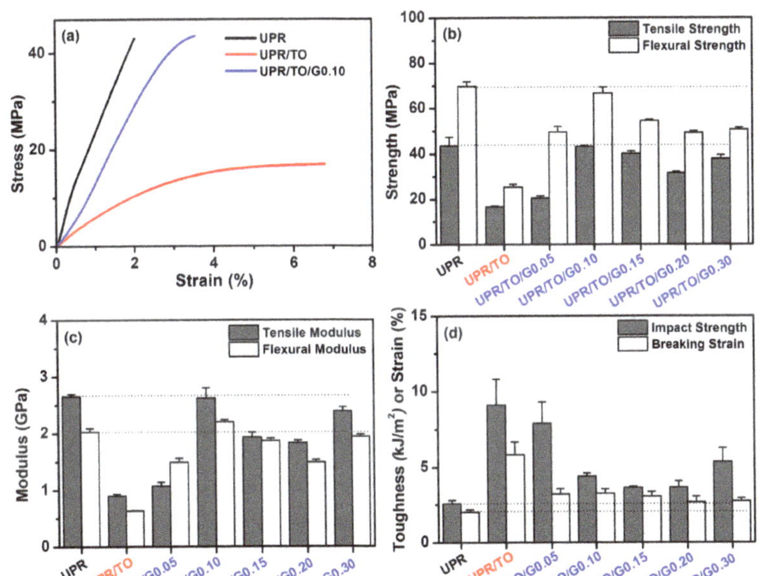

Figure 6. (**a**) Stress–strain curves of neat UPR, UPR/TO, and UPR/TO/G0.10; (**b**) tensile and flexural strengths, (**c**) tensile and flexural moduli, and (**d**) impact strength and tensile breaking strain of UPR, UPR/TO, and UPR/TO/GO composites.

DMA was performed to study the thermo-mechanical properties of neat UPR, UPR/TO, and UPR/TO/GO composites, as shown in Figure 7. The related results are presented in Table 2.

The storage modulus at 25 °C (E'_{25}), which can also reflect the stiffness, was 1.67 GPa for neat UPR. By the incorporation of 10% TO, the data decreased to 0.83 GPa due to the phase separation. As the increase of GO content, the value also experienced a firstly-increasing and subsequently-decreasing process, and showed a rebound at 0.30% of GO. The optimal value occurred at 0.10% of GO and was 1.76 GPa, which was 5.39% and 112% higher than those of UPR and UPR/TO, respectively. Glass transition temperature (T_g) can be determined from the peaks of loss factors (tan δ). The variation of T_g demonstrated an analogous trend as E'_{25}. The T_g of UPR/TO/G0.10 was 105.2 °C, which was only 4.88% lower than neat UPR's (110.6°C) but 49.4% higher than that of UPR/TO (70.4 °C). The improvements of E'_{25} and T_g for UPR/TO/G0.10 are ascribed to the good exfoliation of FGS and the strong FGS–polymer interaction too.

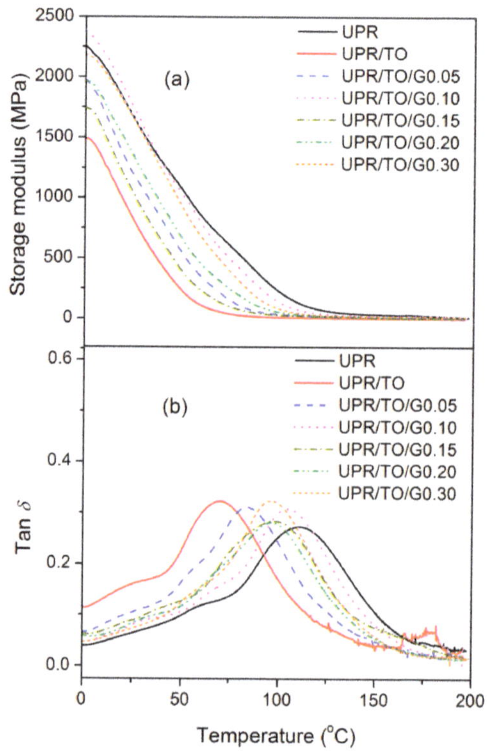

Figure 7. (**a**) Storage modulus and (**b**) Loss factor of neat UPR, UPR/TO, and UPR/TO/GO composites.

Table 2. Results of DMA and TGA for neat UPR, UPR/TO, and UPR/TO/GO composites.

Sample	E'_{25} [a] (GPa)	T_g [b] (°C)	T_5 [c] (°C)	T_p [d] (°C)	w_{char} [e] (%)
UPR	1.67	110.6	212.2	388.1	5.70
UPR/TO	0.83	70.4	229.6	385.5	8.61
UPR/TO/G0.05	1.21	84.3	241.7	391.3	8.08
UPR/TO/G0.10	1.76	105.2	270.6	392.9	8.37
UPR/TO/G0.15	1.04	98.4	259.4	392.9	8.76
UPR/TO/G0.20	1.36	96.4	246.1	392.9	9.16
UPR/TO/G0.30	1.67	97.1	229.1	397.8	8.55

[a] Storage modulus at 25 °C. [b] Glass transition temperature. [c] 5% Weight–loss temperature. [d] Peak temperature at the curves of weight-loss rate. [e] Char yield.

TGA was carried out to investigate thermal stability of neat UPR, UPR/TO, and UPR/TO/GO composites, as indicated in Figure 8. The thermal-property data, including 5% weight-loss temperature (T_5), peak temperature at the curve of weight–loss rate (T_p), and char yield (w_{char}) are summarized in Table 2. All the properties possessed similar trends of variation as E'_{25} or T_g when the GO content increased. For the neat UPR they were 212.2 °C, 388.1 °C, and 5.70%, respectively. And for the UPR/TO/G0.10 composite they were 270.6 °C, 392.9°C, and 8.37%, which were 27.5%, 1.24%, and 46.8% larger than those of neat UPR, respectively. Therefore, the thermal stability of UPR/TO/G0.10 composite was apparently superior to that of neat UPR.

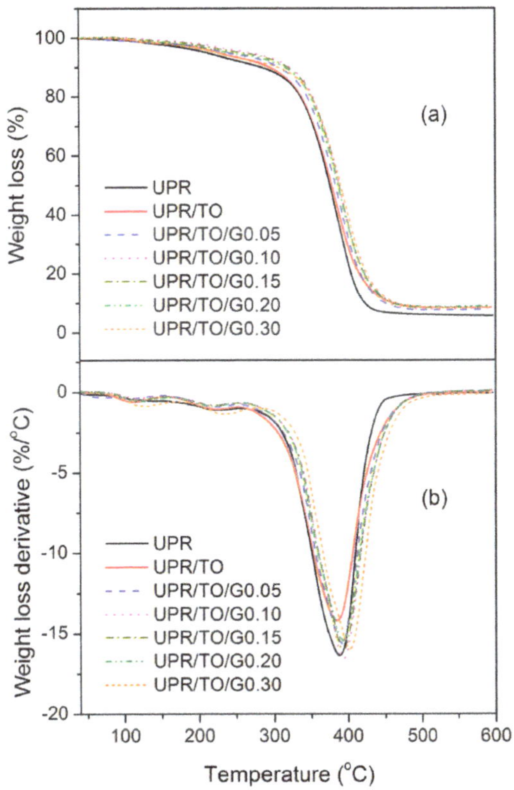

Figure 8. (**a**) TGA curves and (**b**) their derivatives of neat UPR, UPR/TO, and UPR/TO/GO composites.

4. Conclusions

We sucessfully prepared the novel graphene-reinforced biobased UPR nanocomposites via in situ melt polycondensation intergrated with D–A addition. The fabricating method can strengthen the interfical interactions between the graphene, polymer, and biomodifier and does not involve any solvent, which is efficient and ecofriendly. Notably, the stiffness of TO-based UPR was greatly enhanced by only incoporating 0.10% of GO, and the comprehensive properties of the resulting biobased UPR nanocomposite were even superior to the petroleum-based UPR. The excellent performance can be attributed to the well exfoliation of FGS and the strong interactions between FGS and polymeric matrix. Consequently, the developed UPR/TO/GO composites show great potential in the UPR industry and the fabricating method can be employed for other biobased polyester nanocomposites.

Author Contributions: Conceptualization & Writing-Original Draft Preparation, C.L.; Investigation, C.W.; Data Curation & Formal Analysis, J.T.; Data Curation & Formal Analysis, J.Z.; Data Curation, Q.S.; Data Curation,

Y.H.; Data Curation, H.W.; Data Curation, Q.W.; Writing-Review & Editing, Y.Z.; Writing-Review & Editing, W.L.; Writing-Review & Editing, Z.L.

Funding: This research was funded by the Fundamental Research Funds of Jiangsu Province Biomass Energy and Materials Laboratory [JSBEM-S-201501], the Fundamental Research Funds of CAF [CAFYBB2017QB006], and the Natural Science Foundation of Jiangsu Province [BK20161122].

Acknowledgments: This work was supported by the Fundamental Research Funds of Jiangsu Province Biomass Energy and Materials Laboratory (JSBEM-S-201501), the Fundamental Research Funds of CAF (CAFYBB2017QB006), and the Natural Science Foundation of Jiangsu Province (BK20161122).

Conflicts of Interest: The authors declare no conflict of interest.

References

1. Mighani, H. Synthesis of Thermally Stable Polyesters. In *Polyester*; Saleh, H.E.-D.M., Ed.; Intech Open: Rijeka, Croatia, 2012; pp. 3–17.
2. Penczek, P.; Czub, P.; Pielichowski, J. Unsaturated polyester resins: Chemistry and technology. In *Crosslinking in Materials Science*; Ameduri, B., Ed.; Springer: Berlin, Germany, 2005; pp. 1–95.
3. Raquez, J.M.; Deléglise, M.; Lacrampe, M.F.; Krawczak, P. Thermosetting (bio)materials derived from renewable resources: A critical review. *Prog. Polym. Sci.* **2010**, *35*, 487–509. [CrossRef]
4. Fertier, L.; Koleilat, H.; Stemmelen, M.; Giani, O.; Joly-Duhamel, C.; Lapinte, V.; Robin, J.-J. The use of renewable feedstock in UV-curable materials—A new age for polymers and green chemistry. *Prog. Polym. Sci.* **2013**, *38*, 932–962. [CrossRef]
5. Zhang, C.Q.; Garrison, T.F.; Madbouly, S.A.; Kessler, M.R. Recent advances in vegetable oil-based polymers and their composites. *Prog. Polym. Sci.* **2017**, *71*, 91–143. [CrossRef]
6. Meier, M.A.R.; Metzger, J.O.; Schubert, U.S. Plant oil renewable resources as green alternatives in polymer science. *Chem. Soc. Rev.* **2007**, *36*, 1788–1802. [CrossRef] [PubMed]
7. Biermann, U.; Bornscheuer, U.; Meier, M.A.R.; Metzger, J.O.; Schäfer, H.J. Oils and fats as renewable raw materials in chemistry. *Angew. Chem. Inter. Edit.* **2011**, *50*, 3854–3871. [CrossRef] [PubMed]
8. Mehta, G.; Mohanty, A.K.; Misra, M.; Drzal, L.T. Biobased resin as a toughening agent for biocomposites. *Green Chem.* **2004**, *6*, 254–258. [CrossRef]
9. Miyagawa, H.; Mohanty, A.K.; Burgueno, R.; Drzal, L.T.; Misra, M. Development of biobased unsaturated polyester containing functionalized linseed oil. *Ind. Eng. Chem. Res.* **2006**, *45*, 1014–1018. [CrossRef]
10. Ghorui, S.; Bandyopadhyay, N.R.; Ray, D.; Sengupta, S.; Kar, T. Use of maleated castor oil as biomodifier in unsaturated polyester resin/fly ash composites. *Ind. Crop. Prod.* **2011**, *34*, 893–899. [CrossRef]
11. Haq, M.; Burgueno, R.; Mohanty, A.; Misra, M. Hybrid bio-based composites from blends of unsaturated polyester and soybean oil reinforced with nanoclay and natural fibers. *Compos. Sci. Technol.* **2008**, *68*, 3344–3351. [CrossRef]
12. Haq, M.; Burgueno, R.; Mohanty, A.K.; Misra, M. Bio-based unsaturated polyester/layered silicate nanocomposites: Characterization and thermo-physical properties. *Compos. Part A* **2009**, *40*, 540–547. [CrossRef]
13. Haq, M.; Burgueno, R.; Mohanty, A.K.; Misra, M. Bio-based polymer nanocomposites from UPE/EML blends and nanoclay: Development, experimental characterization and limits to synergistic performance. *Compos. Part A* **2011**, *42*, 41–49. [CrossRef]
14. Liu, C.G.; Li, J.; Lei, W.; Zhou, Y.H. Development of biobased unsaturated polyester resin containinghighly functionalized castor oil. *Ind. Crop. Prod.* **2014**, *52*, 329–337. [CrossRef]
15. Lu, J.; Hong, C.K.; Wool, R.P. Bio-based nanocomposites from functionalized plant oils and layered silicate. *J. Polym. Sci. Polym. Phys.* **2004**, *42*, 1441–1450. [CrossRef]
16. Lu, J.; Wool, R.P. Additive toughening effects on new bio-based thermosetting resins from plant oils. *Compos. Sci. Technol.* **2008**, *68*, 1025–1033. [CrossRef]
17. Novoselov, K.S.; Geim, A.K.; Morozov, S.V.; Jiang, D.; Zhang, Y.; Dubonos, S.V.; Grigorieva, I.V.; Firsov, A.A. Electric field effect in atomically thin carbon films. *Science* **2004**, *306*, 666–669. [CrossRef] [PubMed]
18. Stankovich, S.; Dikin, D.A.; Dommett, G.H.B.; Kohlhaas, K.M.; Zimney, E.J.; Stach, E.A.; Piner, R.D.; Nguyen, S.T.; Ruoff, R.S. Graphene-based composite materials. *Nature* **2006**, *442*, 282–286. [CrossRef] [PubMed]

19. Huang, X.; Qi, X.; Boey, F.; Zhang, H. Graphene-based composites. *Chem. Soc. Rev.* **2012**, *41*, 666–686. [CrossRef] [PubMed]
20. Cui, Y.B.; Kundalwal, S.I.; Kumar, S. Gas barrier performance of graphene/polymer nanocomposites. *Carbon* **2016**, *98*, 313–333. [CrossRef]
21. Li, X.G.; McKenna, G.B. Ultrathin Polymer films: Rubbery stiffening, fragility, and T_g reduction. *Macromolecules* **2015**, *48*, 6329–6336. [CrossRef]
22. Kim, H.; Abdala, A.A.; Macosko, C.W. Graphene/polymer nanocomposites. *Macromolecules* **2010**, *43*, 6515–6530. [CrossRef]
23. Hu, H.T.; Wang, X.B.; Wang, J.C.; Wan, L.; Liu, F.M.; Zheng, H.; Chen, R.; Xu, C. Preparation and properties of graphene nanosheets-polystyrene nanocomposites via in situ emulsion polymerization. *Chem. Phys. Lett.* **2010**, *484*, 247–253. [CrossRef]
24. Patole, A.S.; Patole, S.P.; Kang, H.; Yoo, J.B.; Kim, T.H.; Ahn, J.H. A facile approach to the fabrication of graphene/polystyrene nanocomposite by in situ microemulsion polymerization. *J. Colloid. Interface Sci.* **2010**, *350*, 530–537. [CrossRef] [PubMed]
25. Rafiee, M.A.; Rafiee, J.; Srivastava, I.; Wang, Z.; Song, H.H.; Yu, Z.Z.; Koratkar, N. Fracture and fatigue in graphene nanocomposites. *Small* **2010**, *6*, 179–183. [CrossRef] [PubMed]
26. Bao, C.L.; Guo, Y.Q.; Song, L.; Kan, Y.C.; Qian, X.D.; Hu, Y. In situ preparation of functionalized graphene oxide/epoxy nanocomposites with effective reinforcements. *J. Mater. Chem.* **2011**, *21*, 13290–13298. [CrossRef]
27. Yavari, F.; Rafiee, M.A.; Rafiee, J.; Yu, Z.Z.; Koratkar, N. Dramatic increase in fatigue life in hierarchical graphene composites. *ACS Appl. Mater. Inter.* **2010**, *2*, 2738–2743. [CrossRef] [PubMed]
28. Liu, C.; Wang, Z.; Huang, Y.; Xie, H.; Liu, Z.; Chen, Y.; Lei, W.; Hu, L.; Zhou, Y.; Cheng, R. One-pot preparation of unsaturated polyester nanocomposites containing functionalized graphene sheets via a novel solvent-exchange method. *RSC Adv.* **2013**, *3*, 22380–22388. [CrossRef]
29. Bora, C.; Bharali, P.; Baglari, S.; Dolui, S.K.; Konwar, B.K. Strong and conductive reduced graphene oxide/polyester resin composite films with improved mechanical strength, thermal stability and its antibacterial activity. *Compos. Sci. Technol.* **2013**, *87*, 1–7. [CrossRef]
30. He, S.Y.; Petkovich, N.D.; Liu, K.W.; Qian, Y.Q.; Macosko, C.W.; Stein, A. Unsaturated polyester resin toughening with very low loadings of GO derivatives. *Polymer* **2017**, *110*, 149–157. [CrossRef]
31. Liu, C.G.; Lei, W.; Cai, Z.C.; Chen, J.Q.; Hu, L.H.; Dai, Y.; Zhou, Y. Use of tung oil as a reactive toughening agent in dicyclopentadiene-terminated unsaturated polyester resins. *Ind. Crop. Prod.* **2013**, *49*, 412–418. [CrossRef]
32. Xu, Z.; Gao, C. In situ polymerization approach to graphene-reinforced nylon-6 composites. *Macromolecules* **2010**, *43*, 6716–6723. [CrossRef]
33. Liu, K.; Chen, L.; Chen, Y.; Wu, J.L.; Zhang, W.Y.; Chen, F.; Fu, Q. Preparation of polyester/reduced graphene oxide composites via in situ melt polycondensation and simultaneous thermo-reduction of graphene oxide. *J. Mater. Chem.* **2011**, *21*, 8612–8617. [CrossRef]
34. Tang, Z.H.; Kang, H.L.; Shen, Z.L.; Guo, B.C.; Zhang, L.Q.; Jia, D.M. Grafting of polyester onto graphene for electrically and thermally conductive composites. *Macromolecules* **2012**, *45*, 3444–3451. [CrossRef]
35. Wei, H.X.; Li, Y.Y.; Chen, J.P.; Zeng, Y.; Yang, G.Q.; Li, Y. Dispersion of reduced graphene oxide in multiple solvents with an imidazolium-modified hexa-peri-hexabenzocoronene. *Chem. Asian J.* **2012**, *7*, 2683–2689. [CrossRef] [PubMed]
36. Dreyer, D.R.; Park, S.; Bielawski, C.W.; Ruoff, R.S. The chemistry of graphene oxide. *Chem. Soc. Rev.* **2010**, *39*, 228–240. [CrossRef] [PubMed]
37. Sahoo, S.K.; Ray, B.C.; Mallik, A. Role of electrochemically in-house synthesized and functionalized graphene nanofillers in the structural performance of epoxy matrix composites. *Phys. Chem. Chem. Phys.* **2017**, *19*, 16219–16230. [CrossRef] [PubMed]
38. Ramanathan, T.; Abdala, A.A.; Stankovich, S.; Dikin, D.A.; Herrera-Alonso, M.; Piner, R.D.; Adamson, D.H.; Schniepp, H.C.; Chen, X.; Ruoff, R.S.; et al. Functionalized graphene sheets for polymer nanocomposites. *Nat. Nanotechnol.* **2008**, *3*, 327–331. [CrossRef] [PubMed]
39. Das, K.; Ray, D.; Banerjee, C.; Bandyopadhyay, N.R.; Mohanty, A.K.; Misra, M. Novel materials from unsaturated polyester resin/styrene/tung oil blends with high impact strengths and enhanced mechanical properties. *J. Appl. Polym. Sci.* **2011**, *119*, 2174–2182. [CrossRef]

40. Chen, D.; Zhu, H.; Liu, T. In situ thermal preparation of polyimide nanocomposite films containing functionalized graphene sheets. *ACS Appl. Mater. Inter.* **2010**, *2*, 3702–3708. [CrossRef] [PubMed]
41. Putz, K.W.; Compton, O.C.; Palmeri, M.J.; Nguyen, S.T.; Brinson, L.C. High-nanofiller-content graphene oxide-polymer nanocomposites via vacuum-assisted self-assembly. *Adv. Funct. Mater.* **2010**, *20*, 3322–3329. [CrossRef]

polymers

MDPI

Article

Study on Graphene/CNC-Coated Bamboo Pulp Fabric Preparation of Fabrics with Thermal Conductivity

Feng Yang [1,*], Cuiqin Lan [1], Haiming Zhang [1], Jian Guan [1], Fan Zhang [1], Benhua Fei [2,*] and Jilei Zhang [3]

[1] Fashion Accessory Art and Engineering College, Beijing Institute of Fashion Technology, Beijing 100029, China
[2] Key Laboratory of Bamboo and Rattan Science and Technology of the State Forestry Administration, Department of Bio-materials, International Centre for Bamboo and Rattan, Futong Dong Dajie, Chaoyang District, Beijing 100102, China
[3] Department of Sustainable Bioproducts, Mississippi State University, Mississippi State, MS 39762, USA
* Correspondence: yangfeng@bift.edu.cn (F.Y.); feibenhua@icbr.ac.cn (B.F.)

Received: 5 June 2019; Accepted: 9 July 2019; Published: 31 July 2019

Abstract: Functional fabrics have gained attention as an environmentally-friendly synthesis route. In the current study, novelty bamboo pulp fabrics with thermal conductivity properties were prepared by coating the fabric with graphene and cellulose nanocrystal (G/CNC) solutions. The influences of G and CNC concentrations on properties of fabrics were studied. The viscosities of the G/CNC solutions increased with an increase of G contents. G had an obvious thickening effect. Furthermore, compounded fabrics with different G and CNC contents (GCBPFs) were prepared and extensively characterized in terms of thermal and mechanical properties, and morphology. The ultimate thermal conductivity, bursting strength, and tensile strength of the GCBPF were 0.136 W/m·K, 1.514 MPa, and 25.8 MPa, with 4 wt.% CNC and 3 wt.% G contents, respectively. The results demonstrated that the as-fabricated GCBPFs with favorable thermal conductivity could be applied as a novel fast cooling textile for the clothing industry.

Keywords: fabric; graphene; cellulose nanocrystal; thermal conductivity

---

1. Introduction

Energy consumption was one of the major issues for humans in the 21st century [1,2]. To regulate indoor temperature, energy consumption for space heating and cooling in residential and commercial buildings were dominant, which contributed to 12.3% of total U.S. energy consumption [3]. The concept of "personal thermal management" has emerged as a promising alternative for reducing the demand for indoor temperature regulation. A personal cooling technique could provide thermal comfort by locally cooling the temperature of an individual in a low-cost and energy-saving way [4].

First efforts for personal thermal management techniques to regulate the thermal conductivity of fabrics focused on thickness [5], aerial density [6], porosity [7], and weave structure of the fabric matrix [8]. The combination of personal thermal management with wearable devices was subsequently regarded as one of the most promising strategies, with technologies including cold pack textiles with phase change materials [9], air-adjusted textiles [10], and liquid-adjusted textiles [11]. All the above technologies have their limitations, such as unremarkable performance improvement, inconvenience from the bulky size of the devices, massive consumption of energy, complex craft-work, and high-cost. To address these issues, directly incorporating thermal management materials into textiles for effective personal thermal management has received tremendous attention recently. Hsu et al. [3] reported a good IR-transparent textile made from nanoporous polyethylene, with interconnected pores from 50 to 1000 nm in diameter, for efficient human body cooling. The same group further developed a wearable face mask based on their nanoporous system. Yang et al. [12] demonstrated a scalable randomized glass-polymer hybrid metamaterial for daytime individual thermal regulation.

Bamboo has emerged as the ultimate green, cost-efficient textile material that meets the definition of a renewable and sustainable raw material, with little harmful impact to the environment due to its fast growing speed (3–4 years renewal rate) and abundant quantity in China [13–15]. Compared to traditional fibers, the closed-loop system production process from bamboo pulp can be eco-friendly without any harmful substances going into the eco-system [16,17]. Therefore, bamboo fiber as an alternative to traditional fibers has been employed in the production of apparel, sanitary and hygiene products, geo-textiles, composites, and filtration fabrics [18,19].

For fibrous textile materials, conduction, convection, and radiation are the three main routes to processing heat transfer, among which the most significant way is conduction [20]. Therefore, heat conductive textiles for individual thermal regulation can be attractive [21–23]. The conductive textiles for personal cooling lead to a satisfactory body feeling due to a decrease in body-generated heat escaping easily into the external environment. Based on the conductive textiles, the concept of direct coating conductive materials onto the conventional textile has been seen as one of the more effective and simple options for cooling textiles, which can remain flexible and wearable [24]. Further, dip-coating methods for cost-effective thermal conductive textiles are attractive because of their similar dying processes to large scale applications in the textile industry [25]. Concerning these issues, various conductive materials, such as iron [26], steel, copper [27] and carbon materials [28] have been utilized to coat textile fibers for making personal cooling conductive textiles [29]. Among these coating materials, graphene (G) has emerged as a revolutionary functional material due to its amazing high thermal conductivity properties (above 3000 $W{\cdot}m{\cdot}K^{-1}$) [30]. Coated graphene materials include mainly G, oxidation graphene (GO) and oxidation-reduction graphene (RGO). When graphene is employed in aqueous solutions for coating fabrics, it tends to precipitate due to its no charged surface compared to GO and RGO. For aiding the deposition, dispersants should be used to help G dispersing in stable solutions. Chitosan [31], polyether-imide (PEI) [32], sodium dodecyl sulfate (SDS) [33], sodium cholate surfactant [34], and cellulose fibers [35] have been employed as a dispersant in an attempt to bound G with treated fabrics or textile fibers.

In this study, G material was dispersed in a cellulose nanocrystalline (CNC) aqueous solution to make a G/CNC solution. The solution was coated onto bamboo pulp fabrics to the compound thermal conductive textile by a dip-coating method. The thermal and mechanical properties of the G/CNC-coated bamboo pulp fabrics (GCBPFs) were evaluated by the thermal conductivity, tensile strength, bursting strength, viscosity, and crystallinity.

2. Experimental Section

2.1. Materials

Plain weave fabric made from 100% bamboo viscose (14.8 gsm, 0.5 mm of thickness) was purchased from Jilin Chemical Fiber Co. Ltd. (Jilin, China). The bamboo fabric was bleached using a standard process to remove inherent impurities and additives. All treated fabrics were subjected to the same scoring and beaching process before coating.

The multilayer graphene-based carbon nano-materials were provided by the Department of Sustainable Bio-products, Mississippi State University (Starkville, Mississippi, US). The graphene nanoplates consisted of several sheets (with a range of 2–30 layers) of graphene with an overall thickness of approximately 1–10 nm [36]. The morphology of graphene material is shown in Figure 1. Microcrystalline cellulose (column chromatography 97%) was obtained from Shanghai Jinsui Biological Inc. (Shanghai, China), and sulfuric acid (analytical reagent 98%) from Nanjing Chemical Reagents Co. (Nanjing, China). The deionized water was used exclusively.

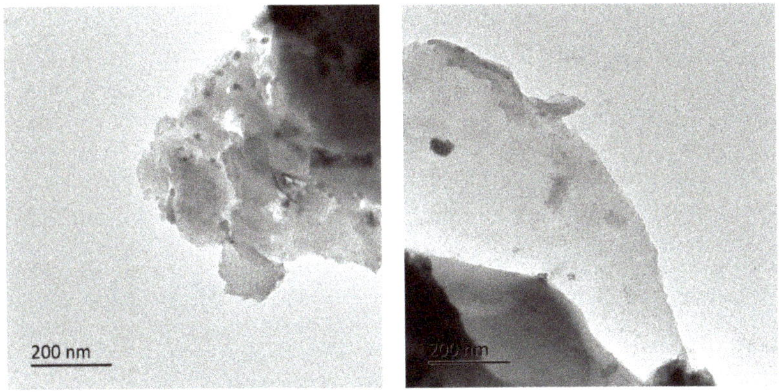

Figure 1. TEM images of graphene materials.

2.2. *Preparation of Coating Solution*

The cellulose nano-crystalline (CNC) suspension was prepared through the sulfuric acid hydrolysis of microcrystalline celluloses [37]. The obtained CNC suspension had the typical acicular structure, with dimensions ranging from 100 to 200 nm in length and 5 to 10 nm in diameter, as shown in Figure 2. There were three concentration levels of CNC suspensions (1, 2, and 4 wt.%) used to disperse G for evaluating the effect of CNC concentrations on the degree of G dispersion. The G and CNC mixing procedure started by adding G materials with 1, 2, and 3 wt.% into CNC suspensions, followed by ultrasonic treatment for 2 min, respectively. Then the G/CNC solutions were obtained with different G and CNC ratios. The additions of G were the ratios to the weight of CNC suspension.

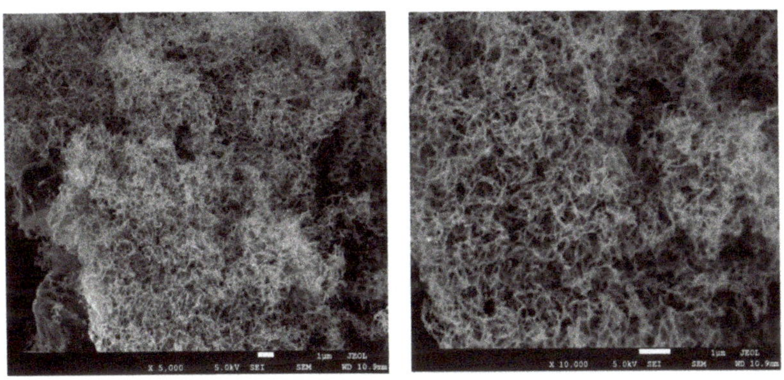

Figure 2. SEM images of cellulose nanocrystalline (CNC).

2.3. *Coating Procedure*

The prepared G/CNC solutions were applied to impregnate the bamboo pulp fabrics simply by a dip-coating method. The fabrics were first dipped into the G/CNC solutions and then padded through a rolling device to remove the excessive solutions. The above procedure was repeated for several times until the resistance of the fabric became constant. The obtained G/CNC coated bamboo pulp fabrics (GCBPFs) were dried at 45 °C for 2 h before being tested. For comparison, CNC suspensions coated fabric samples (CBPFs) were also prepared.

2.4. Characterizations

For measurement of the viscosity of dipping solution before and after the addition of G, an NDJ-79 rotational viscometer was used with a rotational speed of 750 r·min^{-1}. The surface morphologies of G, CNC, and fabrics were tested by Quanta 200 scanning electron microscope (SEM, FEI company, Hillsboro, OR, US) and transmission electron microscope (TEM, FEI company, Hillsboro, OR, US), respectively.

The X-ray diffraction (XRD) was performed using Cu kα radiation source (step size 0.02°, scanning speed 6°/min, voltage 40 kV, current 100 mA) in an Ultima IV XRD instrument (Rigaku company, Tokyo, Japan). The tensile strength of fabrics was determined according to ASTM D2256 by an HY-932CS micro-computer control electron universal testing machine, with a tensile rate of 1 mm/min. The bursting strength of samples was evaluated according to ASTM D3787-07 using a bursting strength tester (HY-953, Hengyu company, Dongguan, China). Five duplicates were repeated in each mechanical test group.

A heat conduction coefficient tester (ISOMET 2104, Judeng company, Beijing, China) was employed to record the heat conductivity of samples. The test standard was according to GB/T 11048-2018. The thermal conductivity (λ) was calculated by using the following law:

$$\lambda = mc\frac{\Delta T}{\Delta t}\Big|_{T=T_2} \bullet \frac{h}{T_1 - T_2} \bullet \frac{1}{\pi R^2} \quad (1)$$

where λ is the thermal coefficient (W/m·k), m is the quality of copper cooling plate (kg), c is the specific heat capacity of copper cooling plate, 385 J/(kg·K), $\frac{\Delta T}{\Delta t}\big|_{T=T_2}$ is the heat dissipation rate of copper cooling plate at T_2 (mV/s), h is the thickness of tested sample (m), $T_1 - T_2$ is the temperature difference between upper and lower sample surfaces (K), and πR^2 is the area of copper cooling plate (m^2).

3. Results and Discussion

3.1. Viscosity of G/CNC Solution

The viscosity, a key feature for the fabric impregnation solution, depends on the solid content, pH value, and temperature of finishing liquid [38]. For the textile industry, it is difficult to impregnate finishing liquid into fiber bundles because of the extremely high viscosity of impregnation solutions or resins. Figure 3a shows the viscosities of G/CNC solutions with different ratios of G to CNC suspension. For the pure CNC suspensions, viscosity increased with increasing CNC contents in the suspensions. Specifically, the viscosity of 4 wt.% CNC suspension was 226.7 mPa s, which was much higher than that of the 1 wt.% CNC suspension (20.1 mPa s). A three-dimensional network structure was formed through hydrogen bonding between each of the other groups (hydroxyl, carbonyl or carboxyl) on CNC or the groups with water. The network structure was constantly improved and strengthened with an increasing amount of CNC [39]. Moreover, the viscosities of the G/CNC solutions increased with an increase of G contents. It was found that G had an obvious thickening effect. The viscosity of the G/CNC solution appeared to have a maximum value of 312.6 mPa s with the concentrations of 3 wt.% G and 4 wt.% CNC. There were many factors that affected the rheological properties of polymers, such as polymer chain structure, relative molecular weight, and solution concentration. The viscosity of the majority of polymers increased with an increase of relative molecular weight or solution concentration. Figure 3b shows the viscosity increase rate of the G/CNC solution, which was calculated as the viscosity difference between G/CNC solution and pure CNC suspension divided by the viscosity of pure CNC suspension. The G/CNC solution with 3 wt.% G and 1 wt.% CNC suspension had the highest viscosity increase rate of 108.5%, and the lowest value appeared at the solution with 1 wt.% G and 4 wt.% CNC suspension.

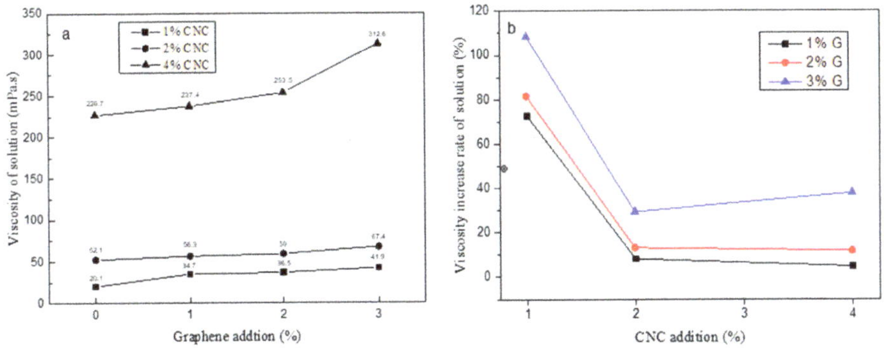

Figure 3. Viscosity (**a**) and viscosity increase rate (**b**) of the G/CNC solution.

3.2. XRD Analysis

The XRD test was carried out to investigate the degree of crystallinity (*Cr*) for the pure CNC suspension and G/CNC solutions. The *Cr* of pure CNC suspension was calculated at 83%, and three concentration levels of CNC suspensions (1, 2, and 4 wt.%) had the same *Cr* values as shown in Figure 4. Simultaneously, it was found that the *Cr* values of the G/CNC solutions were lower than that of pure CNC, and decreased with an increase of G contents. The lower the CNC suspension concentrations, the faster the *Cr* values descend in range of G/CNC solutions with the increase of G contents. The lowest *Cr* value of the G/CNC solution was 19% with 3 wt.% G and 1 wt.% CNC. This phenomenon may be the continuous crystallization of CNC in the G/CNC solution which was organized by graphene sheets [40], resulting in a significant decrease in the peak value of *Cr* in the G/CNC dispersion. In addition, the interaction between graphene and CNC may also lead to the disappearance of the periodic interval between graphene layers, thus greatly improving the dispersion of graphene in CNC suspension [41], and decreasing the *Cr* of G/CNC solution. From the TEM images (Figure 5a), it can be seen that graphene sheets were homogeneously coated well on the surface of CNC via a hydrogen bonding interaction and the rod-like morphology, and the size of CNC was retained well after the reaction [42]. Uniform and stable G/CNC solutions were formed, which was left at room temperature for 3 months without stratification or precipitation as shown in Figure 5b.

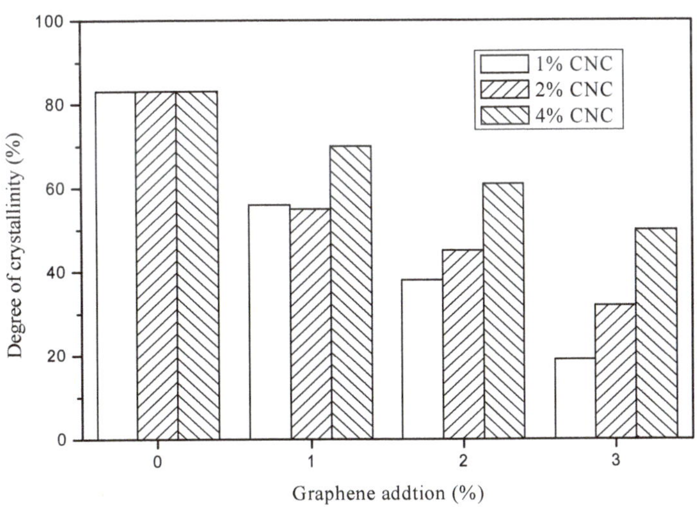

Figure 4. Degree of crystallinity of G/CNC solutions.

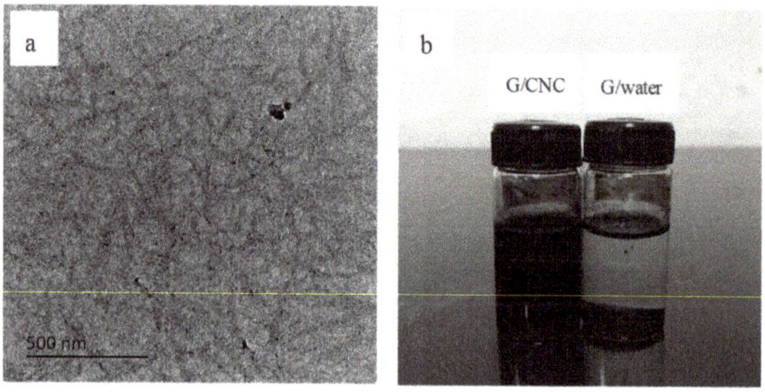

Figure 5. TEM image of G/CNC solution with 3 wt.% G and 4 wt.% CNC (**a**) and photo of graphene/CNC suspension and water after 3 months at room temperature (**b**).

Figure 6 shows the XRD pattern of CNC and G/CNC-coated fabrics with the lowest and highest thermal conductivities. It can be observed in the curve of CBPF that the diffraction peaks appeared around $2\theta = 12.2°$ and $21.7°$, respectively, which is the typical cellulose II crystal structure. The diffraction peaks decreased to $2\theta = 21.2°$ and $21.5°$ in GCBPF-L and GCBPF-H, respectively. The GCBPFs had very similar characteristic peaks with CBPF. As shown in Figure 6, the addition of graphene resulted in the decrease of *Cr* for the G/CNC solutions, therefore the obtained composite fabrics (GCBPFs) also had the lower diffraction peak values than the CBPF one.

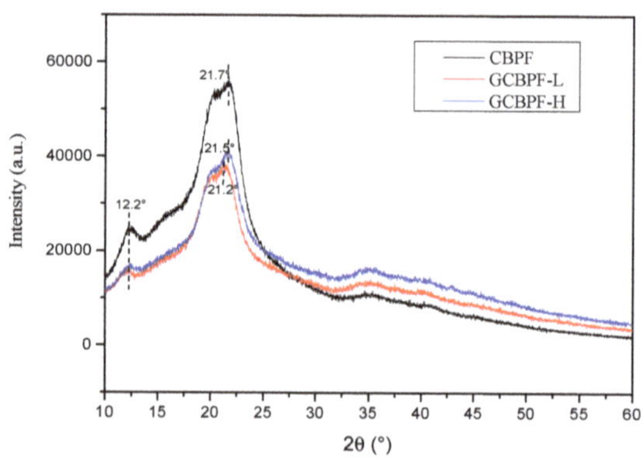

Figure 6. Degrees of crystallinity of CNC-coated bamboo pulp fabric (CBPF), and G/CNC-coated fabrics with the lowest (GCBPF-L) and highest conductivities (GCBPF-H).

3.3. Thermal Conductivity

The thermal conductivities of CBPFs ranged from 0.046 to 0.049 W/m·K tested by the thermal conductivity meter using a steady plate method as shown in Figure 7. The G/CNC-coated fabrics exhibited better conductive behavior than the pure CNC-coated samples. Due to the existence of the graphene in the G/CNC solutions, the solution coated bamboo pulp fabric turned into thermal conductive functional fabrics (GCBPFs). During the test, the thermal conductivity of all GCBPF samples had been improved, indicating that the addition of graphene can further improve the thermal conductivity of fabric to a certain extent. The maximum thermal conductivity of the GCBPF was 0.136 W/m·K, which was 178–196% higher than the CBPF sample, and as high as some previously prepared graphene coated conductive viscose fabrics [18,19]. After dispersing by CNC suspension uniformly, the lamellar structure of G can be maintained by van der Waals or restoration π-π stacking between adjacent G layers.

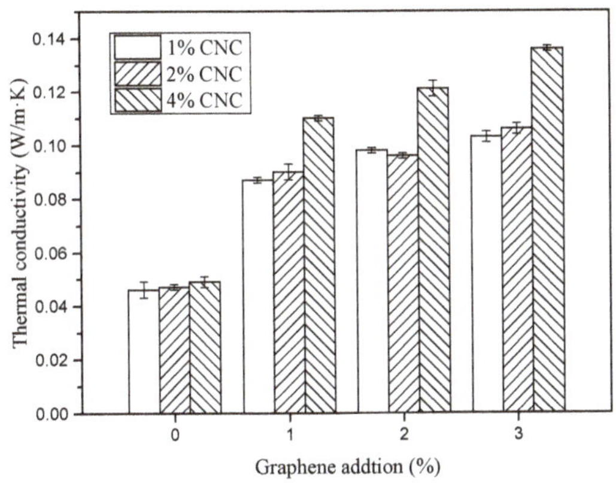

Figure 7. Thermal conductivities of CBPFs and GCBPFs.

The GCBPF sample with a higher CNC concentration of 4 wt.% had higher thermal conductivity with different graphene contents from 1 wt.% to 3 wt.%. Lower thermal conductivity for low contents of CNC might be due to situations such as the coating solution (to some extent) blocking the interstices and pores, and water droplets spreading in/on the fabric [43]. Furthermore, energy came from the effective dissipation of heat through the different coated layer referring irradiative, and it was difficult to be measured. To clear the thermal conductive property systematically, the investigation of air-permeability, wettability, and color appearance of the graphene-coated fabric had been arranged for further study.

3.4. Tensile and Bursting Properties

The results of tensile and elongation at break were plotted in Figures 8 and 9, respectively. When impregnated with pure CNC suspension, the elongation at break decreased a little with an increase of CNC contents from 1 to 4 wt.%. The addition of graphene improved the elongation at break of G/CNC solution coated fabrics, and with the increase of graphene content the elongation at break increased. The maximum elongation at break for GCBPF appears in 3 wt.% G and 2 wt.% CNC, with the value of 12.5%. As shown in Figure 8, the elongation at break increased with an increased content of graphene. The addition of graphene destroyed the crystallization of the CNC solution, and the flake graphene increased the flexibility of the G/CNC impregnating solution, which resulted in the decrease of *Cr* (Figure 4) and increase of elongation at break.

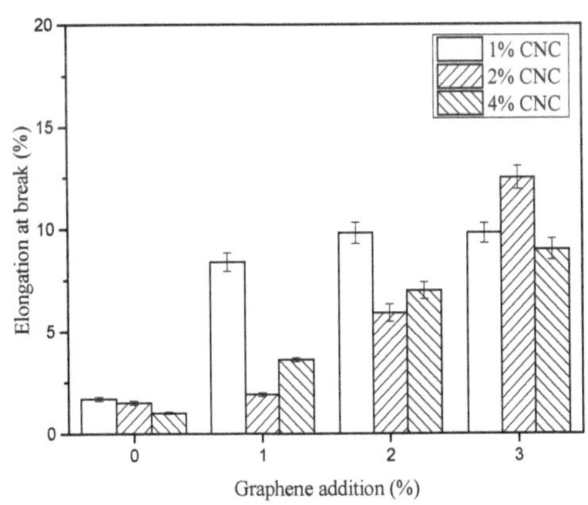

Figure 8. Elongation at break of CBPFs and GCBPFs.

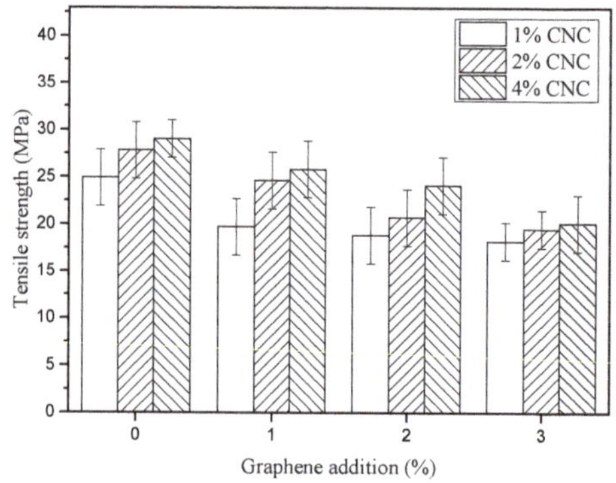

Figure 9. Tensile strength of CBPFs and GCBPFs.

A gradual increase in the tensile strength (TS) of CPBFs was observed (from 24.9 MPa to 29.0 MPa) in Figure 9 when the CNC concentration was increased from 1 wt.% to 4 wt.%. Moreover, the TS of GCBPFs decreased significantly ($p < 0.05$) with the addition of graphene from 1 wt.% to 3 wt.% compared with CPBF, with the values of TS in GCBPFs ranging from 18.2 MPa to 25.8 MPa. The larger decrease of TS in GCBPFs than CBPFs might be contribute to the addition of graphene destroying the complete self-assembly of CNC and the stronger electron scattering by graphene resulting in GCBPFs yielding brighter structures than CBPFs [44].

The bursting strength (BS) of CPBFs and GCPBFs is shown in Figure 10. The values of bursting strength in CPBFs increased with an increase of CNC contents, while the addition of graphene made the values decrease significantly ($p < 0.05$). The ultimate strength of the GCPBF was up to 1.514 MPa, which was slightly lower than the CPBF (1.696 MPa). This indicates that GCPBFs inherited the excellent mechanical properties of CPBFs, and the dip coating treatment did not change the inner structure of fibers; the complex of graphene and hydroxyl groups in cellulose makes cellulose molecules cross-linked [45]. Meanwhile, the GCPBFs also had excellent thermal conductivity mentioned by the former discussion in this paper.

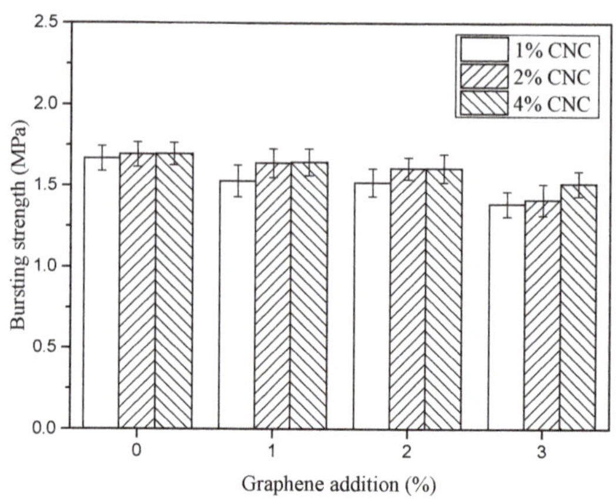

Figure 10. Bursting strength of CBPFs and GCBPFs.

3.5. Morphology of Coated Bamboo Pulp Fabrics

The structures of the fabrics and G/CNC-coated bamboo viscose fabrics (GCBPFs) were investigated by SEM as shown in Figure 11. It can be seen from Figure 11a,b that pristine bamboo viscose fabric

(BPF) had smooth surfaces, while the GCBPFs had much coarser surfaces as shown in Figure 11c–f, indicating the G/CNC solutions had successfully coated the bamboo viscose fiber surface.

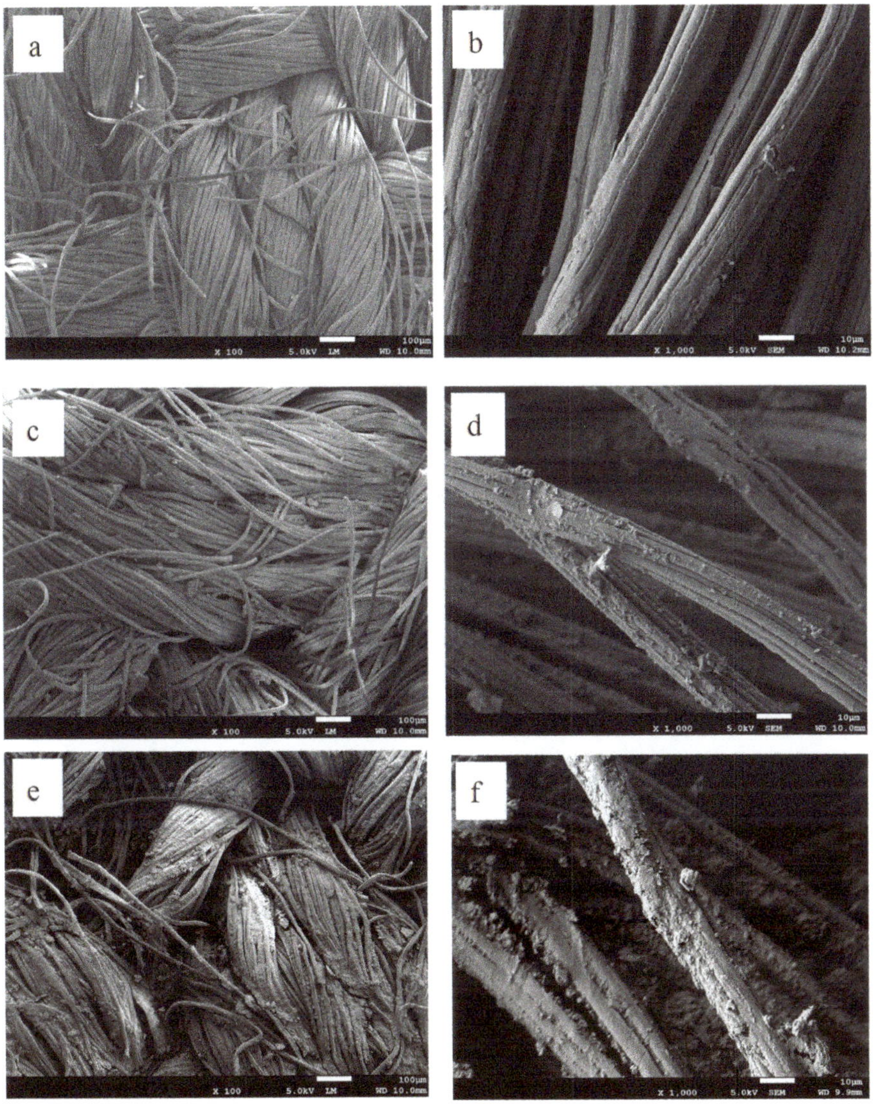

Figure 11. SEM images of CBPFs and GCBPFs ((**a**,**b**) CNC-coated bamboo pulp fabric; (**c**,**d**) G/CNC-coated fabric with lowest thermal conductivity; (**e**,**f**) G/CNC-coated fabric with highest thermal conductivity).

Figure 11c,d are images of GCBPFs with 1 wt.% CNC and 1 wt.% G, which had the lowest thermal conductivity. With increasing contents of CNC and G by 4 wt.% and 3 wt.%, respectively, the graphene can be seen to be uniformly distributed on the surface of the fabrics (Figure 11e,f) and interacting with the bamboo fibers through hydrogen interactions. The GCBPFs are filled more tightly by the G/CNC fillers, which also correspond to highest thermal conductivity [46].

4. Conclusions

In summary, the GCBPFs were synthesized by impregnating BPFs into the G/CNC solutions through a wet coating process in this study. The structure, and mechanical and thermal properties were studied subsequently. The graphene was found to uniformly coat bamboo pulp fiber surfaces due to the dispersion of CNC suspension, which enhanced the thermal and mechanical properties of the fabric at the same time. The GCBPF sample with a higher CNC concentration of 4 wt.% and G of 3 wt.% had excellent thermal conductivity and mechanical properties. The G/CNC coatings formed a firmly linked thermal conductive network, which made the originally insulating BPF thermally conductive. The results of this study indicate that the prepared GCBPFs have great application potentials for cooling textiles, conductive fabrics, wearable electronics, etc. Moreover, the durability of the G/CNC coating, i.e., the washing experiment, color fastness, and wear test, will be evaluated in order to validate the practical interest of this solution in our future work.

Author Contributions: F.Y., C.L., H.Z., J.G., and F.Z. contributed to synthesis, testing, data analysis, and wrote the manuscript. B.F. and J.Z. suggested and supervised the work and revised the manuscript.

Funding: The authors gratefully acknowledge the financial support of the project which was funded by the Special Scientific Research Fund of Construction of High-level teachers Project of Beijing Institute of Fashion Technology (BIFTQG201805), the Special Scientific Research Fund of Construction of High-level teachers Project of Beijing municipal university (IDHT20180511), the National Natural Science Foundation of China (31770599), and the specially invited Professor Support Program of Beijing Municipal University (IDHT20190217).

Acknowledgments: The authors gratefully acknowledgement the financial support from China and would like to thank Wu Yan for helping us prepare for the samples.

Conflicts of Interest: The authors declare no conflicts of interest.

References

1. Pérez, L.L.; Ortiz, J.; Pout, C. A review on buildings energy consumption information. *Energy Build.* **2008**, *40*, 394–398. [CrossRef]
2. Soytas, U.; Sari, R. Energy consumption and GDP: Causality relationship in G-7 countries and emerging markets. *Energy Econ.* **2003**, *25*, 33–37. [CrossRef]
3. Hsu, P.C.; Song, A.Y.; Catrysse, P.B.; Liu, C.; Peng, Y.; Xie, J.; Fan, S.; Cui, Y. Radiative human body cooling by nanoporous polyethylene textile. *Science* **2016**, *353*, 1019–1023. [CrossRef] [PubMed]
4. Gao, T.; Yang, Z.; Chen, C.; Li, Y.; Fu, K.; Dai, J.; Hitz, E.M.; Xie, H.; Liu, B.; Song, J.; et al. Three-dimensional printed thermal regulation textiles. *ACS Nano* **2017**, *11*, 11513–11520. [CrossRef] [PubMed]
5. Etheridge, M.A.; Wall, V.J.; Vernon, R.H. The role of the fluid phase during regional metamorphism and deformation. *J. Metamorph. Geol.* **2010**, *1*, 205–226. [CrossRef]
6. Huang, C.; Juan He Du, L.; Min, D.; Yong, Q. Structural characterization of the lignins from the green and yellow bamboo of bamboo culm (*Phyllostachys pubescens*). *J. Wood Chem. Technol.* **2016**, *36*, 157–172. [CrossRef]
7. Ming, L.; Wood, J.R.; Cathles, L.M. Prediction of thermal conductivity in reservoir rocks using fabric theory. *J. Appl. Geophys.* **1994**, *32*, 321–334.
8. Che, J.; Cagin, T.; Goddard, W.A.I. Thermal conductivity of carbon nanotubes. *Nanotechnology* **2000**, *11*, 2083–2085. [CrossRef]
9. Zalba, B.; Marín, J.M.; Cabeza, L.F.; Mehling, H. Review on thermal energy storage with phase change: materials, heat transfer analysis and applications. *Appl. Therm. Eng.* **2003**, *23*, 251–283. [CrossRef]
10. Majumdar, A.; Mal, P.; Ghosh, A.; Banerjee, D. Multi-objective optimization of air permeability and thermal conductivity of knitted fabrics with desired ultraviolet protection. *J. Text. Inst.* **2016**, *108*, 7. [CrossRef]
11. Rezaei, B.; Askari, M.; Shoushtari, A.M.; Malek, R.A.M. The effect of diameter on the thermal properties of the modeled shape-stabilized phase change nanofibers (PCNs). *J. Therm. Anal. Calorim.* **2014**, *118*, 1619–1629. [CrossRef]
12. Yang, A.; Cai, L.; Zhang, R.; Wang, J.; Hsu, P.C.; Wang, H.; Zhou, G.; Xu, J.; Cui, Y. Thermal Management in Nanofiber-Based Face Mask. *Nano Lett.* **2017**, *17*, 3506–3510. [CrossRef] [PubMed]

13. Huang, C.; Lin, W.; Lai, C.; Li, X.; Jin, Y.; Yong, Q. Coupling the post-extraction process to remove residual lignin and alter the recalcitrant structures for improving the enzymatic digestibility of acid-pretreated bamboo residues. *Bioresour. Technol.* **2019**, *258*, 121355. [CrossRef] [PubMed]
14. Huang, C.; He, J.; Wang, Y.; Min, D.; Yong, Q. Associating cooking additives with sodium hydroxide to pretreat bamboo residues for improving the enzymatic saccharification and monosaccharides production. *Bioresour. Technol.* **2015**, *193*, 142–149. [CrossRef] [PubMed]
15. Wang, Y.; Ge Cheng, H.; Tian, G.; Zheng, L.; Feng, X. Structures of bamboo fiber for textiles. *Text. Res. J.* **2010**, *80*, 334–343. [CrossRef]
16. Zhang, G.; Lin, H.; Morikawa, H.; Miura, M. In-situ growth of ZnO particles on bamboo pulp fabric and its anti-UV property. *Fibers Polym.* **2013**, *14*, 1803–1807. [CrossRef]
17. Qin, Z.Y.; Chen, Y.; Zhang, P.; Zhang, G.; Liu, Y. Structure and properties of Cu (II) complex bamboo pulp fabrics. *J. Appl. Polym. Sci.* **2010**, *117*, 1843–1850. [CrossRef]
18. Majumdar, A.; Mukhopadhyay, S.; Yadav, R. Thermal properties of knitted fabrics made from cotton and regenerated bamboo cellulosic fibres. *Int. J. Therm. Sci.* **2010**, *49*, 2042–2048. [CrossRef]
19. Xu, Y.; Lu, Z.; Tang, R. Structure and thermal properties of bamboo viscose, tencel and conventional viscose fiber. *J. Therm. Anal. Calorim.* **2007**, *89*, 197–201. [CrossRef]
20. Mondal, S. Phase change materials for smart textiles—An overview. *Appl. Therm. Eng.* **2008**, *28*, 1536–1550. [CrossRef]
21. Maity, S.; Chatterjee, A.; Singh, B.; Pal, S.A. Polypyrrole based electro-conductive textiles for heat generation. *J. Text. Inst. Proc. Abstr.* **2014**, *105*, 887–893. [CrossRef]
22. Bashir, T.; Skrifvars, M.; Persson, N.K. Production of highly conductive textile viscose yarns by chemical vapor deposition technique: A route to continuous process. *Polym. Adv. Technol.* **2011**, *22*, 2214–2221. [CrossRef]
23. Jang, J.; Park, H.C.; Lee, H.S.; Khil, M.S.; Kim, S.Y. Electrically and thermally conductive carbon fibre fabric reinforced polymer composites based on nanocarbons and an in-situ polymerizable cyclic oligoester. *Sci. Rep.* **2018**, *8*, 7659. [CrossRef] [PubMed]
24. Cai, G.; Xu, Z.; Yang, M.; Tang, B.; Wang, X. Functionalization of cotton fabrics through thermal reduction of graphene oxide. *Appl. Surf. Sci.* **2017**, *393*, 441–448. [CrossRef]
25. Zhang, J.; Li, B.C.; Wu, L.; Wang, A.Q. Facile preparation of durable and robust superhydrophobic textiles by dip coating in nanocomposite solution of organosilanes. *Chem. Commun.* **2013**, *49*, 11509–11511. [CrossRef] [PubMed]
26. Yue, P.; Wang, S.; Li, X.; Ge, M. Preparation of polyaniline/Ag composite conductive fabric via one-step oxidation-reduction reaction. *J. Text. Res.* **2014**, *60*, 33–42.
27. Yu, S.; Park, B.I.; Park, C.; Hong, S.M.; Han, T.H.; Koo, C.M. RTA-treated carbon fiber/copper core/shell hybrid for thermally conductive composites. *ACS Appl. Mater. Interfaces* **2014**, *6*, 7498–7503. [CrossRef] [PubMed]
28. Mengal, N.; Sahito, I.A.; Arbab, A.A.; Sun, K.C.; Qadir, M.B.; Memon, A.A. Fabrication of a flexible and conductive lyocell fabric decorated with graphene nanosheets as a stable electrode material. *Carbohydr. Polym.* **2016**, *152*, 19–25. [CrossRef] [PubMed]
29. Kongahge, D.; Foroughi, J.; Gambhir, S.; Spinks, G.M.; Wallace, G.G. Fabrication of a graphene coated nonwoven textile for industrial applications. *RSC Adv.* **2016**, *6*, 73203–73209. [CrossRef]
30. Gan, L.; Shang, S.; Yuen, C.W.M.; Jiang, S.X. Graphene nanoribbon coated flexible and conductive cotton fabric. *Compos. Sci. Technol.* **2015**, *117*, 208–214. [CrossRef]
31. Zhang, L.; Tang, P.; Zhang, W.; Xu, M.; Wang, Y. Effect of chitosan as a dispersant on collagen-hydroxyapatite composite matrices. *Tissue Eng. Part C Methods* **2010**, *16*, 71–79. [CrossRef] [PubMed]
32. Posudievsky, O.Y.; Khazieieva, O.A.; Cherepanov, V.V.; Koshechko, V.G.; Pokhodenko, V.D. High yield of graphene by dispersant-free liquid exfoliation of mechanochemically delaminated graphite. *J. Nanoparticle Res.* **2013**, *15*, 2046. [CrossRef]
33. Bin, W.U.; Yang, X. Molecular Simulation of Electrolyte-Induced Interfacial Interaction between SDS/Graphene Assemblies. *J. Phys. Chem. C* **2013**, *117*, 23216–23223.
34. Lotya, M.; King, P.J.; Khan, U.; De, S.; Coleman, J.N. High-concentration, surfactant-stabilized graphene dispersions. *ACS Nano* **2010**, *4*, 3155–3162. [CrossRef] [PubMed]

35. Wu, Y.; Wang, S.; Zhou, D.; Zhang, Y.; Wang, X.; Yang, R. Biodegradable polyvinyl alcohol nanocomposites made from rice straw fibrils: Mechanical and thermal properties. *J. Comp. Mater.* **2013**, *47*, 1449–1459. [CrossRef]
36. Yan, Q.; Zhang, X.; Li, J.; Hassan, E.B.; Wang, C.; Zhang, J.; Cai, Z. Catalytic conversion of Kraft lignin to bio-multilayer graphene materials under different atmospheres. *J. Mater. Sci.* **2018**, *53*, 8020. [CrossRef]
37. Huang, C.; Chu, Q.; Xie, Y.; Li, X.; Jin, Y.; Min, D.; Yong, Q. Effect of kraft pulping pretreatment on the chemical composition, enzymatic digestibility, and sugar release of moso bamboo residues. *Bioresources* **2015**, *10*, 240–255. [CrossRef]
38. Tondi, G.; Thevenon, M.F.; Mies, B.; Standfest, G.; Petutschnigg, A.; Wieland, S. Impregnation of Scots pine and beech with tannin solutions: Effect of viscosity and wood anatomy in wood infiltration. *Wood Sci. Technol.* **2013**, *47*, 615–626. [CrossRef]
39. Montes, S.; Carrasco, P.M.; Ruiz, V.; Germán, C.; Odriozola, I. Synergistic reinforcement of poly(vinyl alcohol) nanocomposites with cellulose nanocrystal-stabilized graphene. *Comp. Sci. Technol.* **2015**, *117*, 26–31. [CrossRef]
40. Carrasco, P.M.; Montes, S.; García, I.; Borghei, M.; Jiang, H.; Odriozola, I. High-concentration aqueous dispersions of graphene produced by exfoliation of graphite using cellulose nanocrystals. *Carbon* **2014**, *70*, 157–163. [CrossRef]
41. Wang, Y.; Zhang, H.; Lin, X.; Chen, S.; Jiang, Z.; Wang, J.; Huang, J.; Zhang, F.; Li, H. Naked Au nanoparticles monodispersed onto multifunctional cellulose nanocrystals-graphene hybrid sheets: Towards efficient and sustainable heterogeneous catalysts. *New J. Chem.* **2018**, *42*, 2197–2203. [CrossRef]
42. Tao, X. *Preparation and Thermal Conductivity of Modified Wood Impregnated by Graphene*; Nanjing Forestry University: Nanjing, China, 2019; p. 84.
43. Abbas, A.; Zhao, Y.; Zhou, J.; Wang, X.; Lin, T. Improving thermal conductivity of cotton fabrics using composite coatings containing graphene, multiwall carbon nanotube or boron nitride fine particles. *Fibers Polym.* **2013**, *14*, 1641–1649. [CrossRef]
44. Zhang, L.; Becton, M.; Wang, X. Mechanical analysis of graphene-based woven nano-fabric. *Mater. Sci. Eng. A* **2015**, *620*, 367–374. [CrossRef]
45. Fernandes, M.; Souto, A.P.; Esteves, M.F.; Bonastre, J.; Cases, F. Plasma treatment of polyester fabrics to increase the adhesion of reduced graphene oxide. *Synth. Metals* **2015**, *202*, 110–122.
46. Kim, H.; Lee, S. Characteristics of Electrical Heating Elements Coated with Graphene Nanocomposite on Polyester Fabric: Effect of Different Graphene Contents and Annealing Temperatures. *Fibers Polym.* **2018**, *19*, 965–976. [CrossRef]

polymers

MDPI

Article

Effects of Graphene Nanoplatelets and Cellular Structure on the Thermal Conductivity of Polysulfone Nanocomposite Foams

Hooman Abbasi, Marcelo Antunes and José Ignacio Velasco *

Department of Materials Science and Engineering, Technical University of Catalonia (UPC BarcelonaTech), ESEIAAT, C/Colom 11, E-08222 Terrassa, Spain; hooman.abbasi@upc.edu (H.A.); marcelo.antunes@upc.edu (M.A.)

* Correspondence: jose.ignacio.velasco@upc.edu; Tel.: +34-937398152

Received: 21 November 2019; Accepted: 18 December 2019; Published: 20 December 2019

Abstract: Polysulfone (PSU) foams containing 0–10 wt% graphene nanoplatelets (GnP) were prepared using two foaming methods. Alongside the analysis of the cellular structure, their thermal conductivity was measured and analyzed. The results showed that the presence of GnP can affect the cellular structure of the foams prepared by both water vapor induced phase separation (WVIPS) and supercritical CO_2 ($scCO_2$) dissolution; however, the impact is greater in the case of foams prepared by WVIPS. In terms of thermal conductivity, the analysis showed an increasing trend by incrementing the amount of GnP and increasing relative density, with the tortuosity of the cellular structure, dependent on the used foaming method, relative density, and amount of GnP, playing a key role in the final value of thermal conductivity. The combination of all these factors showed the possibility of preparing PSU-GnP foams with enhanced thermal conductivity at lower GnP amount by carefully controlling the cellular structure and relative density, opening up their use in lightweight heat dissipators.

Keywords: polysulfone foams; graphene; thermal conductivity; tortuosity; water vapor induced phase separation; $scCO_2$

1. Introduction

Polysulfone (PSU) is a high performance thermoplastic with high thermal and chemical stability, excellent strength and toughness, good environmental stress-crack resistance, and inherent fire resistance [1,2]. Additionally, it is resistant to gamma and e-beam radiation due to its high aromatic content [3]. PSU foaming methods such as carbon dioxide dissolution foaming [4–8], extrusion of microcellular polysulfone using chemical blowing agents [9] and PSU membrane formation [10–13] have been vastly studied. Nevertheless, the addition of thermally conductive nanofillers such as graphene is still incipient [14].

The high aspect ratio of carbon-based nanofillers allows the preparation of polymer-based nanocomposites with high performance and multifunctionality [15]. Their addition to polymers can provide thermal and electrical conductivity at low nanofiller content, overcoming one of the major technological barriers of polymers and enabling their use in applications such as heat sinks [16] and electronic packaging [17]. Additionally, these nanocomposites are suitable candidates as alternatives to conductive polymers that lack thermal stability and proper mechanical performance [18].

The thermal conductivity of nanocomposites is highly dependent on the filler and polymer type [19]. Other factors such as purity and dimension of the filler, polymer crystallinity and measuring methods, explain the scattered data reported for the thermal conductivity of nanocomposites [18,20]. Previous studies of such materials are focused mainly on various types of carbon-based fillers, owing to their intrinsically high thermal conductivity [21–23]. Among them, graphene nanoplatelets (GnP) have been some of the most considered nanofillers in recent studies for enhancing the thermal conductivity of polymer-based nanocomposites [24–26]. However, results suggest certain constraints regarding the improvement of thermal conductivity, as nanocomposites prepared with GnP have shown limited thermal conductivity even at high filler content [27].

Foaming of these nanocomposites has been the center of attention with the goal of creating novel foams with improved specific properties. In this sense, we have previously investigated thoroughly the preparation and properties of foams based on other high performance thermoplastics like polyetherimide (PEI) reinforced with GnP and carbon nanotubes [28–31]. The results presented in these works indicate that foaming provided further enhancement of the electrical conductivity by promoting a better dispersion of the nanofillers through the continuous phase of the nanocomposite foams [29].

Likewise, a great interest has appeared very recently regarding the possibility of tuning and enhancing the thermal conductivity of polymer-based foams and hence extend their applicability by generating a more effective thermal conduction network throughout the polymer cell walls by means of guaranteeing a proper distribution and dispersion of thermally conductive nanoparticles, such as GnP. Foaming could hence provide a viable strategy for developing tailored structures to enhance the heat dissipation efficiency of novel lightweight devices. In this sense, PSU nanocomposite foams containing variable concentrations of GnP (up until 10 wt%) were prepared using two foaming methods: water vapor induced phase separation (WVIPS) and $scCO_2$ dissolution. The resulting foams were characterized regarding their cellular structure and thermal conductivity.

The novelties of the present research are (1) the comparison between two very different foaming methods as is the common method of $scCO_2$ dissolution foaming with the less common method of water vapor induced phase separation, usually considered for preparing polymeric membranes [32], with obvious advantages of the second one such as the non-requirement of having to melt-compound the material or the possibility of controlling in an easy way the density and cellular structure of the resulting foams; and (2) the consideration for the first time of how the addition of variable amounts of GnP and the developed cellular structure affect the thermal conductivity values through a tortuosity factor that takes into account the complexity of the cellular structure. By considering this tortuosity factor and the relative density as fundamental parameters in the final thermal conductivity, our work shows the possibility of optimizing the thermal conductivity of PSU-GnP foams through the control of their density and cellular structure, ultimately related to the possibility of obtaining conductive foams at lower GnP amounts, for instance for lightweight heat dissipation components.

2. Materials and Methods

2.1. Materials

PSU pellets (UDEL P-1700) were purchased from Solvay (Brussels, Belgium) with a density of 1.24 g/cm^3 and a glass transition temperature of 185 °C. The GnP used in this study was acquired from XG Science Inc. (Lansing, MI, USA) with the commercial name of xGnP-Grade M15. These nanoplatelets have a reported thickness of 6–8 nm, an average platelet diameter of 15 µm, a surface area of 120–150 m^2/g, a density of 2.2 g/cm^3, and a thermal conductivity of 3000 and 6 W·m^{-1}·K^{-1} parallel and perpendicular to the surface, respectively. *N*-methyl pyrrolidone (NMP) used in this study was obtained from Panreac Química SA (Barcelona, Spain) with 99% purity and a boiling and flash points of 202 and 95 °C, respectively.

2.2. Foaming Methods

Three series of foams were prepared in this study, the first two using the WVIPS method. In this method, graphene nanoplatelets were initially dispersed in NMP for 30 minutes using a FB-705 ultrasonic processor (Fisher Scientific, Hampton, NH, USA) at maximum amplitude using a 12 mm solid tip probe and 20 kHz, applying a total amount of energy of 90 kJ at 95–130 W. The temperature of this NMP-GnP solution was maintained at 50 °C using an ice-bath. PSU was then dissolved in the NMP-GnP solution at 15 wt% PSU concentration for the first foam series (series 1) and 25 wt% PSU for the second one (series 2) at 50 °C and kept stirring at 450 rpm for 24 h. Foamed samples were respectively coded as "15 PSU x GnP" and "25 PSU x GnP", with x representing GnP's weight percentage. In the following step, the solutions were poured on a flat glass and kept at room temperature exposed to air with an average humidity of 75% for 6 days. Foams with variable GnP content (0, 1, 2, 5 and 10 wt%) were prepared for both series. The phase inversion between the solution and water led to the formation of the cellular structure. The resulting foams were then washed in water and ethanol in order to remove the residual solvent and afterwards dried in a vacuum dryer. The process is shown in Figure 1.

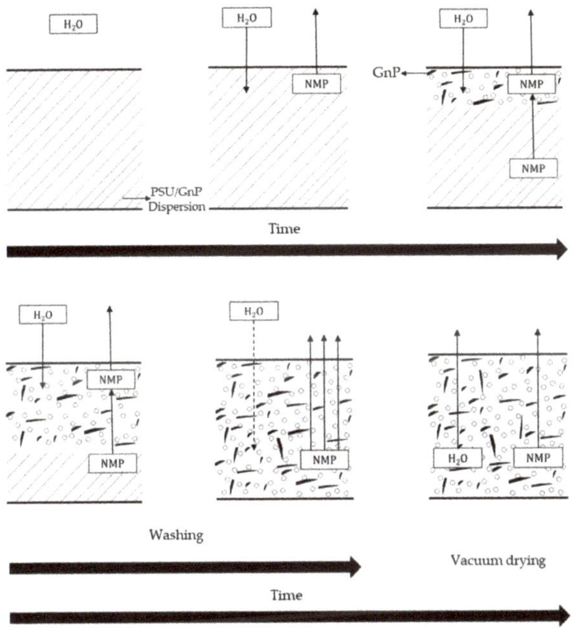

Figure 1. Scheme of the water vapor induced phase separation (WVIPS) process used to prepare PSU-GnP foams from series 1 (15 wt% PSU) and series 2 (25 wt% PSU).

A third series of PSU-GnP nanocomposite foams (series 3) were prepared by initially melt-compounding PSU pellets and GnP powder using a Brabender Plastic-Corder (Brabender GmbH and Co., Duisburg, Germany) and foamed using $scCO_2$ dissolution foaming inside a high pressure vessel. In order to prepare foams with variable GnP content (0, 0.4, 0.7, 1.0, 1.5 and 2 wt%), a masterbatch containing 50 wt% of ultrasonicated GnP in PSU was initially prepared in an NMP solution and then washed and dried. This masterbatch was then melt-mixed with PSU pellets in the Brabender in order to obtain the desired compositions. Subsequently, the nanocomposites were removed from the Brabender mixing chamber and molded into circular-shaped disks (foam precursors) using a hot-plate press (PL15, IQAP LAP, IQAP Masterbatch Group S.L., Barcelona, Spain) at 250 °C and 80 bar for 14 min. The resulting disks had a nominal thickness of 3 mm and a diameter of 74 mm.

Foaming took place in the high-pressure vessel (CH-8610 Uster/Schweiz, Büchiglasuster, Switzerland) by dissolving $scCO_2$ at 185 °C and 180–210 bar for 5 h, followed by a sudden depressurization at ~0.3 MPa/s and controlled cooling of the vessel using circulating water. Both steps of $scCO_2$ dissolution/heating and depressurization/cooling are shown in Figure 2. Foams from this series were coded as "PSU x GnP" (x representing the weight percentage of GnP).

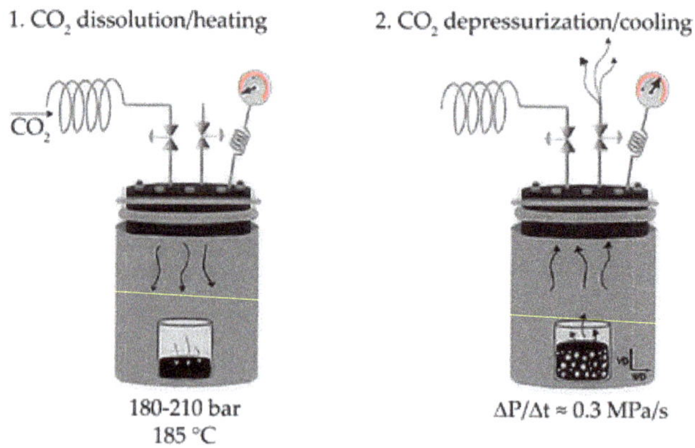

Figure 2. Scheme of the $scCO_2$ dissolution foaming process used to prepare PSU-GnP foams from series 3.

2.3. Testing Procedure

The foam's density values were measured using the ISO-845 standard procedure. The cellular structure of the foams was analyzed using a JEOL JSM-5610 (Tokyo, Japan) scanning electron microscope (SEM) applying a voltage of 10 kV and a working distance of 40 mm. Samples were brittle-fractured using liquid nitrogen and later coated with a thin layer of gold by sputter deposition using a BAL-TEC SCD005 (Los Angeles, CA, USA) sputter coater under an argon atmosphere. The values of the average cell size (Φ) were measured using the intercept counting method, explained in detail in [33]. Five ×300 magnification SEM micrographs were used for each foam. Cell nucleation density (N_0) and cell density (N_f), which respectively represent the number of cells per volume of unfoamed material and the number of cells per volume of foamed material, were calculated assuming an isotropic distribution of spherical cells according to:

$$N_0 = \left(\frac{n}{A}\right)^{\frac{3}{2}}\left(\frac{\rho_s}{\rho}\right), \tag{1}$$

$$N_f = \frac{6}{\pi\Phi^2}\left(1 - \frac{\rho}{\rho_s}\right), \tag{2}$$

where n is the number of cells in the micrograph, A is its area in cm^2, and ρ_s and ρ are the solid and foam densities, respectively.

The thermal conductivity of PSU and PSU-GnP nanocomposites and foams was measured using a C-Therm TCi Thermal Conductivity Analyzer, which employs the Modified Transient Plane Source technique, with a sensor radius of 3.189 mm, optimizing both the power output and measuring time according to the thermal characteristics of each sample (0.005–0.015 W and 15–80 s, respectively). The samples were prepared with the following dimensions: 20 mm × 20 mm × 2 mm (thickness), cutting directly from the obtained foams and flattening the surface using sandpaper.

3. Results

3.1. Cellular Structure of the Foams

The composition of PSU-GnP nanocomposite foams prepared by WVIPS (series 1 and series 2) and $scCO_2$ dissolution (series 3), their respective relative density (ρ/ρ_s), and main cellular structure characteristics are presented in Table 1.

Table 1. Composition, relative densities and cellular structure characteristics of PSU and PSU-GnP nanocomposite foams.

Foam Series	Foam Code	Density (g/cm³)	Relative Density	V_g	V_{PSU}	V_{GnP}	Φ (μm)	N_0 (cells/cm³)	N_f (cells/cm³)
Series 1	15 PSU	0.299	0.24	0.759	0.241	0.000	29.7	1.9×10^8	5.5×10^7
	15 PSU 1 GnP	0.351	0.28	0.719	0.280	0.002	55.5	3.3×10^7	8.1×10^6
	15 PSU 2 GnP	0.370	0.30	0.705	0.292	0.003	56.7	3.2×10^7	7.5×10^6
	15 PSU 5 GnP	0.402	0.32	0.683	0.308	0.009	50.1	5.1×10^7	1.0×10^7
	15 PSU 10 GnP	0.500	0.39	0.615	0.362	0.023	Open cell	-	-
Series 2	25 PSU	0.435	0.35	0.649	0.351	0.000	19.9	5.2×10^8	1.5×10^8
	25 PSU 1 GnP	0.402	0.32	0.677	0.321	0.002	21.2	3.7×10^8	1.3×10^8
	25 PSU 2 GnP	0.507	0.41	0.595	0.401	0.005	33.9	1.0×10^8	2.9×10^7
	25 PSU 5 GnP	0.532	0.42	0.580	0.408	0.012	Open cell	-	-
	25 PSU 10 GnP	0.588	0.45	0.547	0.426	0.027	Open cell	-	-
Series 3	PSU	0.443	0.36	0.643	0.357	0.000	19.1	2.1×10^8	1.8×10^8
	PSU 0.1 GnP	0.428	0.35	0.655	0.345	0.000	17.3	2.5×10^8	2.4×10^8
	PSU 0.4 GnP	0.457	0.37	0.632	0.367	0.001	13.9	3.4×10^8	4.5×10^8
	PSU 0.7 GnP	0.472	0.38	0.620	0.378	0.002	15.1	2.7×10^8	3.5×10^8
	PSU 1 GnP	0.510	0.41	0.590	0.407	0.002	13.1	4.1×10^8	5.1×10^8
	PSU 1.5 GnP	0.575	0.46	0.539	0.457	0.004	13.8	3.7×10^8	4.0×10^8
	PSU 2 GnP	0.561	0.45	0.552	0.443	0.005	14.9	2.1×10^8	3.1×10^8

As can be seen in Table 1, the addition of GnP resulted in a general increase in the average cell size for foams prepared using the WVIPS method when compared to the unfilled PSU foams. The addition of GnP in the first two foam series, that is, those prepared by WVIPS, seemed to affect the kinetics of cell formation due to GnP's affinity for NMP, which slowed down the process by hindering the phase exchange with water. The greater amounts of GnP in these foams (10 wt% GnP in series 1 and 5 and 10 wt% GnP in series 2) resulted in the formation of foams with open interconnected pores, which could be the result of the mentioned GnP affinity for NMP (see micrographs presented in Figures 3 and 4).

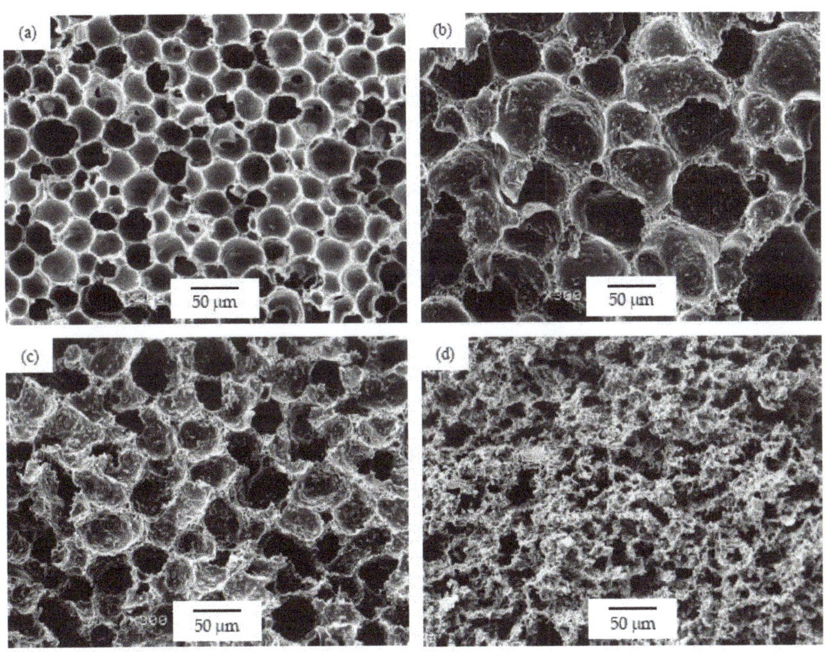

Figure 3. Micrographs at ×300 magnification illustrating the cellular structure of PSU-GnP foams from series 1: (**a**) 15 PSU; (**b**) 15 PSU 2 GnP; (**c**) 15 PSU 5 GnP; and (**d**) 15 PSU 10 GnP.

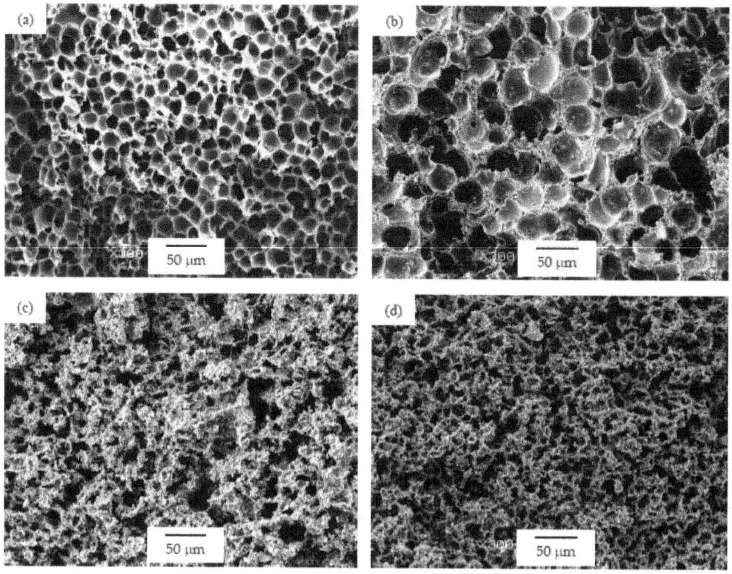

Figure 4. Micrographs at ×300 magnification illustrating the cellular structure of PSU-GnP foams from series 2: (**a**) 25 PSU; (**b**) 25 PSU 2 GnP; (**c**) 25 PSU 5 GnP; and (**d**) 25 PSU 10 GnP.

Regarding the cellular structure of foams prepared via $scCO_2$ dissolution (series 3), which displayed slightly smaller cell sizes than foams from series 1 and series 2, the addition of GnP resulted in a slight decrease of the average cell size when compared to the unfilled PSU foam (see Figure 5), which could be related to a barrier effect of the platelet-like GnP to the diffusion of $scCO_2$. No cell size reduction was observed with further increasing the amount of GnP.

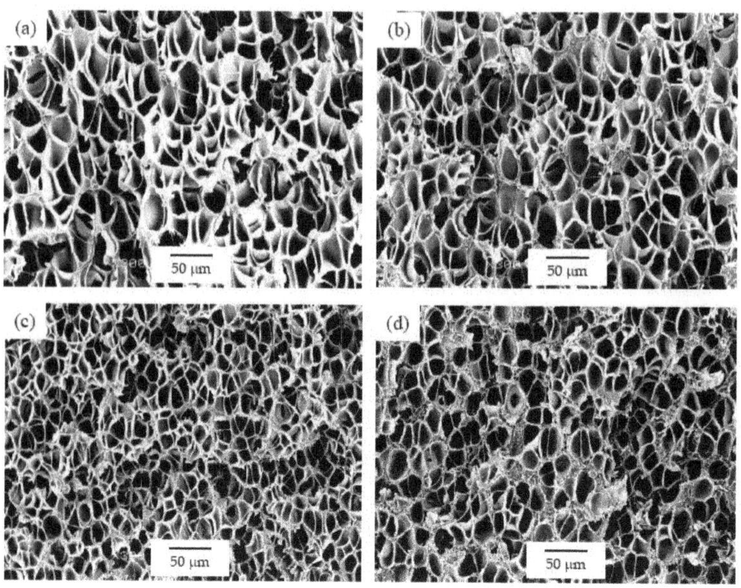

Figure 5. Micrographs at ×300 magnification illustrating the cellular structure of PSU-GnP foams from series 3: (**a**) PSU; (**b**) PSU 0.4 GnP; (**c**) PSU 1 GnP; and (**d**) PSU 2 GnP.

Cell density and cell nucleation density decreased for the first two series of foams with the addition of GnP when compared to the respective unfilled PSU foams. However, by increasing the amount of GnP, these values did not follow a generic pattern. In a similar way, foams prepared by $scCO_2$ dissolution displayed a general increase in cell density and cell nucleation density with adding GnP. Nevertheless, no clear trend was observed within these samples related to the increment in the amount of GnP.

3.2. Thermal Conductivity

The experimental thermal conductivity (λ_{exp}) of all foams and their relative density values are presented in Table 2.

Table 2. Experimental thermal conductivity values of PSU and PSU-GnP nanocomposite foams.

Foam Series	Foam Code	Relative Density	λ_{exp} (W·m^{-1}·K^{-1})
Series 1	15 PSU	0.24	0.056
	15 PSU 1 GnP	0.28	0.073
	15 PSU 2 GnP	0.30	0.073
	15 PSU 5 GnP	0.32	0.111
	15 PSU 10 GnP	0.39	0.193
Series 2	25 PSU	0.35	0.066
	25 PSU 1 GnP	0.32	0.090
	25 PSU 2 GnP	0.41	0.116
	25 PSU 5 GnP	0.42	0.221
	25 PSU 10 GnP	0.45	0.510
Series 3	PSU	0.36	0.077
	PSU 0.1 GnP	0.35	0.072
	PSU 0.4 GnP	0.37	0.088
	PSU 0.7 GnP	0.38	0.087
	PSU 1 GnP	0.41	0.091
	PSU 1.5 GnP	0.46	0.102
	PSU 2 GnP	0.45	0.108

As can be seen, there was a direct relation between the thermal conductivity and the GnP amount, as all PSU-GnP nanocomposite foams displayed an increasingly higher thermal conductivity with incrementing the amount of GnP (see Figure 6), owing to the inherently high thermal conductivity of GnP and the higher probability of physical contact between graphene nanoplatelets (see the ×1500 magnification micrographs presented in Figures 7–9 and especially the comparative high magnification micrographs presented in Figure 10 showing the distribution and dispersion of GnP throughout the cell walls of PSU-GnP foams). No significant differences were observed in terms of GnP dispersion between foams with 2 wt% GnP prepared by the two foaming methods (Figure 10). The experimental thermal conductivity of each foam series followed a linear trend with the volume percentage of GnP ($R^2 \geq 0.93$) with different slopes for each of the series (Figure 6).

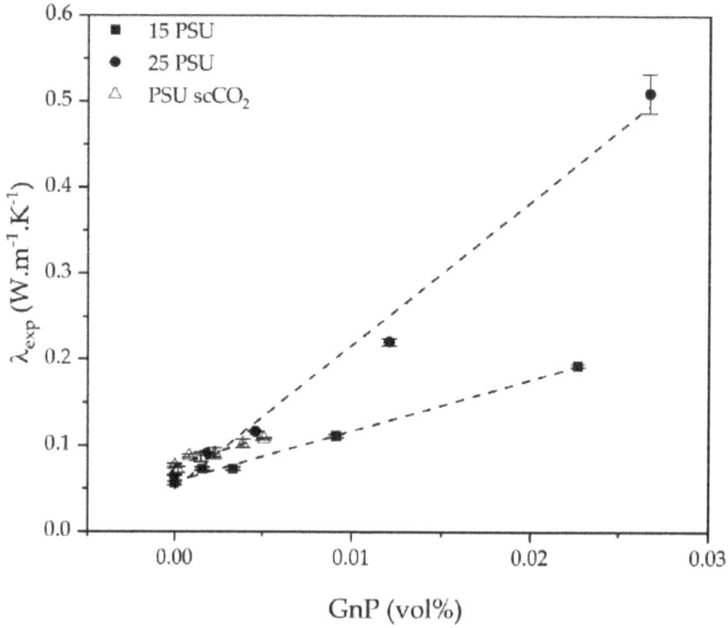

Figure 6. Experimental thermal conductivity enhancement of PSU and PSU-GnP nanocomposite foams with increasing GnP amount.

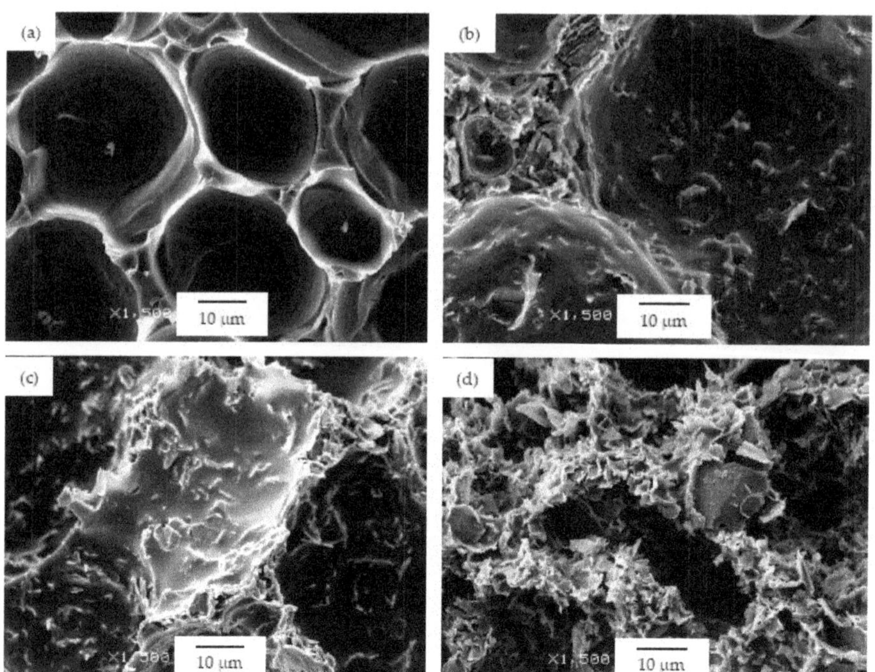

Figure 7. Micrographs at ×1500 magnification illustrating the cell walls of PSU-GnP foams from series 1: (**a**) 15 PSU; (**b**) 15 PSU 2 GnP; (**c**) 15 PSU 5 GnP; and (**d**) 15 PSU 10 GnP.

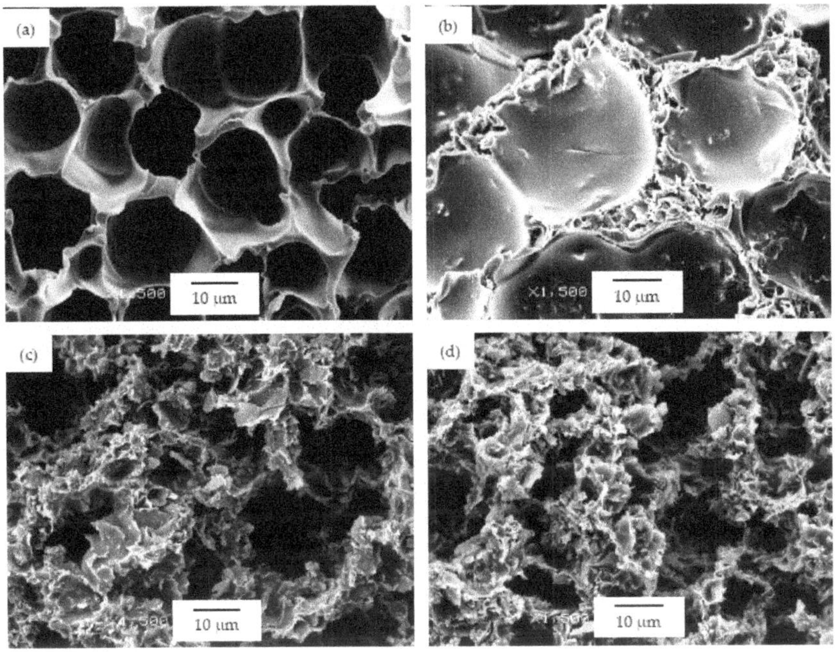

Figure 8. Micrographs at ×1500 magnification illustrating the cell walls of PSU-GnP foams from series 2: (**a**) 25 PSU; (**b**) 25 PSU 2 GnP; (**c**) 25 PSU 5 GnP; and (**d**) 25 PSU 10 GnP.

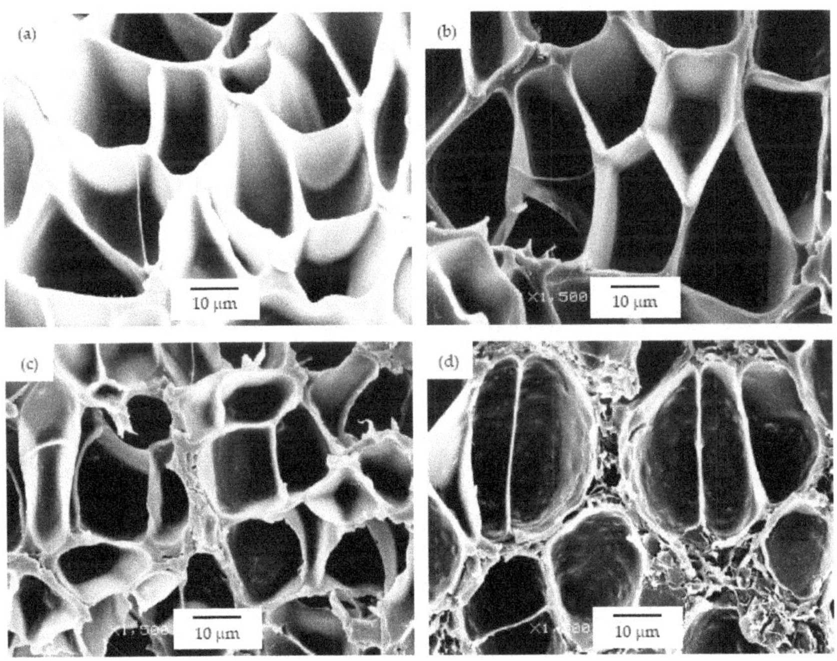

Figure 9. Micrographs at ×1500 magnification illustrating the cell walls of PSU-GnP foams from series 3: (**a**) PSU; (**b**) PSU 0.4 GnP; (**c**) PSU 1 GnP; and (**d**) PSU 2 GnP.

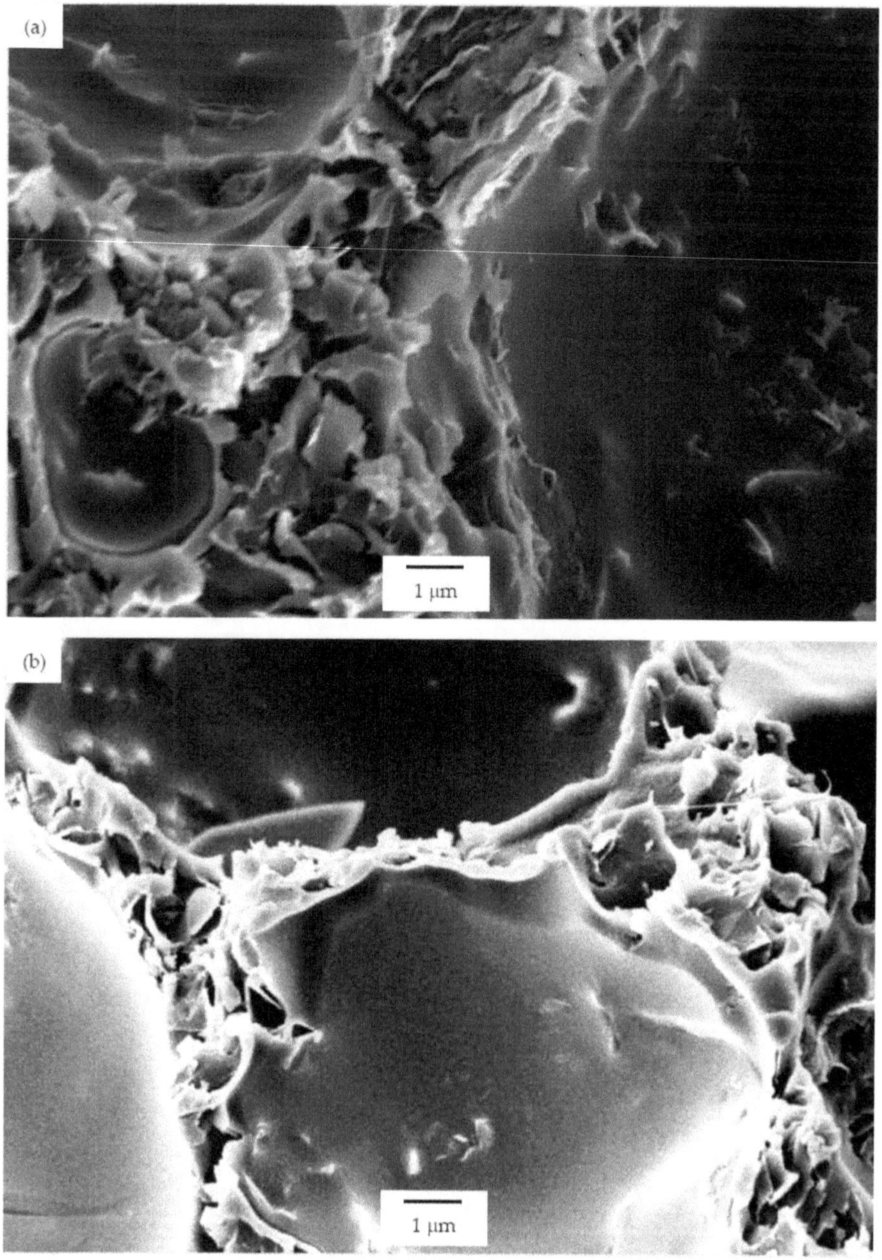

Figure 10. *Cont.*

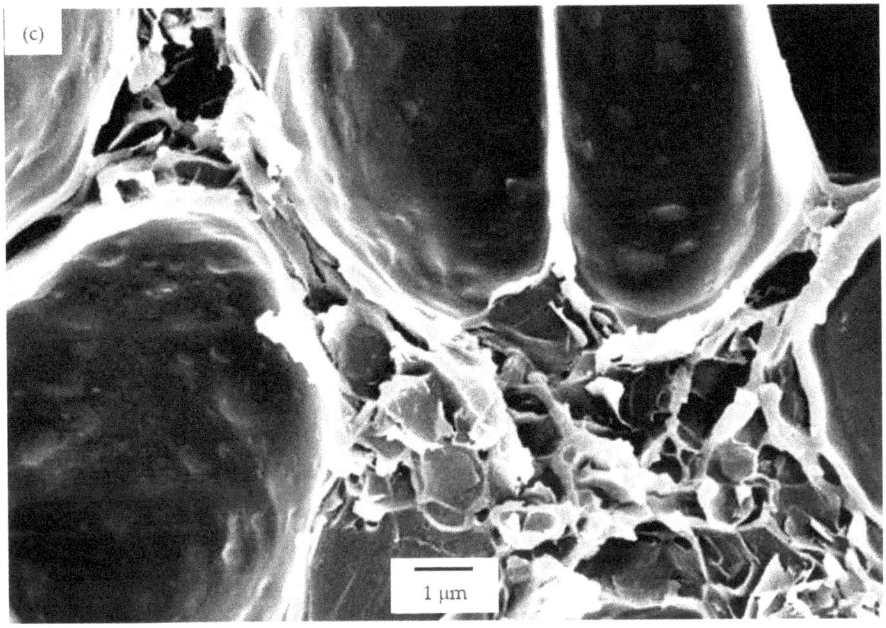

Figure 10. High-magnification micrographs (×5000) illustrating the distribution and dispersion of GnP throughout the cell walls of PSU-GnP foams with 2 wt% GnP: (**a**) 15 PSU 2 GnP; (**b**) 25 PSU 2 GnP; and (**c**) PSU 2 GnP.

Relative density also played a significant role, as thermal conductivity augmented with increasing relative density (Figure 11), especially in the case of PSU-GnP foams from series 2. This increment was related to the increasingly higher importance of thermal conduction through the solid phase, formed by a combination of PSU and thermally conductive GnP.

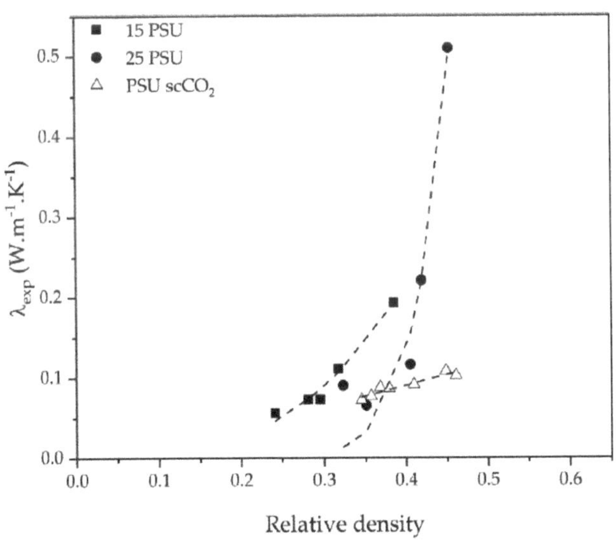

Figure 11. Experimental thermal conductivity enhancement of PSU and PSU-GnP nanocomposite foams with increasing relative density.

The effect of GnP influence on the thermal conductivity of the foams regardless of the changes in relative density can be observed in Figure 12 by representing the specific experimental thermal conductivity (λ_{spec}), defined as the quotient between the experimental thermal conductivity and the density of the foam, as a function of GnP amount.

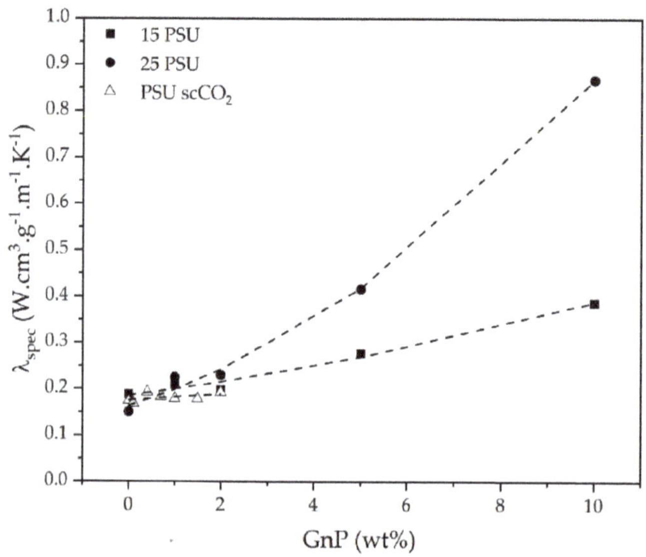

Figure 12. Specific experimental thermal conductivity evolution of PSU and PSU-GnP nanocomposite foams with increasing GnP amount.

The thermal conductivity of nanocomposite foams (λ_f) can be assumed as a contribution of four factors [34]: thermal conduction through the solid nanocomposite (λ_s), conduction through the gas phase (λ_g), convection through the cells (λ_{cv}) and radiation through the cell struts and across the cell voids (λ_r):

$$\lambda_f = \lambda_s + \lambda_g + \lambda_{cv} + \lambda_r. \tag{3}$$

The thermal conductivity due to convection (λ_{cv}) is only considered significant when the Grashof number is greater than 1000, which requires a minimum cell size of 10 mm [35,36]; therefore, it can be ignored for these foams.

Regarding the contribution of radiation (λ_r), the heat flux going through the foam in radiation form can be modeled as radiation across a series of parallel opaque surfaces with a separation equal to the average cell size (Φ) according to [34]:

$$\lambda_r = 4\frac{\varepsilon}{2-\varepsilon}\sigma T^3 \Phi, \tag{4}$$

where ε is the cell wall emissivity and σ is the Stefan–Boltzmann constant. The radiation contribution can also be disregarded for the foams presented in this work, as all of them are black due to the presence of GnP and have been prepared with relatively high thicknesses. Since the conductive GnP particles play a key role in the final thermal conduction behavior of the foams, a two-phase model was suggested for predicting thermal conductivity:

$$\lambda_f = \lambda_g V_g + \xi(\lambda_c V_c), \tag{5}$$

in which λ_f represents the thermal conductivity of the foam, λ_g corresponds to the conductivity of the gas in the cells ($\lambda_{air} = 0.026$ W·m^{-1}·K^{-1} [37]), λ_c represents the conductivity of the solid nanocomposite phase (PSU-GnP, experimentally determined to be between 0.210 W·m^{-1}·K^{-1} for unfilled PSU and 0.282 W·m^{-1}·K^{-1} for PSU-GnP foams containing the highest amount of GnP) and ξ is a parameter related to the tortuosity, which depends on the complexity of the foam's structure. V_g and V_c correspond to the volume fraction of the gas and solid phase, respectively.

As can be seen when comparing the SEM micrographs presented in Figures 3–5, and in Figures 7–9, and the cellular structure characteristics presented in Table 1, the changes in cellular structure could alter the tortuosity factor ξ, directly affecting the values of thermal conductivity; therefore, the $\xi\lambda_c$ product was calculated from the experimental thermal conductivity values for each foam using the following equation and presented in Table 3:

$$\xi\lambda_c = \frac{\lambda_f - \lambda_g V_g}{V_c}. \tag{6}$$

Table 3. Tortuosity effect ($\xi\lambda_c$) and foaming efficiency (n) influence on the thermal conductivity of PSU and PSU-GnP nanocomposite foams.

Foam Series	GnP (wt%)	$\xi\lambda_c$ (W·m^{-1}·K^{-1})	Relative Density	n
Series 1	0	0.150	0.24	0.692
	1	0.192	0.28	0.764
	2	0.184	0.30	0.760
	5	0.294	0.32	0.847
	10	0.459	0.39	0.909
Series 2	0	0.138	0.35	0.712
	1	0.225	0.32	0.806
	2	0.248	0.41	0.840
	5	0.490	0.42	0.918
	10	1.093	0.45	0.964
Series 3	0	0.168	0.36	0.760
	0.1	0.159	0.35	0.744
	0.4	0.194	0.37	0.791
	0.7	0.185	0.38	0.785
	1.0	0.185	0.41	0.791
	1.5	0.192	0.46	0.808
	2.0	0.209	0.45	0.821

The effect of changes in tortuosity could be observed in foams with similar composition. In the case of foams prepared using the WVIPS method, the ones with an open-cell structure (15 PSU 10 GnP, 25 PSU 5 GnP and 25 PSU 10 GnP) showed considerably lower tortuosity (higher $\xi\lambda_c$ values) and, as a consequence, displayed higher values of thermal conductivity. For foams with closed-cell structure, the value of the specific thermal conductivity decreased with decreasing cell size and hence increasing cell nucleation density, directly related to a higher tortuosity.

The drawback in the two-phase model presented in Equation (5) is that the effect of foam's density on the thermal conductivity of the foam is not clearly introduced. The already shown increase in thermal conductivity of the foam by incrementing relative density (Figure 11) can be explained by a power law expression similar to the one suggested by Gibson and Ashby for predicting the mechanical performance of cellular solids [35]:

$$\frac{\lambda_f}{\lambda_c} = K\left(\frac{\rho}{\rho_s}\right)^n. \tag{7}$$

In this equation, if we assume K as being a variable equivalent to the tortuosity parameter ξ, exponent n can be seen as representing the foaming efficiency in the final thermal conductivity, being related to the relative density according to:

$$n = \frac{[\ln(\lambda_f) - \ln(\xi\lambda_c)]}{\left[\ln\left(\frac{\rho}{\rho_s}\right)\right]}. \quad (8)$$

The values of n have been calculated for each foam and are presented in the final column of Table 3. As the values of n suggest, the highest efficiency corresponds to the foams from series 2, i.e., foams prepared by the WVIPS method with a 25% of PSU.

These results, combined with the previous analysis on the tortuosity and cellular structure of the foams, could lead to an optimization of the thermal conductivity of PSU-GnP nanocomposite foams by controlling their density and cellular structure, and hence to the achievement of highly conductive foams at lower amounts of GnP. This can be more clearly seen in Figure 13 by representing the normalized thermal conductivity of the foams (λ_{norm}), calculated by dividing λ_f by ($\xi\lambda_c$), as a function of relative density.

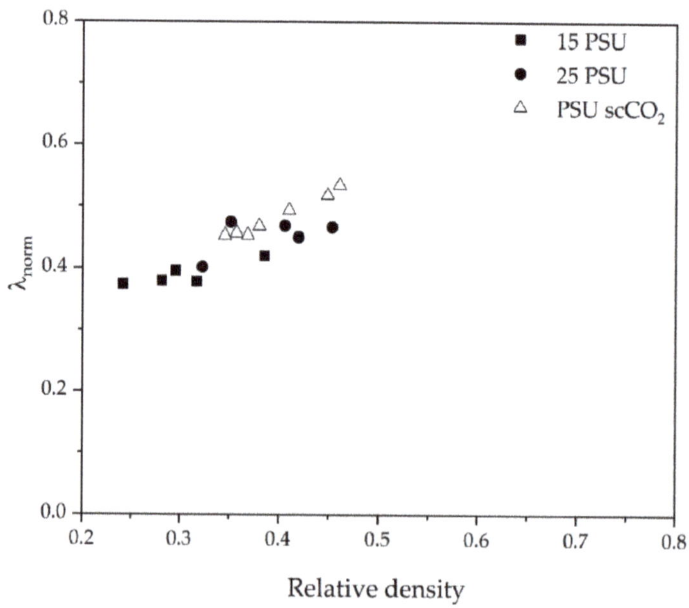

Figure 13. Normalized thermal conductivity evolution with relative density for PSU and PSU-GnP nanocomposite foams.

4. Conclusions

Three series of PSU-GnP nanocomposite foams with variable amounts of GnP (between 0 and 10 wt%) were prepared using two foaming methods: WVIPS and $scCO_2$ dissolution. Foams prepared by the WVIPS method presented relative densities between 0.24–0.39 for series 1 (foams prepared with 15% PSU) and between 0.35–0.45 for series 2 (foams prepared with 25% PSU), while foams prepared by $scCO_2$ dissolution (series 3) displayed relative densities between 0.36–0.45. In all series, relative density increased with incrementing the amount of GnP.

In terms of cellular structure, foams from series 1 and series 2 showed higher average cell sizes and hence lower cell densities and cell nucleation densities with the addition of GnP. Among these foams, those with lower amounts of GnP displayed a homogenous closed-cell structure, whereas those with a higher GnP concentration (10 wt% GnP in series 1 and 5 and 10 wt% GnP in series 2) showed an open-cell interconnected structure. These results suggest that the presence of GnP affected the kinetics during the phase separation exchange process by slowing it down due to the affinity of GnP for NMP. In the case of foams prepared via $scCO_2$ dissolution, an opposite trend was observed with only slight changes. The average cell size of foams from series 3 slightly decreased by adding GnP, which could be the result of a physical barrier effect of the platelet-like GnP to the diffusion of CO_2 in PSU.

The thermal conductivity of the foams was affected mainly by the presence of GnP, the cellular structure, and the density of the foams. The thermal conductivity showed a linear increasing trend with increasing GnP volume fraction, as expected due to the intrinsically high thermal conductivity of GnP. Additionally, the tortuosity of the cellular structure, also directly related to the added amount of GnP, influenced the final value of thermal conductivity by affecting the path of conduction. The highest values of thermal conductivity corresponded to foams with an open-cell interconnected structure. Moreover, density played a key role, as thermal conductivity followed a power law relation with relative density. The combination of these factors showed that PSU-GnP foams can be prepared with higher thermal conductivity at a lower amount of GnP by carefully controlling their cellular structure and relative density.

Author Contributions: Formal analysis: H.A., M.A., and J.I.V.; investigation: H.A., M.A., and J.I.V.; methodology: H.A., and M.A.; writing—original draft: H.A., M.A., and J.I.V.; writing—review and editing: H.A., M.A., and J.I.V. All authors have read and agreed to the published version of the manuscript.

Funding: Research funded by the Spanish Ministry of Science, Innovation, and Universities, Government of Spain (project MAT2017-89787-P).

Conflicts of Interest: The authors declare no conflict of interest.

References

1. McKeen, L.W. 12-High Temperature/High Performance Polymers. In *Plastics Design Library*; William Andrew Publishing: Boston, MA, USA, 2012; pp. 315–337. ISBN 978-1-4557-2551-9.
2. Gohil, S.V.; Suhail, S.; Rose, J.; Vella, T.; Nair, L.S. Polymers and composites for orthopedic applications. In *Materials for Bone Disorders*; Elsevier: Amsterdam, The Netherlands, 2017; pp. 349–403.
3. Kyriacos, D. High-Temperature Engineering Thermoplastics. In *Brydson's Plastics Materials*; Elsevier: Amsterdam, The Netherlands, 2017; pp. 545–615.
4. Krause, B.; Diekmann, K.; van der Vegt, N.F.A.; Wessling, M. Open Nanoporous Morphologies from Polymeric Blends by Carbon Dioxide Foaming. *Macromolecules* **2002**, *35*, 1738–1745. [CrossRef]
5. Li, Z.; Jia, Y.; Bai, S. Polysulfone foam with high expansion ratio prepared by supercritical carbon dioxide assisted molding foaming method. *RSC Adv.* **2018**, *8*, 2880–2886. [CrossRef]
6. Guo, H.; Nicolae, A.; Kumar, V. Solid-state microcellular and nanocellular polysulfone foams. *J. Polym. Sci. Part B Polym. Phys.* **2015**, *53*, 975–985. [CrossRef]
7. Sun, H.; Mark, J.E. Preparation, characterization, and mechanical properties of some microcellular polysulfone foams. *J. Appl. Polym. Sci.* **2002**, *86*, 1692–1701. [CrossRef]
8. Krause, B.; Boerrigter, M.E.; van der Vegt, N.F.A.; Strathmann, H.; Wessling, M. Novel open-cellular polysulfone morphologies produced with trace concentrations of solvents as pore opener. *J. Memb. Sci.* **2001**, *187*, 181–192. [CrossRef]
9. Huang, Q.; Klötzer, R.; Seibig, B.; Paul, D. Extrusion of microcellular polysulfone using chemical blowing agents. *J. Appl. Polym. Sci.* **1998**, *69*, 1753–1760. [CrossRef]
10. Wu, H.; Tang, B.; Wu, P. Development of novel SiO_2–GO nanohybrid/polysulfone membrane with enhanced performance. *J. Memb. Sci.* **2014**, *451*, 94–102. [CrossRef]
11. Chakrabarty, B.; Ghoshal, A.K.; Purkait, M.K. Ultrafiltration of stable oil-in-water emulsion by polysulfone membrane. *J. Memb. Sci.* **2008**, *325*, 427–437. [CrossRef]
12. Anadão, P.; Sato, L.F.; Wiebeck, H.; Valenzuela-Díaz, F.R. Montmorillonite as a component of polysulfone nanocomposite membranes. *Appl. Clay Sci.* **2010**, *48*, 127–132. [CrossRef]
13. Devrim, Y.; Erkan, S.; Baç, N.; Eroğlu, I. Preparation and characterization of sulfonated polysulfone/titanium dioxide composite membranes for proton exchange membrane fuel cells. *Int. J. Hydrog. Energy* **2009**, *34*, 3467–3475. [CrossRef]
14. Lawal, A.T. Graphene-based nano composites and their applications: A review. *Biosen. Bioelectron.* **2019**, *141*, 111384. [CrossRef] [PubMed]
15. Antunes, M.; Realinho, V.; Velasco, J.I. Foaming behaviour, structure, and properties of polypropylene nanocomposites foams. *J. Nanomater.* **2010**, *2010*, 4. [CrossRef]

16. King, J.A.; Tucker, K.W.; Vogt, B.D.; Weber, E.H.; Quan, C. Electrically and thermally conductive nylon 6, 6. *Polym. Compos.* **1999**, *20*, 643–654. [CrossRef]
17. Verdejo, R.; Barroso-Bujans, F.; Rodriguez-Perez, M.A.; De Saja, J.A.; Lopez-Manchado, M.A. Functionalized graphene sheet filled silicone foam nanocomposites. *J. Mater. Chem.* **2008**, *18*, 2221–2226. [CrossRef]
18. Han, Z.; Fina, A. Thermal conductivity of carbon nanotubes and their polymer nanocomposites: A review. *Prog. Polym. Sci.* **2011**, *36*, 914–944. [CrossRef]
19. Fraleoni-Morgera, A.; Chhikara, M. Polymer-based nano-composites for thermal insulation. *Adv. Eng. Mater.* **2019**, *21*, 1801162. [CrossRef]
20. Gedler, G.; Antunes, M.; Borca-Tasciuc, T.; Velasco, J.I.; Ozisik, R. Effects of graphene concentration, relative density and cellular morphology on the thermal conductivity of polycarbonate–graphene nanocomposite foams. *Eur. Polym. J.* **2016**, *75*, 190–199. [CrossRef]
21. Fukushima, H.; Drzal, L.; Rook, B.; Rich, M. Thermal conductivity of exfoliated graphite nanocomposites. *J. Therm. Anal. Calorim.* **2006**, *85*, 235–238. [CrossRef]
22. Gonnet, P.; Liang, Z.; Choi, E.S.; Kadambala, R.S.; Zhang, C.; Brooks, J.S.; Wang, B.; Kramer, L. Thermal conductivity of magnetically aligned carbon nanotube buckypapers and nanocomposites. *Curr. Appl. Phys.* **2006**, *6*, 119–122. [CrossRef]
23. Gu, J.; Yang, X.; Lv, Z.; Li, N.; Liang, C.; Zhang, Q. Functionalized graphite nanoplatelets/epoxy resin nanocomposites with high thermal conductivity. *Int. J. Heat Mass Transf.* **2016**, *92*, 15–22. [CrossRef]
24. Shtein, M.; Nadiv, R.; Buzaglo, M.; Kahil, K.; Regev, O. Thermally conductive graphene-polymer composites: Size, percolation, and synergy effects. *Chem. Mater.* **2015**, *27*, 2100–2106. [CrossRef]
25. Yang, J.; Zhang, E.; Li, X.; Zhang, Y.; Qu, J.; Yu, Z.-Z. Cellulose/graphene aerogel supported phase change composites with high thermal conductivity and good shape stability for thermal energy storage. *Carbon* **2016**, *98*, 50–57. [CrossRef]
26. Wang, F.; Drzal, L.T.; Qin, Y.; Huang, Z. Mechanical properties and thermal conductivity of graphene nanoplatelet/epoxy composites. *J. Mater. Sci.* **2015**, *50*, 1082–1093. [CrossRef]
27. Hu, L.; Desai, T.; Keblinski, P. Thermal transport in graphene-based nanocomposite. *J. Appl. Phys.* **2011**, *110*, 33517. [CrossRef]
28. Abbasi, H.; Antunes, M.; Velasco, J.I.I. Graphene nanoplatelets-reinforced polyetherimide foams prepared by water vapor-induced phase separation. *eXPRESS Polym. Lett.* **2015**, *9*, 412–423. [CrossRef]
29. Abbasi, H.; Antunes, M.; Velasco, J.I.I. Enhancing the electrical conductivity of polyetherimide-based foams by simultaneously increasing the porosity and graphene nanoplatelets dispersion. *Polym. Compos.* **2018**, *40*, E1416–E1425. [CrossRef]
30. Abbasi, H.; Antunes, M.; Velasco, J.I. Polyetherimide foams filled with low content of graphene nanoplatelets prepared by $scCO_2$ dissolution. *Polymers* **2019**, *11*, 328. [CrossRef]
31. Abbasi, H.; Antunes, M.; Velasco, J.I. Effects of Carbon Nanotubes/Graphene Nanoplatelets Hybrid Systems on the Structure and Properties of Polyetherimide-Based Foams. *Polymers* **2018**, *10*, 348. [CrossRef]
32. Liang, C.Z.; Chung, T.-S.; Lai, J.-Y. A review of polymeric composite membranes for gas separation and energy production. *Prog. Polym. Sci.* **2019**, *97*, 101141. [CrossRef]
33. Sims, G.L.A.; Khunniteekool, C. Cell-size measurement of polymeric foams. *Cell. Polym.* **1994**, *13*, 137–146.
34. Glicksman, L.R. Heat transfer in foams. In *Low Density Cellular Plastics. Physical Basis of Behaviour*; Chapman & Hall: London, UK, 1994; pp. 104–152.
35. Gibson, L.J.; Ashby, M.F. *Cellular Solids: Structure and Properties*; Cambridge University Press: Cambridge, UK, 1999; ISBN 131602542X.
36. Holman, J.P. *Heat Transfer*; McGraw-Hill: New York, NY, USA, 2010; ISBN 0071267697.
37. Weast, R.C.; Astle, M.J.; Beyer, W.H. *Handbook of Chemistry and Physics*; CRC Press: Boca Raton, FL, USA, 1988; Volume 69.

polymers

MDPI

Article

Anisotropic Cellulose Nanofibers/Polyvinyl Alcohol/Graphene Aerogels Fabricated by Directional Freeze-drying as Effective Oil Adsorbents

Lijie Zhou, Shengcheng Zhai, Yiming Chen and Zhaoyang Xu *

College of Materials Science and Engineering, Nanjing Forestry University, Nanjing 210037, China; zhoulijie678@163.com (L.Z.); zhai_sc@njfu.edu.cn (S.Z.); 18362983199@163.com (Y.C.)

* Correspondence: zhaoyangxunjfu@hotmail.com; Tel./Fax: +86-25-8542-7517

Received: 19 March 2019; Accepted: 16 April 2019; Published: 18 April 2019

Abstract: Under the current situation of frequent oil spills, the development of green and recyclable high-efficiency oil-absorbing aerogel materials has attracted wide attention from researchers. In this study, we report a high-strength, three-dimensional hydrophobic cellulose nanofiber (CNF)/polyvinyl alcohol (PVA)/graphene oxide (GO) composite aerogel with an anisotropic porous structure, which was fabricated by directional freeze-drying technology using anisotropically grown ice crystals as a template, followed by hydrophobic treatment with a simple dip coating process. The prepared composite aerogel presented anisotropic multi-level pore microstructures, low density (17.95 mg/cm^3) and high porosity (98.8%), good hydrophobicity (water contact angle of 142°) and great adsorption capacity (oil absorption reaching 96 times its own weight). More importantly, the oriented aerogel had high strength, whose compressive stress at 80% strain reached 0.22 MPa and could bear more than 22,123 times its own weight without deformation. Therefore, the CNF/PVA/GO composite aerogel prepared by a simple and easy-to-operate directional freeze-drying method is a promising absorbent for oil-water separation.

Keywords: cellulose nanofibers; polyvinyl alcohol; graphene; directional freeze-drying; oil absorption

---

1. Introduction

Water pollution caused by oil and chemical spills seriously harms the ecological environment and human health [1–3]. Therefore, effectively solving the problem of oil pollution has become an urgent and arduous task. Traditional methods for dealing with petroleum pollution include the following methods: physical and mechanical methods [4], direct combustion methods [5], bioremediation methods [6], chemical treatment methods [7] and dosing materials [8], in which the adsorption of adsorbent materials is simple and environmentally friendly. The oil-absorbing materials reported in the literature can be summarized into three categories according to the nature and source of the materials: (1) porous inorganic minerals such as clay [9], silica [10], perlite [11], fly ash [12], etc. These materials come from a wide range of sources, which have low oil absorption, poor oil holding capacity, and difficulty in recycling; (2) synthetic polymers such as polyurethane [13], polystyrene foam [14]. Although these materials have good oil absorption ability and reusability, they have poor biodegradability; (3) natural plant fibers such as cotton fiber [15], straw [16], etc. These materials are provided with a wide range of sources, low cost and biodegradability but lack a good oil absorption rate. Therefore, novel, environmentally friendly, highly efficient and sustainable materials are urgently needed.

Nanocellulose has a wide range of sources, good biodegradability and outstanding mechanical properties [17]. Cellulose aerogel not only possesses the advantages of low density and high porosity compared with traditional aerogel but also has good biocompatibility that makes it an environmentally friendly adsorbent and has attracted widespread attention [18]. Moreover, hydrophobic cellulose-based aerogels can be prepared by coating metal oxides and hydrophobic polymers. It has been studied to modify cellulose-based aerogels with methyltrimethoxysilane and trimethylchlorosilane to make them hydrophobic and lipophilic [19,20]. Studies have shown that hydrophobically modified cellulose-based aerogels are potential oil-absorbing materials, which are used to remove marine oil spills [21,22].

Despite the excellent properties of cellulose-based aerogels, their low strength and modulus limit their application compared to inorganic polymer materials. The poor compressive strength of oil-absorbing aerogels will cause them to be broken during the oil absorption process, so it is important to increase their strength. To enhance the compressive properties of the aerogel, we prepared cellulose nanofiber (CNF) / polyvinyl alcohol (PVA) / graphene oxide (GO) composite aerogels. The low cost PVA solution has excellent water solubility, biodegradability and biocompatibility. The long polymer chains of PVA lead to high-density hydrogen bonding with CNFs and GO, which can enhance the mechanical properties of the cellulose aerogel [23–26]. Meanwhile, GO forms a strong interaction with CNFs and PVA through hydrogen bonding due to its excellent mechanical properties and large number of oxygen atoms on the surface. However, most of the aerogels reported have a random internal structure, leading to relatively poor compressive strength. In recent years, orientation techniques have been used to prepare porous ceramic materials and have attracted the attention of researchers as a special method for preparing aerogels [27–29]. Lee and Deng reported the fabrication of layered cellulose foams by directional freezing, emphasizing the differences in the microstructure and mechanical properties of microfiber foams [30]. Therefore, in order to further increase compressive strength, designing anisotropic porous structures in aerogels by directional freezing may be a viable solution.

"Directional freeze-drying" (Figure 1a) means that the bottom of the container contacts with liquid nitrogen, causing the solvent to freeze from the bottom of the container and that the ice crystals grow in one direction. When the vessel containing the mixed aqueous suspension is frozen, phase separation causes the particles and polymer molecules to be expelled from the formed ice crystals. Then, the molecules are accumulated between the growing ice crystals, and the freezing process can be carried out in a more controlled manner. In this way, a defined alignment structure is formed. After sublimation of the ice crystals after lyophilization, the resulting solids can form a continuous 3D network to form an anisotropic porous aerogel [31]. In addition, containers containing aqueous suspensions are directly frozen in liquid nitrogen and a refrigerator to produce a relatively disordered porous structure (we refer to them as "non-directional freeze-drying" and "refrigerator freeze-drying") [32–34]. "Non-directional freeze-drying" (Figure 1b) means that the bottom of the container and the bottle body are in direct contact with liquid nitrogen, then the ice crystal grows from the outside of the suspension to the inside in the direction of the bottom and the bottle body, followed by lyophilization. "Refrigerator freeze-drying" (Figure 1c) means that when water suspension is frozen by refrigerator, the container is surrounded by cold air, so the ice crystal grows from the outside of the suspension to the inside in all directions. Due to the different growth patterns of ice crystals during freezing, the three composite aerogels prepared by different freezing methods have different structures and properties.

In this study, a novel, simple, safe and reliable "directional freezing" method was used to prepare a CNF/PVA/GO composite aerogel with an anisotropic alignment structure. Then, CPGA was hydrophobically modified with methyltrichlorosilane (TMCS) by simple thermal chemical vapor deposition to obtain a hydrophobic CNF/PVA/GO composite aerogel (MCPGA). To confirm the properties of the resulting aerogel, characterization tests were performed by scanning electron microscopy (SEM), Fourier transform infrared (FTIR) spectroscopy, mechanical testing machine and water contact angle (WCA). The results show that the obtained composite aerogel has a low density, high porosity, aligned porous structure and high compressive strength. It also exhibited high oil-water selectivity and excellent oil absorption capacities for various oils and organic solvents.

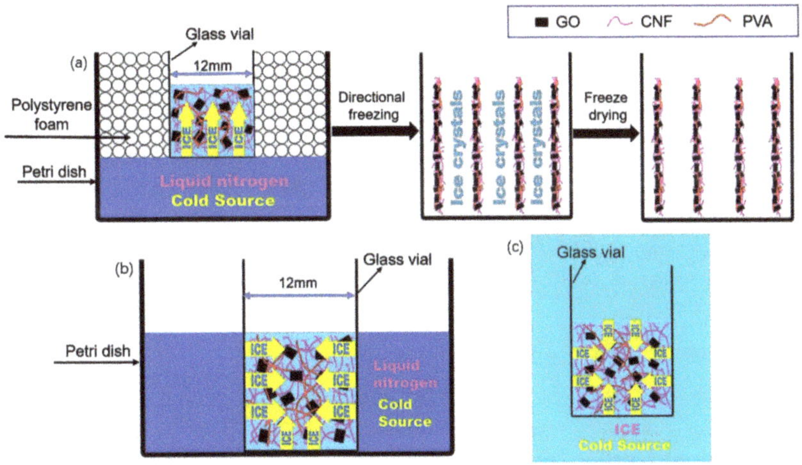

Figure 1. Schematic representation of (**a**) the directional freeze-drying method; (**b**) the non-directional freeze-drying method and (**c**) the refrigerator freeze-drying method. Cellulose nanofiber (CNF); polyvinyl alcohol (PVA); graphene oxide (GO).

2. Experimental

2.1. Materials

Bamboo powder (60 mesh) was purchased from Chongqing Dongyi Xiabu Co., Ltd. (Chongqing, China) and was the raw material for preparing nanocellulose. Graphite powder was provided by Qingdao Henglide Graphite Co., Ltd. (Qingdao, China). Analytical grade PVA (Mw: 95,000 g/mol), glutaraldehyde (GA, Crosslinker, 25 wt % in H_2O), potassium hydroxide (KOH), Sudan III, TMCS (99 wt %), acetic acid (CH_3COOH), hydrogen peroxide (H_2O_2, 30%), sodium nitrate ($NaNO_3$), potassium permanganate ($KMnO_4$), concentrated sulfuric acid (H_2SO_4, 98%), phosphorus pentoxide (P_2O_5), hydrochloric acid (HCl), sodium chlorite ($NaClO_2$) and potassium persulfate ($K_2S_2O_8$) were purchased from Aladdin Industrial Co (Shanghai, China). All materials were analytical and used without further purification. Deionized water was self-made in our laboratory.

2.2. Preparation of Cellulose Nanofibers

The chemical treatment of bamboo powder referred to the treatment method is described in the document [35]. The cellulose suspensions used in this study were prepared according to the methods reported in the previously reported literature [36]. The prepared cellulose suspensions were stored at 4 °C before future utilization.

2.3. Preparation of Polyvinyl Alcohol Solution

PVA (10.0 g, Mw: 95,000 g/mol) was dissolved in deionized water (100 mL) and stirred in a water bath (HH-1, Nanjing Xian'ou Instrument Manufacturing Co., Ltd., Nanjing, China) at 85 °C for 12 h until the PVA was completely dissolved. The PVA solution was then stored at room temperature for further use.

2.4. Preparation of Graphene Oxide

GO was prepared according to a modified Hummers' method [37]. Natural graphite powder (3 g), $K_2S_2O_8$ (4 g) and P_2O_5 (4 g) were mixed and placed into a water-bath (80 °C) with concentrated H_2SO_4 (25 mL). After treatment for 5 h under strong mixing by magneton, the mixture was washed with distilled water until the solution was neutral, then placed in a drying oven (60 °C) (DZF-6090, Shanghai

Jinghong Laboratory Equipment Co., Ltd., Shanghai, China). The mixture, $NaNO_3$ (1 g) and $KMnO_4$ (10 g) were slowly added into a beaker pre-filled with concentrated H_2SO_4 (100 mL) in the ice-bath under strong stirring for 2 h. Then, the mixture was warmed to 35 °C and kept for 4 h, heated to 90 °C with adding distilled water (800 mL) and held for 1 h, and then mixed with H_2O_2 (20 mL) to a bright yellow solution. The solution was filtered with HCl (5 wt %) for several times and the undissolved substance was washed with deionized water until it was neutral. In order to remove the residual salts and acids, the resulting graphite oxide colloid was dialyzed with a molecular weight cut-off membrane (MW 3500 Da) and deionized water for 2 weeks at a room temperature. Finally, the solution was diluted to the concentration of 1.5 wt %; then, exfoliation of graphite oxide to graphene oxide sheets was performed by sonicating for 0.5 h under the condition of 960 W, and a homogeneous GO solution was obtained.

2.5. Preparation of CNF/PVA/GO (CPGA) by Directional Freeze-drying

The CNF suspension (3 g, 1.275 wt %), PVA solution (0.76 mL, 0.1 g/mL), GO solution (3.825 g, 1.5 wt %) and water (2 mL) were mixed and vigorously stirred in a beaker for 1 h. Sulfuric acid (8 μL, 1.0% by volume) was added to the above suspension to adjust the pH of the mixture from 4 to 6. Then, the GA solution (80 μL, 25 wt %) was added to the resulting CNF/PVA/GO mixture. The above mixture was mechanically stirred for 1 h and sonicated in an ultrasonic bath for 30 min, and finally placed in a vacuum oven to remove any residual air bubbles. At the final stage, the obtained mixture was transferred to a cylindrical mold and then crosslinked in an oven at 75 °C for 3 h. The crosslinked aqueous gel was stored overnight in a 4 °C refrigerator for pre-cooling to avoid macroscopic cracking during the freezing step. A cylindrical mold with a mixed aqueous gel was placed in a directional freezer for a few minutes to ensure it was completely frozen; then, the mold was transferred to a freeze dryer and lyophilized for 48 h at −50 °C, and then the directional CNF/PVA/GO aerogel (d-CPGA) was obtained. In addition, refrigerator freeze-dried CNF/PVA/GO aerogel (r-CPGA) and non-directional freeze-dried CNF/PVA/GO aerogel (n-CPGA) were prepared for comparison.

2.6. Modification of CPGA to Prepare Hydrophobic CPGA Aerogel (MCPGA)

In order to prepare the hydrophobic composite aerogel, TMCS was used as a modification agent. The CPGA (d-CPGA, r-CPGA, n-CPGA) was placed in a sealed container together with 8 mL of TMCS in an oven at 40 °C for 10 h to produce a hydrophobic and oleophilic MCPGA (d-MCPGA, r-MCPGA, n-MCPGA) by a silanization reaction. All experimental steps of the sample were shown in Figure 2.

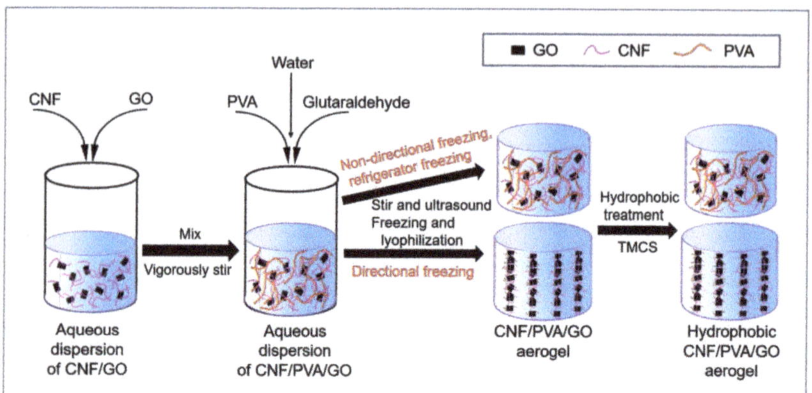

Figure 2. Schematic showing experimental process of the samples.

2.7. Density and Porosity Measurements

The density (ρ) of aerogels was calculated according to Equation (1):

$$\rho = m/v \quad (1)$$

where ρ is the density of the aerogel, m is the mass of the aerogel, and v is the volume of the aerogel.

The density (ρ_s) of the solid material is calculated according to Equation (2) based on the solid density of each component and their weight ratios:

$$\rho_s = 1/(W_{CNFs}/\rho_{CNFs} + W_{PVA}/\rho_{PVA} + W_{GO}/\rho_{GO} + W_{silane}/\rho_{silane}) \quad (2)$$

where W is the weight percentage of the different components. ρ_{CNFs}, ρ_{PVA}, ρ_{GO} and ρ_{silane} are the solid densities of CNFs, PVA, GO and silane, respectively. The densities of the CNFs, PVA, GO and silane used for this study are 1460, 1269, 2100 and 1273 kg/m^3, respectively, according to the manufacturer's data sheet.

The porosity (P) of aerogels was calculated according to Equation (3)

$$P(\%) = (1 - \rho/\rho_S) \times 100\% \quad (3)$$

where P, ρ and ρ_s are the porosity, actual density and relative density of the material, respectively.

2.8. Sample Characterization

For each test described below, each sample was tested at least three times and the average results were reported. Field emission scanning electron microscopy (FE-SEM, Hitachi S-4800, Tokyo, Japan) was used to investigate the microscopic morphology of different types of aerogels. The Fourier transform infrared spectroscopy (FTIR) (Nicolet iS10, Thermo Electron Corp., Madison, WI, USA) data was recorded in the range of 500–4000 cm^{-1}. X-ray diffraction (XRD) pattern was measured on Ultima IV multipurpose XRD system (Ultima IV, Rigaku, Akishima, Japan) with Cu Kα radiation at a scanning rate of 5°/min. The water contact angle was measured at room temperature using a contact angle goniometer (OCA 15/20, Future Digital Scientific Corp., Natick, MA, USA) with 4 μL of water droplets. The static drop method (angle measurement method) was used for WCA testing. Three aerogels of the same type were taken and the droplets were separately dropped onto the surface of the aerogel for 2 min with deionized water. The data was read at 1 min and 2 min and the arithmetic mean was taken, respectively. The data was obtained after multiple measurements. The compression test was carried out at room temperature using Shimadzu CMT4204 (Kyoto, Japan). The test sample had a diameter of 12 mm and a height of 8 mm and the loading rate was set to 2 mm/min. The oil absorption capacity of MCPGA, Qt, was measured using the equation Qt = $(m_2 - m_1)/m_1$, where m_1 and m_2 are the weights of the aerogels before and after absorption. The oil absorption capacity is the average of all absorption tests repeated three times.

3. Results and Discussion

3.1. Morphology and Microstructure

Figure 3a,d,g are digital photographs of d-MCPGA, r-MCPGA, n-MCPGA, respectively. From the pictures, we could not find the difference in macroscopic appearance. In order to explore the difference in the internal structure of the aerogel, we observed the microstructure by SEM (Figure 3b–f,h–i). As shown in the c, e and i of Figure 3, d-MCPGA, r-MCPGA, and n-MCPGA all had a three-dimensional interpenetrating porous structure. Previous studies have shown that solidification has an important influence on the microstructure and pore morphology of porous materials [38,39]. D-MCPGA produces an aligned layered structure in the vertical direction. This particular structure strongly indicates that the ice crystals grow mainly in the vertical direction, and the sheet fibers of d-MCPGA grow parallel to the freezing direction, which indicates the d-MCPGA had anisotropic porous structure and the pore size was about 60 μm (Figure 3b–c). Furthermore, as indicated by the red markings in Figure 3c, there were a plurality of connecting channels between the holes which formed an open porous structure that allowed the oil to easily reach the interior of the aerogel to enhance oil absorption. Freezing at −40 °C had the slowest freezing rate and produced the largest ice crystals, so r-MCPGA had a maximum pore size of nearly 70 μm and a disordered three-dimensional porous structure (Figure 3e–f). For n-MCPGA, the freezing rate of the frozen slurry in liquid nitrogen was too fast to produce a dense structure with only a small pore size of 10 μm (Figure 3i). Under such rapid freezing conditions, the thin wall was thin and the space between the sheets was small, which was not conducive to the entry of oil and organic solvents.

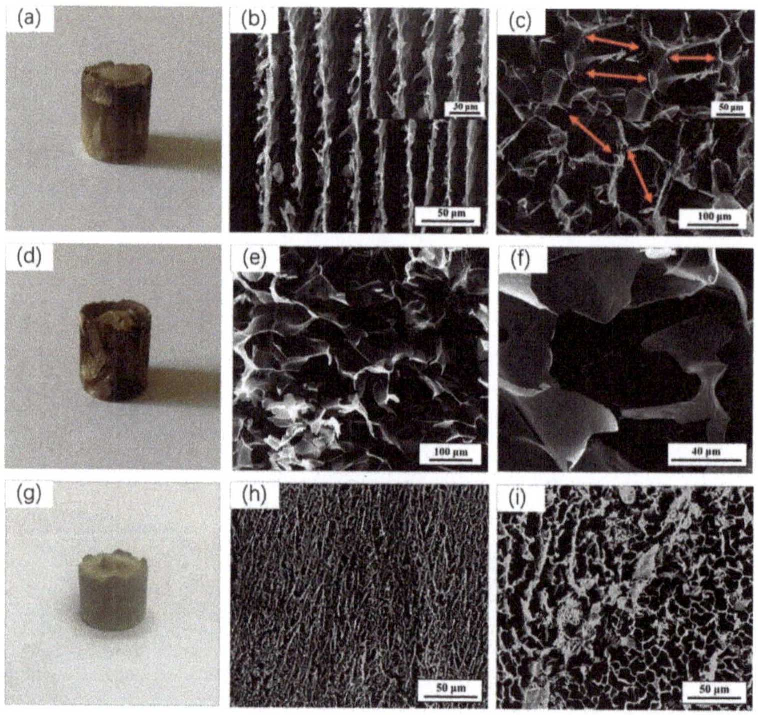

Figure 3. Macroscopic pictures and microstructures of aerogels formed by different freeze-drying methods. (**a**), (**d**) and (**g**) are macro photographs of d-MCPGA, r-MCPGA, n-MCPGA, respectively; (**b**) vertical section of d-MCPGA; (**c**) cross section of d-MCPGA; (**e–f**) internal SEM image of r-MCPGA; (**h**) vertical section of n-MCPGA; (**i**) cross section of n-MCPGA.

The MCPGA were ultra-light and had high porosity. The density and porosity of MCPGA (d-MCPGA, r-MCPGA, n-MCPGA) are summarized in Table 1.

Table 1. Physical properties of MCPGA (d-MCPGA, r-MCPGA, n-MCPGA).

Sample	Density (kg/m^3)	Porosity (%)
d-MCPGA	17.95	98.8
r- MCPGA	15.41	99.0
n-MCPGA	18.04	98.8

3.2. Chemical Properties

To evaluate the formation of TMCS/CNF/PVA/GO composites, we used Fourier transform infrared spectrometer and an X-Ray diffractomer. Figure 4 showed FTIR spectra of (a) CNFs, (b) GO, (c) PVA, (d) CNF/PVA/GO and (e) TMCS/CNF/PVA/GO aerogels.

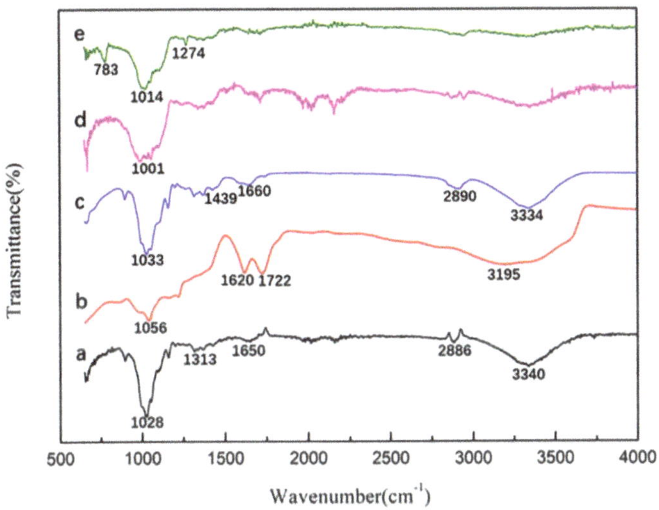

Figure 4. FTIR spectra of (**a**) CNFs, (**b**) GO, (**c**) PVA, (**d**) CNF/PVA/GO and (**e**) TMCS /CNF/PVA/GO aerogels.

The spectrum of CNFs (Figure 4a) showed that 3340 cm^{-1} and 1650 cm^{-1} were the stretching vibration and bending vibration of -OH, 2886 cm^{-1} was the stretching vibration of CH, 1313 cm^{-1} was the bending vibration of -OH, 1028 cm^{-1} was the stretching vibration absorption peak of C-O, whose vicinity had many weak shoulder peaks [35,40]. The spectrum of GO (Figure 4b) showed a strong peak corresponding to an oxygen-containing group. A broad peak near 3195 cm^{-1} was a hydroxyl group, 1722 cm^{-1} was a C=O stretching vibration from a carbonyl group and a carboxyl group, 1620 cm^{-1} was C=C from an aromatic ring, 1225 and 1056 cm^{-1} were C-O-C and C-O stretching vibrations, respectively [41]. In the infrared spectrum of polyvinyl alcohol (Figure 4c), 3334 cm^{-1} and 2890 cm^{-1} were stretching vibrations of -OH and -CH_2, respectively [42], 1660 cm^{-1} was H-O-H bending vibration, 1439 cm^{-1} was CH stretching vibration and 1033 cm^{-1} was the C-O-C stretching vibration absorption peak [43]. In the CNF/PVA/GO spectrum (Figure 4d), we could find that the diffraction peaks of CNFs, PVA, and GO all appeared, but the peaks were weakened, which might indicate that a strong interaction was forced among the three components of CNFs, PVA, and GO [34]. Figure 4e was the FTIR spectrum of the TMCS/CNF/PVA/GO aerogel. There was no significant change in the characteristic peak of the composite aerogel before and after the TMCS hydrophobic treatment. However, new diffraction peaks appeared at 783 and 1274 cm^{-1}, which were ascribed to characteristic vibrations of Si-O-Si and C-Si asymmetric stretching, respectively. In the region of 1000–1130 cm^{-1}, the Si-O-Si bond absorption band in the siloxane compound overlaps with the C-O bond of cellulose [28], which indicated that the silylation reaction proceeded successfully and only reacted on the surface of the aerogel so that the skeletal structure and mechanical properties of the aerogel were not affected.

The XRD patterns of (a) CNFs, (b) PVA, (c) GO, (d) CNF/PVA/GO and (e) TMCS/CNF/PVA/GO aerogels are shown in Figure 5. The distinct peak at $2\theta = 22.5°$ observed from CNFs (Figure 5a) was attributed to the typical reflection planes (002), and a broad peak at around 15.6° was assigned to the (101) and (10$\bar{1}$) lattice planes of the cellulose I crystalline structure [44]. The XRD pattern of the PVA (Figure 5b) exhibited a broad peak at $2\theta = 19.6°$, which was ascribed to the orthorhombic lattice structure of semi-crystalline PVA [45]. GO exhibited a diffraction peak at 2θ = 10.4° (Figure 5c), corresponding to the (002) crystal plane [46]. The diffraction peaks of the CNFs, PVA and GO all appeared in the CNF/PVA/GO aerogel (Figure 5d). A broad peak was observed at 2θ = 20.7° and the peak intensity decreased, corresponding to the diffraction peaks of the CNFs and PVA. Moreover, the peak of GO sharply weakened, probably due to the relatively small amount of GO added to the sample. The TMCS/CNF/PVA/GO aerogel (Figure 5e) did not show any different diffraction peaks of CNF/PVA/GO aerogel as the TMCS silylation reaction only occurred on the aerogel surface without changing its chemical structure.

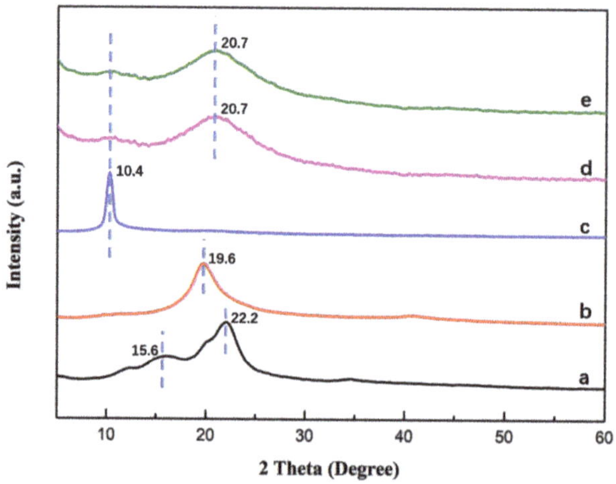

Figure 5. XRD patterns of (**a**) CNFs, (**b**) PVA, (**c**) GO, (**d**) CNF/PVA/GO and (**e**) TMCS/CNF/PVA/GO aerogels.

3.3. Mechanical Properties

The mechanical properties of aerogels were influenced by many factors, such as raw materials and ratios, material properties of each component, and microstructure. Under the same raw materials and ratios, different preparation methods have produced aerogels with different microstructures. This structural difference played a key role in influencing the mechanical properties of aerogels. The compression behavior of aerogels (d-MCPGA, r-MCPGA, n-MCPGA) made by CNFs, PVA and GO using the directional freeze-drying method is shown in Figure 6, and the compression behavior of aerogels prepared by conventional refrigerator freeze-drying and non-directional freeze-drying are used as contrast. The curve characteristics of this three-characteristic deformation region of Figure 6 are consistent with previous studies [47]. The compressive stress of r-MCPGA at 80% strain reached 0.11 MPa, slightly higher than other cellulose-based aerogels described in previous studies [48]. This result might be attributed to the strong interaction between the CNFs and GO through hydrogen bonding and the addition of PVA with a long polymeric chain that could bond with the high density hydrogen of the CNFs and GO. At 80% strain, the ultimate compressive stress of d-MCPGA (0.22 MPa) was two times higher than that of r-MCPGA (0.11 MPa) and about 1.5 times higher than that of n-MCPGA (0.15 MPa). This significant increase might be due to the microstructure of the radially aligned directional aerogel (Figure 3b). Moreover, the d-MCPGA had a higher compressive capacity than most of the reported oil-absorbing materials. Figure 7 compares the compressive strength of d-MCPGA and other oil-absorbing materials. The directional freezing process caused the aerogel to have an axially ordered vertical pore structure, which contributed to the high strength characteristics of the d-MCPGA, but the extrusion behavior destroyed its internal orientation structure and caused the structure to be unrecoverable; thus, d-MCPGA had high strength but was relatively inferior in flexibility and repeatability.

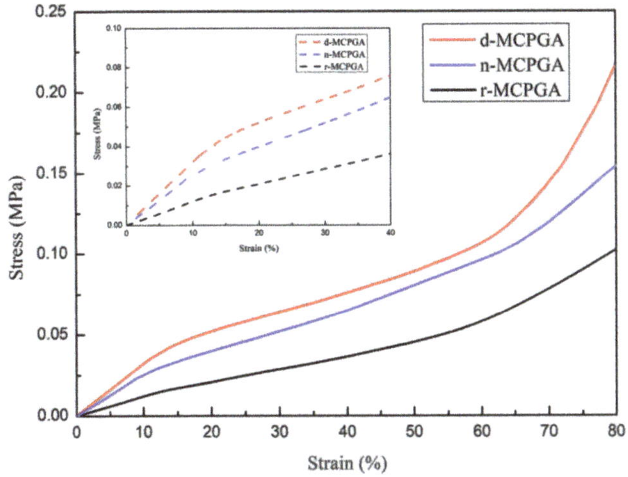

Figure 6. Compressive behavior of aerogels (d-MCPGA, r-MCPGA, n-MCPGA) prepared by different freeze-drying methods consisting of CNFs, PVA and GO.

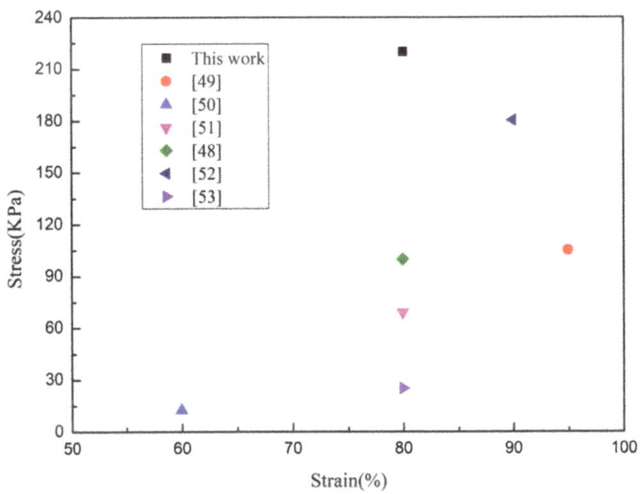

Figure 7. Comparison of the compressive strength of different materials [49–53].

As shown in Figure 8a–d, 0.0226 g of lightweight d-MCPGA were placed on the young leaves and carried a weight of 500 g without any deformation (bearing more than 22123 times its own weight), which far exceeded the previous the study. This high strength was mainly attributed to the excellent entanglement of CNFs with PVA and GO, and the directional freeze-drying method formed a highly oriented microstructure.

3.4. Surface Wettability

Hydrophobicity was one of the most important criteria for evaluating oil-absorbing materials. The surface wettability of CPGA, MCPGA (d-MCPGA, r-MCPGA, n-MCPGA) was investigated by contact angle measurement. Due to the presence of abundant hydroxyl groups hydroxyl groups on the surface of the CNFs, PVA and GO, CPGA exhibited strong hydrophilicity. As shown in Figure 9a, for the original CPGA prior to silane coating, 4 μL of water droplets were completely absorbed by in

0.2 s and the WCA of 0°. However, the silane-coated d-MCPGA, r-MCPGA, and n-MCPGA were all hydrophobic; the WCA was as high as 142, 138 and 139 (Figure 9b–d). After 120 s, the water droplets retained the initial contact angle and were not absorbed by the aerogel. The water contact angle of the outer surface did not change significantly with time. Various MCPGAs had a microporous structure, which could promote the complete filling of high porosity aerogels by TMCS. These results indicate that the aerogel had a highly hydrophobic surface after silane treatment.

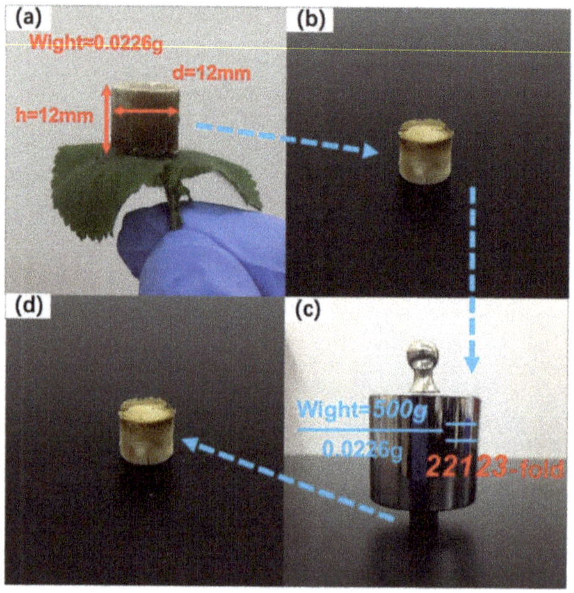

Figure 8. (**a**) Ultra-light cylindrical d-MCPGA supported by young leaves; (**b–d**) the pre, middle and post processes of the d-MCPGA supporting a 500 g mass load.

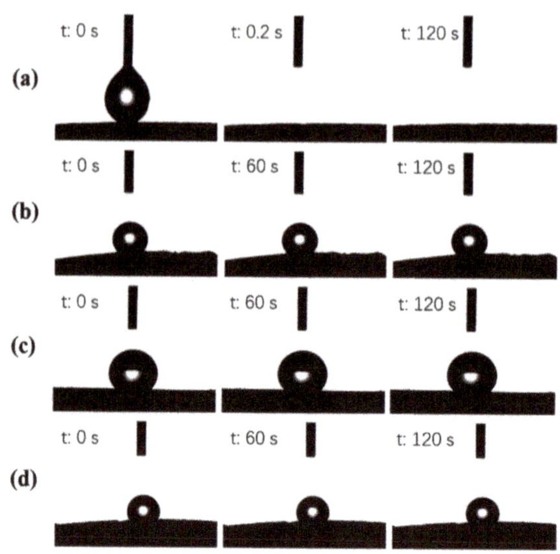

Figure 9. Water contact angle of (**a**) CPGA, (**b**) d-MCPGA, (**c**) r-MCPGA and (**d**) n-MCPGA.

3.5. Absorption Capacity

Silane-coated hydrophobic lipophilic MCPGA was an ideal absorbent material for the removal of oils and organic solvents from water. Due to its porous structure and oil-water selectivity, MCPGA has a strong ability to selectively absorb oil and organic liquids in water. Figure 10 showes the process of MCPGA removal of corn oil (stained with Sudan Red). MCPGA and dyed corn oil didn't touch and float on the surface of deionized water. When the two substances were in contact, the oil was quickly absorbed by MCPGA. After removing MCPGA, the remaining was clean water without any oil. In order to further study the oil absorption capacity of the MCPGA, we measured the adsorption capacity of different oils and organic solvents by weighing the samples before and after oil absorption. Figure 11 showed the absorption capacity of the MCPGA (d-MCPGA, r-MCPGA, n-MCPGA) for different oils and organic liquids. The oils and organic liquids tested included diesel, corn oil, N,N-dimethylformamide, pump oil, secondary pump oil, engine oil and ethanol.

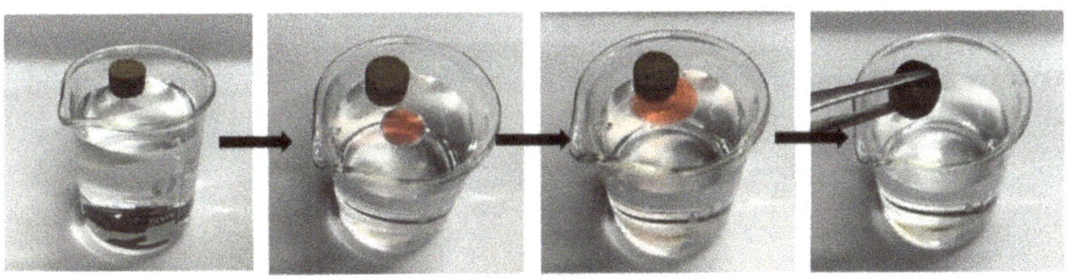

Figure 10. Removal of corn oil (dyed with Sudan red) from the water surface using MCPGA.

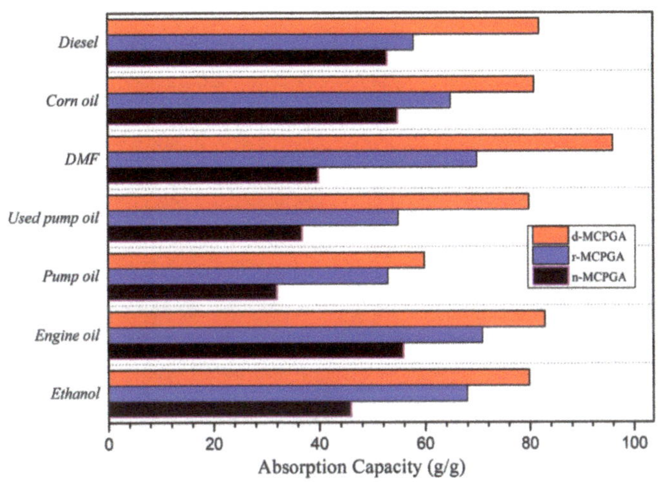

Figure 11. Absorption capacity of MCPGA (d-MCPGA, r-MCPGA, n-MCPGA) for different oils and organic liquids.

The absorption ratios of d-MCPGA, r-MCPGA and n-MCPGA were different. The oil absorption capacity of d-MCPGA was higher (60~96 times), the oil absorption capacity of r-MCPGA was second (53~71 times), while n-MCPGA had the worst oil absorption capacity (32~56 times). The oil was mainly stored in the macropores of the aerogel, so the difference in the absorption capacity between the various organic solvents depended on the surface tension and density of the solvent tested, and the different porous structures and porosities inside the three aerogels. Capillary forces were responsible for the adsorption capacity of aerogels. Due to the capillary interaction between the mobile liquid and the aerogel surface, the liquid could flow spontaneously into the narrow network channel inside the aerogel [54]. Since MCPCA was a porous structure, it had massive capillary channels. Compared to n-MCPGA and r-MCPGA, organic solvents could more easily enter the pores of d-MCPGA through capillary tension. The results of the absorption ratios were consistent with the microstructure and porosity results of the three aerogels previously analyzed. In addition, the oil absorption capacity of d-MCPGA was higher than most existing oil absorption materials (Table 2).

Table 2. Comparison of the absorption capacities of different materials.

Absorbent Material	WCA	Absorption Capacity (g/g)	Ref.
TMCS/rGO/CNF aerogel	117	33–39	[36]
TiO_2-coated nanocellulose aerogel	>90	40	[55]
Spongy graphene	95	20–86	[56]
Graphene-based aerogel	>90	28–40	[57]
MTMS-coated cellulose aerogel	135	18–20	[58]
CNF aerogel	Unknown	28	[59]
chitosan-silica aerogel	137	13–30	[60]
Carbon fiber aerogel from bamboo	145	22–80	[61]
Kymene-coated CNF aerogel	144	24–46	[62]
Anisotropic CNF/PVA/GO aerogel	142	60–96	This work

As shown in Figure 12, the absorption capacity of (a) pump oil; (b) engine oil and (c) corn oil on the MCPGA (d-MCPGA, r-MCPGA, n-MCPGA) was plotted as a function of absorption time. MCPGA took about 20–40 s to achieve the absorption balance of oil and pump oil, while in corn oil it only needed 10 s. The greater the viscosity of liquid, the worse the fluidity and the slower the movement of the molecules. Therefore, the pump oil with higher viscosity reached the absorption saturation in a longer time (viscosity: pump oil > engine oil > corn oil). During the adsorption process, initial adsorption rate was faster and slowed down over time until the adsorption stopped, which might be attributed to the pore structure of the porous aerogel providing a large number of channels for the oil. In addition, the surface and interior of the aerogel possessed a large number of available active adsorption sites, which tended to saturate with time [36]. There was no significant difference in the adsorption rate of d-MCPGA and r-MCPGA. As can be seen from Figure 12b,c, the adsorption rate of n-MCPGA was relatively slow, which might be due to its smaller pores and the narrower channels.

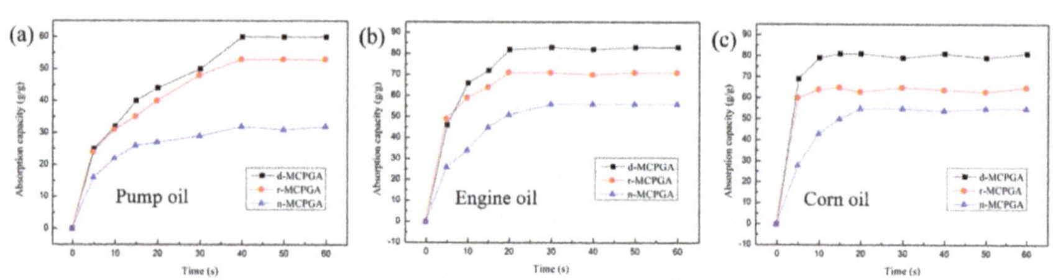

Figure 12. The absorption capacity of MCPGA (d-MCPGA, r-MCPGA, n-MCPGA) to adsorb (**a**) pump oil; (**b**) engine oil and (c) corn oil as a function of time.

4. Conclusions

In summary, we have successfully fabricated hydrophobic anisotropic CNF/PVA/GO aerogel for oil adsorbents using a simple directional freeze-drying process followed by thermal chemical vapor deposition of TMCS. The CNF/PVA/GO aerogel provided a parallel wall porous structure, thereby achieving an excellent oil and organic solvent absorption capacity of up to 96 times its own weight. Further, the assembled anisotropic aerogel delivered low density (17.95 mg/cm^3), high strength (the compressive stress at 80% strain reaching 0.22 MPa and bearing capacity exceeds 22,123 times its own weight). Therefore, this high strength MCPGA is a potential selective oil absorption material.

Author Contributions: Formal analysis, L.Z.; Funding acquisition, S.Z.; Software, Y.C.; Writing—original draft, L.Z.; Writing—review & editing, Z.X.

Acknowledgments: This work was supported by the National Natural Science Foundation of China (31770607), the National Key Research and Development Program of China (2017YFD0600204) and the Natural Science Foundation of Jiangsu Province of China (BK20171450). The authors acknowledge the Advanced Analysis & Testing Center of Nanjing Forestry University.

Conflicts of Interest: The authors declare no conflict of interest.

References

1. Al-Majed, A.A.; Adebayo, A.R.; Hossain, M.E. A sustainable approach to controlling oil spills. *J. Environ. Manag.* **2012**, *113*, 213–227. [CrossRef] [PubMed]
2. Shannon, M.A.; Bohn, P.W.; Elimelech, M.; Georgiadis, J.G.; Marinas, B.J.; Mayes, A.M. Science and technology for water purification in the coming decades. *Nature* **2008**, *452*, 301–310. [CrossRef] [PubMed]
3. Vilcáez, J.; Li, L.; Hubbard, S.S. A new model for the biodegradation kinetics of oil droplets: Application to the Deepwater Horizon oil spill in the Gulf of Mexico. *Geochem. Trans.* **2013**, *14*, 4. [CrossRef] [PubMed]
4. Ge, J.; Ye, Y.D.; Yao, H.B.; Zhu, X.; Wang, X.; Wu, L.; Yu, S.H. Pumping through porous hydrophobic/oleophilic materials: An alternative technology for oil spill remediation. *Angew. Chem. Int. Ed.* **2014**, *126*, 3686–3690. [CrossRef]
5. Liu, H.; Geng, B.; Chen, Y.; Wang, H. A review on the aerogel-type oil sorbents derived from nanocellulose. *ACS Sustain. Chem. Eng.* **2016**, *5*, 49–66. [CrossRef]
6. Prince, R.C. Bioremediation of marine oil spills. *Oil Gas Sci. Technol.* **2003**, *58*, 463–468. [CrossRef]
7. Syed, S.; Alhazzaa, M.I.; Asif, M. Treatment of oily water using hydrophobic nano-silica. *Chem. Eng. J.* **2011**, *167*, 99–103. [CrossRef]
8. Adebajo, M.O.; Frost, R.L.; Kloprogge, J.T.; Carmody, O.; Kokot, S. Porous Materials for Oil Spill Cleanup: A Review of Synthesis and Absorbing Properties. *J. Porous Mater.* **2003**, *10*, 159–170. [CrossRef]
9. Carmody, O.; Frost, R.; Xi, Y.; Kokot, S. Adsorption of hydrocarbons on organo-clays—Implications for oil spill remediation. *J. Colloid Interface Sci.* **2007**, *305*, 17–24. [CrossRef]
10. Wiehn, M.; Levario, T.J.; Staggs, K.; Linneen, N.; Wang, Y.C.; Pfeffer, R. Adsorption of Short-Chain Alcohols by Hydrophobic Silica Aerogels. *Ind. Eng. Chem. Res.* **2013**, *52*, 18379–18385. [CrossRef]
11. Bastani, D.; Safekordi, A.A.; Alihosseini, A.; Taghikhani, V. Study of oil sorption by expanded perlite at 298.15 K. *Sep. Purif. Technol.* **2006**, *52*, 295–300. [CrossRef]
12. Karakasi, O.K.; Moutsatsou, A. Surface modification of high calcium fly ash for its application in oil spill clean up. *Fuel* **2010**, *89*, 3966–3970. [CrossRef]
13. Wu, D.; Wu, W.; Yu, Z.; Zhang, C.; Zhu, H. Facile Preparation and Characterization of Modified Polyurethane Sponge for Oil Absorption. *Ind. Eng. Chem. Res.* **2014**, *53*, 20139–20144. [CrossRef]
14. Lin, J.; Shang, Y.; Ding, B.; Yang, J.; Yu, J.; Al-Deyab, S.S. Nanoporous polystyrene fibers for oil spill cleanup. *Mar. Pollut. Bull.* **2012**, *64*, 347–352. [CrossRef] [PubMed]
15. Singh, V.; Jinka, S.; Hake, K.; Parameswaran, S.; Kendall, R.J.; Ramkumar, S. Novel Natural Sorbent for Oil Spill Cleanup. *Ind. Eng. Chem. Res.* **2014**, *53*, 11954–11961. [CrossRef]
16. Sun, X.F.; Sun, R.; Sun, J.X. Isolation and Characterization of Cellulose Obtained from Ultrasonic Irradiated Sugarcane Bagasse. *J. Agric. Food Chem.* **2002**, *50*, 6428–6433. [CrossRef]
17. Chen, C.; Wang, H.; Li, S.; Fang, L.; Li, D. Reinforcement of cellulose nanofibers in polyacrylamide gels. *Cellulose* **2017**, *24*, 5487–5493. [CrossRef]
18. Feng, J.; Nguyen, S.T.; Fan, Z.; Duong, H.M. Advanced fabrication and oil absorption properties of super-hydrophobic recycled cellulose aerogels. *Chem. Eng. J.* **2015**, *270*, 168–175. [CrossRef]
19. Nguyen, S.T.; Feng, J.; Le, N.T.; Le, A.T.T.; Hoang, N.; Tan, V.B.C. Cellulose Aerogel from Paper Waste for Crude Oil Spill Cleaning. *Ind. Eng. Chem. Res.* **2013**, *52*, 18386–18391. [CrossRef]
20. Jin, C.; Han, S.; Li, J.; Sun, Q. Fabrication of cellulose-based aerogels from waste newspaper without any pretreatment and their use for absorbents. *Carbohydr. Polym.* **2015**, *123*, 150–156. [CrossRef] [PubMed]
21. Rafieian, F.; Hosseini, M.; Jonoobi, M.; Yu, Q. Development of hydrophobic nanocellulose-based aerogel via chemical vapor deposition for oil separation? for water treatment. *Cellulose* **2018**, *25*, 4695–4710. [CrossRef]

22. Xu, Z.; Jiang, X.; Zhou, H.; Li, J. Preparation of magnetic hydrophobic polyvinyl alcohol (PVA)–cellulose nanofiber (CNF) aerogels as effective oil absorbents. *Cellulose* **2017**, *25*, 1217–1227. [CrossRef]
23. Matsumura, S.; Kurita, H.; Shimokobe, H. Anaerobic biodegradability of polyvinyl alcohol. *Biotechnol. Lett.* **1993**, *15*, 749–754. [CrossRef]
24. Finlay, K.; Gawryla, M.D.; Schiraldi, D.A. Biologically Based Fiber-Reinforced/Clay Aerogel Composites. *Ind. Eng. Chem. Res.* **2008**, *47*, 615–619. [CrossRef]
25. Maleki, H.; Durães, L.; Portugal, A. Synthesis of lightweight polymer-reinforced silica aerogels with improved mechanical and thermal insulation properties for space applications. *Microporous Mesoporous Mater.* **2014**, *197*, 116–129. [CrossRef]
26. Parmenter, K.E.; Milstein, F. Mechanical properties of silica aerogels. *J. Non-Cryst. Solids* **1998**, *223*, 179–189. [CrossRef]
27. Bai, H.; Chen, Y.; Delattre, B.; Tomsia, A.P.; Ritchie, R.O. Bioinspired Large-Scale Aligned Porous Materials Assembled with Dual Temperature Gradients. *Sci. Adv.* **2015**, *1*, e1500849. [CrossRef]
28. Zhai, T.; Zheng, Q.; Cai, Z.; Turng, L.S.; Xia, H.; Gong, S. Poly(vinyl alcohol)/cellulose nanofibril hybrid aerogels with an aligned microtubular porous structure and their composites with polydimethylsiloxane. *ACS Appl. Mater. Interfaces* **2015**, *7*, 7436–7444. [CrossRef]
29. Zhang, X.; Wang, H.; Cai, Z.; Yan, N.; Liu, M.; Yu, Y. Highly compressible and hydrophobic anisotropic aerogels for selective oil/organic solvent absorption. *ACS Sustain. Chem. Eng.* **2018**, *7*, 332–340. [CrossRef]
30. Lee, J.; Deng, Y. The morphology and mechanical properties of layer structured cellulose microfibril foams from ice-templating methods. *Soft Matter* **2011**, *7*, 6034. [CrossRef]
31. Wang, C.; Chen, X.; Wang, B.; Huang, M.; Wang, B.; Jiang, Y.; Ruoff, R.S. Freeze-Casting Produces a Graphene Oxide Aerogel with a Radial and Centrosymmetric Structure. *ACS Nano* **2018**, *12*, 5816–5825. [CrossRef] [PubMed]
32. Sun, H.; Xu, Z.; Gao, C. Aerogels: Multifunctional, Ultra-Flyweight, Synergistically Assembled Carbon Aerogels. *Adv. Mater.* **2013**, *25*, 2632. [CrossRef]
33. Wang, C.; He, X.; Shang, Y.; Peng, Q.; Qin, Y.; Shi, E. Multifunctional graphene sheet–nanoribbon hybrid aerogels. *J. Mater. Chem. A* **2014**, *2*, 14994–15000. [CrossRef]
34. Xu, Z.; Zhou, H.; Tan, S.; Jiang, X.; Wu, W.; Shi, J.; Chen, P. Ultralight super-hydrophobic carbon aerogels based on cellulose nanofibers/poly(vinyl alcohol)/graphene oxide (CNFs/PVA/GO) for highly effective oil–water separation. *Beilstein J. Nanotech.* **2018**, *9*, 508–519. [CrossRef]
35. Chen, W.; Yu, H.; Liu, Y. Preparation of millimeter-long cellulose I nanofibers with diameters of 30–80 nm from bamboo fibers. *Carbohydr. Polym.* **2011**, *86*, 453–461. [CrossRef]
36. Xu, Z.; Zhou, H.; Jiang, X.; Li, J.; Huang, F. Facile synthesis of reduced graphene oxide/trimethyl chlorosilane-coated cellulose nanofibres aerogel for oil absorption. *IET Nanobiotechnol.* **2017**, *11*, 929–934. [CrossRef]
37. Hummers, W.S., Jr.; Offeman, R.E. Preparation of Graphitic Oxide. *J. Am. Chem. Soc.* **1958**, *80*, 1339. [CrossRef]
38. Deville, S. Freeze-Casting of Porous Ceramics: A Review of Current Achievements and Issues. *Adv. Eng. Mater.* **2010**, *10*, 155–169. [CrossRef]
39. Deville, S.; Saiz, E.; Tomsia, A.P. Freeze casting of hydroxyapatite scaffolds for bone tissue engineering. *Biomaterials* **2006**, *27*, 5480–5489. [CrossRef]
40. Oularbi, L.; Turmine, M.; El Rhazi, M. Electrochemical determination of traces lead ions using a new nanocomposite of polypyrrole/carbon nanofibers. *J. Solid State Electrochem.* **2017**, *21*, 3289–3300. [CrossRef]
41. Tang, G.; Jiang, Z.G.; Li, X.; Zhang, H.B.; Dasari, A.; Yu, Z.Z. Three dimensional graphene aerogels and their electrically conductive composites. *Carbon* **2014**, *77*, 592–599. [CrossRef]
42. Wang, S.; Ren, J.; Li, W.; Sun, R.; Liu, S. Properties of polyvinyl alcohol/xylan composite films with citric acid. *Carbohydr. Polym.* **2014**, *103*, 94–99. [CrossRef]
43. Wu, L.; Yuan, X.; Sheng, J. Immobilization of cellulase in nanofibrous PVA membranes by electrospinning. *J. Membr. Sci.* **2005**, *250*, 167–173. [CrossRef]
44. Zheng, Q.; Cai, Z.; Ma, Z.; Gong, S. Cellulose nanofibril/reduced graphene oxide/carbon nanotube hybrid aerogels for highly flexible and all-solid-state supercapacitors. *ACS Appl. Mater. Interfaces* **2015**, *7*, 3263–3271. [CrossRef]
45. Liu, D.; Sun, X.; Tian, H.; Maiti, S.; Ma, Z. Effects of cellulose nanofibrils on the structure and properties on PVA nanocomposites. *Cellulose* **2013**, *20*, 2981–2989. [CrossRef]

46. Li, C.; Feng, C.; Peng, Z.; Gong, W.; Kong, L. Ammonium-assisted green fabrication of graphene/natural rubber latex composite. *Polym. Compos.* **2013**, *34*, 88–95. [CrossRef]
47. Zheng, Q.; Javadi, A.; Sabo, R.; Cai, Z.; Gong, S. Polyvinyl alcohol (PVA)–cellulose nanofibril (CNF)–multiwalled carbon nanotube (MWCNT) hybrid organic aerogels with superior mechanical properties. *RSC Adv.* **2013**, *3*, 20816–20823. [CrossRef]
48. Javadi, A.; Zheng, Q.; Payen, F.; Javadi, A.; Altin, Y.; Cai, Z.; Gong, S. Polyvinyl alcohol-cellulose nanofibrils-graphene oxide hybrid organic aerogels. *ACS Appl. Mater. Interfaces* **2013**, *5*, 5969–5975. [CrossRef] [PubMed]
49. Yang, X.; Cranston, E.D. Chemically Cross-Linked Cellulose Nanocrystal Aerogels with Shape Recovery and Superabsorbent Properties. *Chem. Mater.* **2014**, *26*, 6016–6025. [CrossRef]
50. Zhou, S.; Wang, M.; Chen, X.; Xu, F. Facile Template Synthesis of Microfibrillated Cellulose/Polypyrrole/Silver Nanoparticles Hybrid Aerogels with Electrical Conductive and Pressure Responsive Properties. *ACS Sustain. Chem. Eng.* **2015**, *3*, 3346–3354. [CrossRef]
51. Martoïa, F.; Cochereau, T.; Dumont, P.J.J.; Orgéas, L.; Terrien, M.; Belgacem, M.N. Cellulose nanofibril foams: Links between ice-templating conditions, microstructures and mechanical properties. *Mater. Des.* **2016**, *104*, 376–391. [CrossRef]
52. Wicklein, B.; Kocjan, A.; Salazar-Alvarez, G.; Carosio, F.; Camino, G.; Antonietti, M.; Bergström, L. Thermally insulating and fire-retardant lightweight anisotropic foams based on nanocellulose and graphene oxide. *Nat. Nanotechnol.* **2015**, *10*, 277. [CrossRef] [PubMed]
53. Jiang, F.; Hsieh, Y.L. Amphiphilic superabsorbent cellulose nanofibril aerogels. *J. Mater. Chem. A* **2014**, *2*, 6337–6342. [CrossRef]
54. Zhan, W.; Yu, S.; Gao, L.; Wang, F.; Fu, X.; Sui, G.; Yang, X. Bioinspired Assembly of Carbon Nanotube into Graphene Aerogel with "Cabbagelike" Hierarchical Porous Structure for Highly Efficient Organic Pollutants Cleanup. *ACS Appl. Mater. Interfaces* **2017**, *10*, 1093–1103. [CrossRef] [PubMed]
55. Ras, R.H.A.; Ikkala, O.; Korhonen, J.T.; Kettunen, M. Hydrophobic Nanocellulose Aerogels as Floating, Sustainable, Reusable, and Recyclable Oil Absorbents. *ACS Appl. Mater. Interfaces* **2011**, *3*, 1813–1816. [CrossRef]
56. Bi, H.; Xie, X.; Yin, K.; Zhou, Y.; Wan, S.; He, L. Spongy Graphene as a Highly Efficient and Recyclable Sorbent for Oils and Organic Solvents. *Adv. Funct. Mater.* **2012**, *22*, 4421–4425. [CrossRef]
57. Wu, T.; Chen, M.; Zhang, L.; Xu, X.; Gao, J. Three-dimensional graphene-based aerogels prepared by a self-assembly process and its excellent catalytic and absorbing performance. *J. Mater. Chem. A* **2013**, *1*, 7612–7621. [CrossRef]
58. Nguyen, S.T.; Feng, J.; Ng, S.K.; Wong, J.P.W.; Tan, V.B.C.; Duong, H.M. Advanced thermal insulation and absorption properties of recycled cellulose aerogels. *Colloids Surf. A* **2014**, *445*, 128–134. [CrossRef]
59. Chin, S.F.; Binti Romainor, A.N.; Pang, S.C. Fabrication of hydrophobic and magnetic cellulose aerogel with high oil absorption capacity. *Mater. Lett.* **2014**, *115*, 241–243. [CrossRef]
60. Ma, Q.; Liu, Y.; Dong, Z.; Wang, J.; Hou, X. Hydrophobic and nanoporous chitosan-silica composite aerogels for oil absorption. *J. Appl. Polym. Sci.* **2015**, 132. [CrossRef]
61. Yang, S.; Chen, L.; Mu, L.; Hao, B.; Ma, P.C. Low cost carbon fiber aerogel derived from bamboo for the adsorption of oils and organic solvents with excellent performances. *RSC Adv.* **2015**, *5*, 38470–38478. [CrossRef]
62. Mulyadi, A.; Zhang, Z.; Deng, Y. Fluorine-Free Oil Absorbents Made from Cellulose Nanofibril Aerogels. *ACS Appl. Mater. Interfaces* **2016**, *8*, 2732–2740. [CrossRef] [PubMed]

polymers

MDPI

Article

Electrospun Graphene Nanosheet-Filled Poly(Trimethylene Terephthalate) Composite Fibers: Effects of the Graphene Nanosheet Content on Morphologies, Electrical Conductivity, Crystallization Behavior, and Mechanical Properties

Chien-Lin Huang *, Hsuan-Hua Wu, Yung-Ching Jeng and Wei-Zhi Liang

Department of Fiber and Composite Materials, Feng Chia University, Taichung 40724, Taiwan; janis781031@gmail.com (H.-H.W.); rattenfingero8@gmail.com (Y.-C.J.); bobby96124@gmail.com (W.-Z.L.)
* Correspondence: clhuang@mail.fcu.edu.tw; Tel.: +886-4-2451-7250 (ext. 3437)

Received: 19 November 2018; Accepted: 15 January 2019; Published: 17 January 2019

Abstract: In this study the effects of increased graphene nanosheet (GNS) concentration on variations in the structure and properties of electrospun GNS-filled poly(trimethylene terephthalate) (PTT/GNS) composite fiber, such as its morphologies, crystallization behavior, mechanical properties, and electrical conductivity, were investigated. The effects of GNS addition on solution rheology and conductivity were also investigated. GNSs were embedded in the fibers and formed protrusions. The PTT cold crystallization rate of PTT/GNS composite fibers increased with the gradual addition of GNSs. A PTT mesomorphic phase was formed during electrospinning, and GNSs could induce the PTT mesomorphic phase significantly during PTT/GNS composite fiber electrospinning. The PTT/GNS composite fiber mats (CFMs) became ductile with the addition of GNSs. The elastic recoveries of the PTT/GNS CFMs with 170 °C annealing were better than those of the as-spun PTT/GNS CFMs. Percolation scaling laws were applied to the magnitude of conductivity to reveal the percolation network of electrospun PTT/GNS CFMs. The electrical conductivity mechanism of the PTT/GNS CFMs differed from that of the PTT/GNS composite films. Results showed that the porous structure of the PTT CFMs influenced the performance of the mats in terms of electrical conductivity.

Keywords: poly(trimethylene terephthalate); graphene; electrospinning; composite fiber; morphology; crystallization; electrical conductivity; mechanical property; elastic recovery

1. Introduction

Graphene nanosheets (GNSs) that are several of nanometers thick and their related materials, such as carbon nanotubes (CNTs), are promising functional nanofillers for advanced applications because of their outstanding mechanical and electrical properties and large surface areas [1]. GNS-filled and CNT-filled polymer composites can be used in high-conductivity applications, such as electrostatic discharge devices [2], electromagnetic interference (EMI)-shielding materials [3,4], sensors [5], electrical switching [6], ambipolar field-effect transistors [7,8], and electrodes [9]. Furthermore, polymer/GNS composites are widely being explored because GNSs are affordable and effective alternatives to CNTs.

Electrospinning has attracted increasing attention in recent years as a method for fabricating polymeric fibers. Electrospun fiber mats exhibit considerable potential in energy devices, filtration, tissue engineering, and biosensors because of their high surface area-to-volume ratio, high porosity, and diverse nanostructures [10]. Furthermore, electrospun fibers filled with GNS or CNT can increase the functional properties of fibers for advanced applications. Su et al. [11] reported that the uniaxially oriented electrospun polyethylene oxide (PEO)/CNT composite fiber mats (CFMs) have better tensile strength, modulus, and electrical conductivity than randomly oriented electrospun PEO/CNT CFMs. Additionally, the crystallization rate and the perfection of PEO crystal in PEO/CNT fibers can be affected owing to the high extensional force and the bending jet yielded during PEO/CNT electrospinning. Kim et al. [12] found that although graphene oxide (GO) protrudes from the electrospun polyacrylonitrile (PAN) fibers, the carbon composite fiber mats (CFMs) with GNS retain good conductivity following PAN/GO fiber carbonization at 1000 °C. Li et al. [13] reported that the a highly sensitive H_2O_2 biosensor could be prepared via the electrospun poly(vinyl alcohol) (PVA)/GNS-decorated with silver nanoparticles CFMs. Jin et al. [14] reported that the electrospun PAN/CNT-cobalt ferrite CFMs exhibits an EMI shielding efficiency (SE) value of approximately 3.9 dB. Furthermore, Li et al. [15] successfully developed a novel poly(methyl methacrylate) (PMMA)/polyamide 6 (PA6)/GNS nanocomposite which was prepared from electrospun PA6/GNS and PMMA CFMs through hot press molding. The mechanical properties of the PMMA/PA6/GNS nanocomposites were significantly improved. Ramazani et al. [16] showed that the fiber diameter of polycaprolactone (PCL)/GO CFMs decreases as GO content increases because the viscosity of the PCL/GO solution decreases. GNS particles usually protrude when incorporated in electrospun fibers because the lateral dimension of the former is usually larger than the diameter of the latter. Therefore, GNS protrusions on electrospun composite fibers could be in contact with other substances. In our previous work [17], PVA solutions filled with GNSs were prepared for electrospinning, and the PVA/GNS CFMs showed good cell adhesion and proliferation, thereby suggesting their possible use in tissue engineering. This phenonmenon caused by the GNS protrusions could help the proliferation of cells. Gao et al. [18] showed that GNS could be anchored onto thermoplastic polyurethane (TPU) fibers through the combination of electrospinning and ultrasonication. Moreover, the conductivity of the TPU/GNS composite fibers significantly increased. However, limited information is available as to the electrical conductivity (σ) and GNSs dispersion state of polymer/GNS CFMs from electrospinning polymer solutions with various amounts of GNSs, especially with extra high GNS loading. Therefore, the novel composite fibers filled with GNS can be used in the development of functional objects for various of flexible applications, ranging from electrical sensors to tissue engineering.

The electrical property of composites filled with GNSs depends on the microstructural properties, such as GNS dispersion state, GNS–GNS interaction, and polymer–GNS interaction. The percolation scaling law is usually used as a theoretical basis for evaluating the microstructure of composites filled with fillers. The percolation equation is typically expressed as $P \sim (\phi - \phi_c)^{\beta}$ [19,20], where P is the asymptotic behavior of a material property, ϕ is the filler volume fraction, β is the critical scaling exponent, and ϕ_c is the fraction threshold. A filler network is formed that is higher than ϕ_c, based on our previous study [21] where the percolated filler–filler network in composites in which the percolation exponent and threshold change was correlated with the filler geometric structure. Mazinani et al. [22] reported a polystyrene (PS) composite fiber filled with different contents and types of CNT. In addition, the σ of the electrospun PS/multi-walled CNT (MWCNT) CFMs was fitted using the percolation theory. Chien et al. [23] also demonstrated that the σ of the electrospun poly(D,L-lactic acid) (PDLLA)/carbon nanocapsule (CNC) CFMs could fit using the percolation theory. Sreeprasad et al. [5] demonstrated that polyallylamine hydrochloride (PAH) microfibers could be spun from a 40% PAH solution. The as-spun PAH fibers were immersed in a graphene quantum dot (GQD) solution and the covalently anchored GQDs on the PAH fibers were obtained to produce a percolating network to exhibit electron-tunneling transport for humidity and pressure sensor applications. However, to the best of our knowledge, the electrical percolation results of the electrospun polymer/GNS CFMs have not yet been reported.

Poly(trimethylene terephthalate) (PTT), a semi-crystalline thermoplastic polymer, is a member of the polyester family with an odd number of methylene groups. PTT possesses a faster crystallization rate and lower melting temperature than poly(ethylene terephthalate) (PET). Moreover, the PTT fiber also possess a high elastic recovery even at a high elongation of 20% [24]. Using PTT for an electrically conductive application has an advantage over that of other thermoplastic polymers, such as thermoplastic polyurethanes, because of its low water absorption, dimensional stability, better mechanical strength, and chemical resistance of the former [25–27]. PTT composites filled with nanofillers show enhanced properties and performance. Gupta et al. [28] demonstrated that PTT composites filled with 5 wt % MWCNT has an EMI SE value of approximately 20 dB. We previously reported that the PTT crystallization rate increases as GNSs are gradually added [29]. Meanwhile, in the case wherein liquid nitrogen was used to quench the melt, a PTT mesomorphic phase was formed despite the extremely short crystallization time with high GNS loading. The cold crystallization of the PTT composites was retarded by the GNSs. Moreover, the percolation exponent and threshold of the PTT/GNS composites are larger than those of the PTT/CNT composites [21]. In the electrospun PTT/CNT composite fibers, the aspect ratio and functionalized group of CNT affects the morphology of the resulting fibers [30]. However, the σ, morphologies, crystallization behavior, and mechanical properties of PTT/GNS composite fibers were poorly described, especially the elastic recovery of the electrospun PTT/GNS CFMs.

Khil et al. [31] reported that electrospun PTT fibers with diameters from 200–600 nm could be obtained via electrospinning with a trifluoroacetic acid (TFA)-methylene chloride co-solvent. Wang et al. [32–34] showed that electrospun PET, PTT, and polybutylene terephthalate could be prepared from a TFA solvent. Polyesters are soluble in strong polar solvents, such as TFA. However, TFA is not a good dispersion medium for GNSs. Therefore, determining the appropriate GNS dispersion in electrospun fibers remains a challenge. First, GNSs should demonstrate good dispersion in a semidilute polymer solution. Common processing routes for mixing polymers and fillers, such as melt compounding [35] and coagulation [36,37], were reported as a means to incorporate fillers into thermoplastic polymer matrices. Nevertheless, GNSs easily aggregate in polymer solutions with high GNS contents due to the van der Waals forces of attraction. Wu et al. [30] reported that PTT/CNT composites could be prepared via melt compounding for electrospinning. Moreover, PTT composites filled with GNS could be prepared through coagulation by using *ortho*-dichlorobenzene (*o*-DCB)-phenol as a solvent to obtain low electrical percolation thresholds of 0.53 and 0.84 vol % in GNS with different aspect ratios [21,29]. Coagulation should work better than melt compounding for GNS to be well dispersed and incorporated in thermoplastic polymer matrices because the low solution viscosity improves the polymer chain diffusion into the GNS interstices. Thus, fillers are dispersed in a polymer matrix to form composite powders, and the composite powders are re-dissolved in a solvent to obtain a well-dispersed polymer/filler solution. This method may be an appropriate route for preparing polyester/GNS solutions for electrospinning.

Although many achievements have been reported in this area, the polymer chain conformations and crystal structures in semicrystalline polymer/GNS composite nanofibers are incompletely known. Thus, understanding how GNS contents interact with electrospun composite fibers and how the functional properties of the fibers can be controlled for advanced applications, such as mechanical and electrical applications is of great interest. In this study, PTT composite powders filled with well-dispersed GNSs were prepared via coagulation, and the composite powders were re-dissolved in a TFA solvent to obtain a well-dispersed PTT/GNS solution. To clarify the electrified jet manipulation for producing PTT composite fibers filled with GNS, systematically studying the effects of GNS on solution properties and the electrospinning process is necessary. Thus, the rheological and conductive properties of the prepared PTT/GNS solution were measured and correlated with the electrospinnability characteristics of the solution. The morphologies of GNS assemblies in the PTT fiber with extra high GNS loadings were also investigated. GNSs formed protrusions on the PTT/GNS composite fibers, and these protrusions helped the formation of the GNS–GNS network in the PTT/GNS CFMs. The influence of GNS loading on the cold crystallization of PTT/GNS CFMs was analyzed via differential scanning calorimetry (DSC), Fourier transform infrared spectroscopy (FTIR), and wide-angle X-ray diffraction (WAXD). GNSs could induce the PTT mesomorphic phase in PTT/GNS composite fibers significantly during PTT/GNS electrospinning. Moreover, the σ, cold crystallization behavior, and mechanical properties of the PTT/GNS CFMs were studied to reveal the GNS dispersion state, GNS–GNS network, and microstructure in the electrospun PTT composite fibers. The advantages for incorporation of GNS in the electrospun PTT fibers increased the electrical conductivity and ductility of the PTT/GNS CFMs.

2. Materials and Methods

2.1. Materials and Composite Preparation

PTT was purchased from DuPont Co. (Wilmington, DE, USA), and the viscosity-average molecular weight of PTT was determined to be 53,100 g/mol [29]. *o*-DCB was purchased from Acros Organics (Morris Plains, NJ, USA). Phenol was purchased from Showa Chemical Co. (Tokyo, Japan). GNSs were purchased from Enerage Inc. (P-LF10, Yilan, Taiwan), which were prepared via thermal reduction of GO at 1500 °C. According to the manufacturer, the average thickness of the GNS was 3 nm. The PTT and GNSs had densities of 1.28 and 2.20 g/cm^3 [21], respectively. The GNSs were added to the *o*-DCB-phenol (1:1, *v*/*v*) solvent, after which ultrasonic treatment was performed for 6 h. Weighed PTT pellets were then added. Afterward, the mixture, with a 1% (*w*/*v*) solid content was stirred at 140 °C to dissolve the PTT pellets. The uniform suspension solution was precipitated dropwise into a 20-fold excess volume of methanol. The precipitated powders were continuously dried in a vacuum oven at 120 °C for 72 h until the residual solvent was removed. Note that the absence of any *o*-DCB absorbance peaks at 662 cm^{-1} of the FTIR spectra, which were assigned to C–Cl stretching vibration, and phenol peaks at 810, 1160, and 3421 cm^{-1}, which were attributed to the asymmetric stretch of phenolic C–C–OH, C–O, and O–H stretching vibration, respectively, indicated the complete removal of residual solvent [21].

2.2. Poly(trimethylene terephthalate)/Graphene Nanosheet (PTT/GNS) Solution Preparation and Properties

To prepare a homogenous PTT/GNS solution with varying GNS amounts (based on the PTT polymer), pre-weighed PTT powders were added to a TFA solvent and vigorously stirred for several hours. A PTT solution (14 wt %) was used to monitor the effect of carbon nanofillers on the morphology of the as-spun fibers. A PTT solution (14 wt %) filled with different amounts of GNSs (based on the weight of the PTT polymer) was then obtained [17]. In the present study, the samples were designated according to their polymer-to-filler weight ratios, for example, 99/1. The corresponding volume fraction ϕ was obtained from the respective densities of the applied components. The solution conductivity properties (κ) were measured at 25 °C using a Consort conductivity meter (C832, Consort, Turnhout, Belgium). The viscosities of the solutions were measured using a viscometer (DV-II+Pro, spindle 18, and cup 13R, AMETEK Brookfield, Middleboro, MA, USA) at 25 °C. The viscoelastic pr−operties of the solutions filled with GNSs were measured in a strain-controlled rheometer (DHR-1, TA Instruments, New Castle, DE, USA) using a flat plate feature with a 60 mm diameter at 25 °C.

2.3. Electrospinning Process

The prepared solutions were subjected to room temperature electrospinning, wherein the nozzle size was D_i/D_0/length = 0.69 mm/1.09 mm/4 cm, and D_i and D_0 denoted the inner and outer nozzle diameters, respectively. The prepared solutions were delivered by a syringe pump (Cole–Parmer, Vernon Hills, IL, USA) to the nozzle at a controlled flow rate (*Q*). A high electrical voltage (*V*) was applied to the spinneret using a high-voltage source (MECC, HVU-40P100, Fukuoka, Japan) to provide a sufficient electric field for electrospinning. To construct a needle-to-plate electrode configuration, an aluminum board (30 × 30 cm^2) was used as the collector for the electrospun fibers at a fixed tip-to-collector distance (*H*) of 14 cm. A summary of the different carbon nanofillers and concentrations studied here and the resulting fiber morphologies is shown in Table 1.

Table 1. Summary of composite fibers obtained and the resulting fibers morphologies.

Poly(trimethylene terephthalate) (PTT) Concentration (wt %)	Graphene Nanosheet (GNS) Concentration (wt %)	Fiber Morphology
7, 8	-	Beads on fiber
9, 11, 12, 14	-	Smooth fiber with nanofibril
14	1, 3, 5, 7	Irregular fiber structure with nanofibril; irregular fiber structure depend on GNS concentrations
11	10, 12	Irregular fiber structure with nanofibril; irregular fiber structure size depend on GNS concentrations
9	14	Irregular fiber structure with nanofibril and big GNS aggregation particles
7	16	Irregular fiber structure with nanofibril and big GNS aggregation particles

2.4. Morphology and Characterization of Fibers

The morphology of the fibers was observed using a scanning electron microscope (SEM, Hitachi S4100, Krefeld, Germany). Fiber diameters were measured from a collection of approximately 200 fibers, from which the average diameter (d_f) was determined. A transmission electron microscope (TEM, Jeol JEM-1200EX, Peabody, MA, USA) was used to determine the locations of the carbon nanofiller particles within the fibers. These PTT/GNS CFMs were subsequently evaluated on the basis of thermo-gravimetric analysis (Figure S1). There was no significant weight loss before degradation. Thus, no residual solvent was detected in the PTT/GNS CFMs. FTIR measurements were performed using a Perkin-Elmer FTIR spectrometer (Spectrum Two, Waltham, MA, USA) equipped with a Mettler heating stage (HT82) for temperature control. A total of 32 scans with a 2 cm^{-1} resolution were obtained for each spectrum with a holding time of 8 min. The wide-angle X-ray diffraction (WAXD) intensity profiles of the PTT/GNS CFMs were obtained using a Bruker diffractometer (NanoSTAR Universal System, Cu K_α radiation, Billerica, MA, USA). In-situ intensity profiles of samples during heating were obtained using a vacuum-assisted heating device. The heating process was realized by a stepwise increase of temperature with a holding time of 20 min each for data acquisition. The crystallization and melting behavior of the PTT/GNS CFMs was investigated using a TA differential scanning calorimetry (DSC) Q20 under a nitrogen atmosphere. The samples were heated to 270 °C at a rate of 10 °C/min.

The PTT/GNS composite films were prepared by hot pressing the PTT/GNS CFMs into rectangular steel molds (with a thickness of 0.3 mm) at 280 °C for 3 min, followed by air cooling to room temperature. Measurements for the samples with high σ (>10^{-6} S/cm) were performed using a Keithley 2400 Source Meter (Solon, OH, USA). A standard four-probe technique with an applied voltage of 1–20 V was applied to reduce the effects of contact resistance. A Keithley 6487 electrometer (Solon, OH, USA) equipped with a Keithley 8009 resistivity fixture was used for the samples with low σ (<10^{-6} S/cm) based on the ASTM D257 standard. The applied voltage for the sample with low σ was between 100–500 V.

The mechanical properties of the as-spun PTT/GNS CFMs with and without annealing were obtained through tensile strength testing. During annealing, the PTT/GNS CFMs were heated up to 170 °C for 30 min in a vacuum and without any stretching. For example, a sample of PTT/GNS 99/1 CFMs with 170 °C annealing is denoted as 99/1@170. The mechanical properties of the PTT/GNS CFMs were obtained from the stress-strain curve using a universal tensile testing machine (HT-2402, Hung Ta Intsrument Co., Taichung, Taiwan) at a stretching rate of 10 mm/min. Specimens with a dog-bone shape were prepared using a sharp cutter. The mean thickness of each sample was ~50 µm. The gauge length and sample width were 15 and 4 mm respectively. The reported data of mechanical properties represent the average results of the five tests. The sample is drawn 10 mm/min to 20% of its initial gauge length. Thereafter, the tensile load was removed immediately, and the crosshead was allowed to returned to its original position with the same crosshead speed. These measurement protocols were repeated five times. The elastic recovery for the first and five stretch cycle was used in the following equation [24]:

$$\text{elastic recovery} = \frac{L_o - L}{L_o} \times 100\% \quad (1)$$

where L_o is the original length of the filament and L is the irreversible length of the filament after the crosshead it was returned to its original position.

3. Results and Discussion

3.1. Effect of Carbon Nanofillers on Solution Properties

In general, obtaining bead-free fibers is closely correlated to a polymer's solution concentration and molecular weight. However, solvent volatility affects the electrospun fiber morphology. Khil et al. [31] demonstrated that smooth PTT fibers could be electrospun via a 13 wt % PTT/TFA-methylene chloride (MC) solution (TFA-MC 1:1, v/v). Wu et al. [30] showed electrospun PTT fibers using different ratios of TFA/MC solvent. As the MC content decreased, a coalescent fiber structure was observed because of the reduced volatile MC content. Wang et al. [32] reported that a smooth PTT fiber could be electrospun using neat PTT/TFA solution. Therefore, TFA is an excellent solvent in polyester electrospinning for producing fine diameters. Based on the method by McKee et al. [38], a log–log plot of the solution's specific viscosity (η_{sp}) versus the volume fraction ($\phi_{PTT/TFA}$) was constructed to determine the C_e (obtaining a semidilute solution regime with entangled PTT chains) for the present PTT/TFA solutions. Figure 1 shows the plots of the η_{sp} versus the volume percentages of the investigated PTT solutions. The η_{sp} of the PTT/TFA solution increased with the increased PTT content. The rapid viscosity increase at 9.44 vol % (8 wt %) suggests that C_e occurred approximately at this concentration, and η_{sp} and the volume percentage of the PTT/TFA solution had a constant slope of 3.93 above this concentration. The determined exponent is consistent with the theoretical prediction for entangled solutions in a good solvent (~3.9). The overlapping concentration C^* of 0.91 vol % was determined by $1/[\eta]$. Therefore, the C_e/C^* ratio is ~10.4, which is in agreement with that reported ratio for polymers dissolved in good solvents. Moreover, the measured κ for the neat TFA solvent is 1.26 μS/cm. A PTT concentration increase to 14 wt % results in slight increase of the κ for 14 wt % PTT/TFA solution to 8.31 μS/cm.

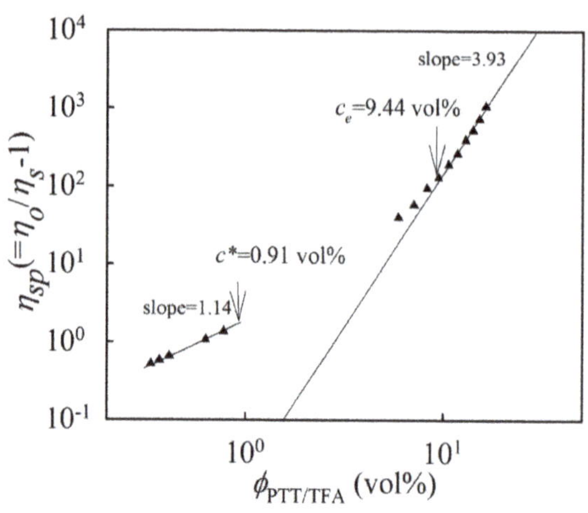

Figure 1. Concentration dependence of the specific viscosity of the PTT solutions.

Figure S2 shows the SEM images of the PTT fiber products collected from the electrospinning of the PTT solutions with different concentrations. The minimum concentration required for the PTT solutions to produce bead-free fibers is 9 wt %. PTT solutions with concentrations higher than 9 wt % possess a stretchable PTT chain network that prevents network fracture during electrospinning, thereby yielding bead-free fibers. In general, electrospinning eventually degenerates to electrospraying in solutions with a concentration lower than the entanglement concentration (C_e). To produce uniform fibers via electrospinning, the solution concentration should be as high as 1.8~2.5 C_e [38,39]. McKee et al. [38] reported that the minimum solution concentration required to electrospin bead-free fibers is 2.0~2.5 C_e for a PET copolymer/chloroform-dimethylformamide solution. However, for the present PTT/TFA solution, the minimum concentration required to obtain smooth fibers is 9 wt %, which is similar to the C_e of 9 wt %. This result revealed that the minimum concentration for producing smooth PTT fibers is approximately 1.0 C_e, which could be attributed to the high TFA volatility. Wang et al. [32,33] reported similar results. Therefore, we selected 14 wt % PTT solution (~1.5 C_e) filled with different amounts of GNS to ensure a completely stable fiber formation during electrospinning and to prevent non-smooth fiber morphologies of PTT/GNS composite fibers. Non-smooth morphologies were identified as bead fibers resulting from the weak network of a PTT solution.

Figure 2 shows the structures of the deposited GNS on the TEM grid prepared from the *o*-DCB solution. GNSs were transparent with folding and wrinkling. In the TEM observation in our previous study [21], the number-average long and short axis lengths of GNS were 1.67 ± 1.48 and 0.65 ± 0.49 µm, respectively. Moreover, the number-average minimum thickness of GNS based on atomic force microscopy (AFM) measurements was 6.5 nm. Numerous studies have incorporated GNSs into polyester via melt compounding; however, fully penetrating the GNS interstices is difficult for the polymer chains of matrices. Based on the above PTT/TFA solution results, TFA is a good solvent for a PTT polymer. Therefore, the PTT/GNS composite powders were prepared via coagulation. Consequently, the PTT/GNS composite powders were re-dissolved in the TFA solvent to prepare the PTT/TFA solution with different amounts of GNS for electrospinning.

Figure 2. Transmission electron microscope (TEM) images of GNS the deposited GNS on the TEM grid prepared from the ortho-dichlorobenzene (*o*-DCB) solution.

A 14 wt % PTT solution filled with various amounts of GNS can be used to obtain PTT/GNS composite fibers via electrospinning. However, the polymer or polymer solution filled with carbon nanofillers could form nanofiller-nanofiller networks [23,40,41]. Thus, to determine the GNS–GNS network effect of GNS addition on a PTT solution, viscoelastic measurements of the 14 wt % PTT solutions filled with 1–7 wt % GNS were conducted. Figure 2a,b show the G' and G'', respectively, of the 14 wt % PTT solutions filled with different GNS amounts. A slight GNS addition of 1 wt % did not change the G' in the low frequency region. When the GNS contents were increased to 3 wt %, the slope in the low frequency region was reduced and a small plateau was observed; however, the increase in GNS content had no effect in the high frequency region. For the PTT solution filled with 5 wt % GNS, a two-order increase of G' in the low frequency region was observed and eventually shifted the whole curve upward. The large plateau increase at the low frequency region indicated a GNS–GNS network formation in the entangled polymer chain network. A mild increase in G'' was observed with the corresponding GNS addition. This finding suggests an increased elasticity of the solution. The rheological properties of the GNS-filled PTT solutions were significantly changed; therefore, their influence on electrospinnability deserves further discussion.

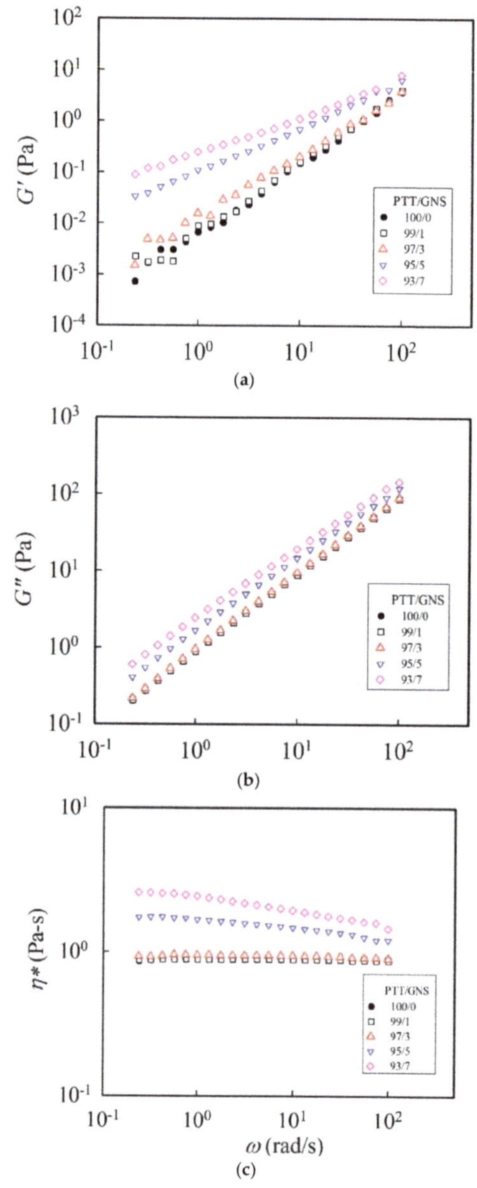

Figure 3. Effect of GNS content on the viscoelastic properties of PTT solutions: (**a**) dynamic storage modulus G', (**b**) dynamic loss modulus G'', and (**c**) complex viscosity η^* at 25 °C. The PTT/trifluoroacetic acid (TFA) concentration is 14 wt %.

Figure 3c shows the complex viscosity (η^*) of the PTT solutions filled with different GNS amounts at 25 °C. Newtonian flow behavior occurred in the neat PTT solution during measurement frequencies, and the value of the zero-shear viscosity (η_o) curves reached 0.88 Pa-S. The η^* value was unchanged when 1 wt % GNS was added to the solution. When the GNS content was increased to 5 wt %, Newtonian flow behavior occurred at low frequencies, but the behavior became non-Newtonian at frequencies higher than 0.75 rad/s. The η^* of the PTT solution with 7 wt % GNS was significantly increased, and η_o reached 2.57 Pa-S. Ramazani et al. [16] examined the viscosity of a PCL/GO solution and concluded that η^* decreases as the GO content increases due to the induced free volume around the GO nanosheets. Our viscosity result showed that η^* increased as the GNS content increased in the PTT/GNS solution, which indicates the formation of a GNS–GNS network, and the absence of a free volume effect around the GNS. Figure S3 shows the conductivity of the PTT solution filled with GNS. GNSs are conductive fillers; therefore, adding GNS into PTT solutions significantly improves the κ values. After adding 7 wt % GNS to the PTT solution, the κ reached 151.1 μS/cm, which is significantly higher (18-fold) than that of the neat PTT solution. This result indicates that adding GNS to PTT solutions could significantly increase the PTT solution conductivity. Thus, the addition of GNS increased the viscosity and conductivity of the PTT solution, and a GNS–GNS network was formed in the 14 wt % PTT solution filled with 5 wt % GNS.

3.2. Effect of GNS Concentration on Electrospinning and As-Spun Fiber Morphology

Figure 4 shows the functioning domain for electrospinning a 14 wt % PTT solution with various GNS contents. Functioning domains [23,42,43] are defined as the operating windows of applied voltage and Q required for a stable cone-jet mode. The lower- and upper-bound applied voltages are denoted as V_s and V_{us}, respectively. An H of 14 cm was used given the volatility of the TFA solvent. At a given Q, the operating windows ($V_{us} - V_s$) were higher for the PTT solution filled with GNS content than those for the neat PTT solution. Moreover, increasing the GNS content increased the operating window at a given Q. Based on the functioning domain for electrospinning PTT/GNS solutions (Figure 4), a common but limited processing window does exist to determine the GNS effect. Therefore, Q and V were determined to be 0.3 mL/h and 11 kV, respectively, in electrospinning the PTT/GNS solution to demonstrate the GNS effects on fiber diameter.

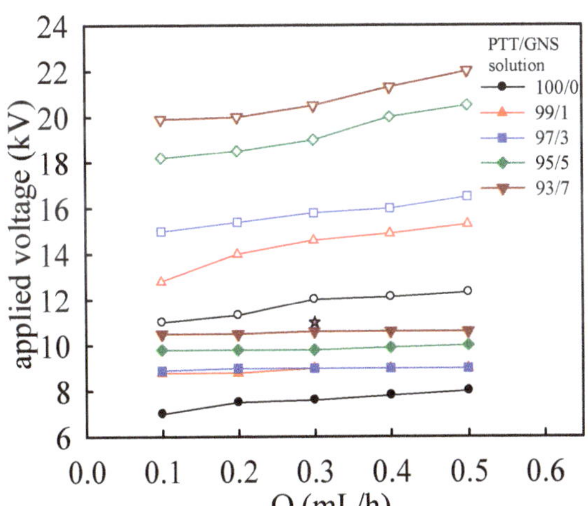

Figure 4. Functional domain for electrospinning of 14 wt % PTT solution with various GNS contents. The domains indicate the range of operating electrical fields required for the stable cone-jet mode. (Filled symbols for lower bond applied voltage and open symbols for upper bond applied voltage).

To determine the effect of GNS additions on fiber morphology and diameter, 14 wt % PTT solutions filled with 1–7 wt % GNSs were electrospun and compared. Figure 5 shows the SEM images of the fibers collected from electrospinning 14 wt % PTT solutions with various GNS amounts with the following conditions: Q = 0.3 mL/h, H = 14 cm, and V = 11 kV. The thick arrows indicate the GNS positions. Based on the SEM micrographs shown in Figure 5, the PTT composite fibers became less smooth and formed an increasingly irregular structure along the fiber as the amount of GNS increased. At 5 wt % GNS content, the PTT/GNS composite fibers with nanofibrils became observable. The thin arrows indicate the positions of the nanofibril whose diameters are 10–30 nm.

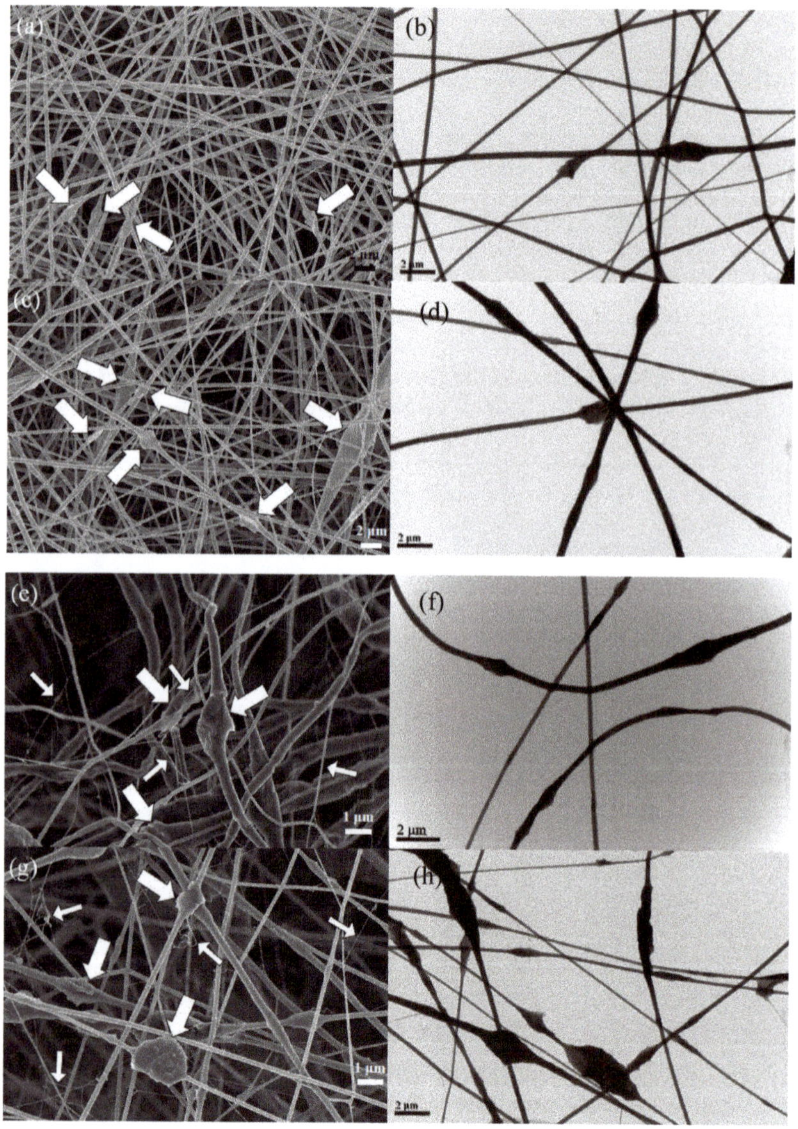

Figure 5. Scanning electron microscope (SEM) and TEM images of electrospun PTT fibers filled with: (**a**,**b**) 1 wt %, (**c**,**d**) 3 wt %, (**e**,**f**) 5 wt %, and (**g**,**h**) 7 wt % GNS. The positions of GNSs and nanofibrils are indicated by the thick and thin arrows, respectively.

The TEM images in Figure 5 show that GNSs are identifiable and that they are embedded in the PTT fibers. The lateral dimension of GNS was larger than the diameter of the electrospun PTT fibers, and the GNS particles protruded from the smooth PTT fiber. At an increased GNS content in the PTT solution, numerous GNS protrusions were observed on the fiber surface, with some GNS appearing curly in the PTT/GNS composite fibers. The GNS protrusions could help the formation of the GNS–GNS network in the PTT/GNS CFMs. The details of this observation will be discussed later. At 7 wt % GNS content, GNSs became distinctly distributed in the PTT fiber (Figure 5h), and the inter-GNS distance decreased (L_{GNS}).

The PTT/GNS composite fiber diameters were measured from a collection of over 200 fibers under the same electrospinning conditions (Q = 0.3 mL/h, H = 14 cm, and V = 11 kV). For a neat PTT fiber, the measured d_f was 256 ± 92 nm. At 1 wt % GNS content, the d_f of the PTT/GNS 99/1 fiber decreased significantly to 141 ± 28 nm. When the GNS content reached 3 wt %, the d_f of the PTT/GNS 97/3 fiber increased to 310 ± 116 nm. Furthermore, when the GNS content was increased to 5 and 7 wt %, the d_f of the PTT/GNS 95/5 and 93/7 fibers slightly increased to 359 ± 133 and 364 ± 114 nm, respectively. The d_f of the PTT/GNS fiber initially decreased and then increased as the GNS content increased. The nanofibrils in the PTT/GNS CFMs increased the standard deviation of d_f. In electrospinning, solution viscosity and conductivity are important factors for determining the d_f of electrospun fibers. In previous studies [44,45], d_f was shown to decrease with a decreasing solution viscosity and increasing solution conductivity. Based on the results shown in Figure S3, the conductivity of the PTT/GNS solutions increased as the GNS content increased. High solution conductivity mainly causes the decrease in the d_f of PTT/GNS composite fibers because it increases the electrostatic force, thereby stretching it during electrospinning. Based on the results shown in Figure 2c, the solution viscosity of the PTT/GNS solution filled with 1 wt % GNS was almost the same as that of the neat PTT solution. However, the conductivity of the PTT/GNS solutions filled with 1 wt % GNS was increase significantly, which resulted in the higher stretched forced in an electrostatic field. This phenomenon resulted in a decrease in the d_f of the PTT/GNS 99/1 composite fiber. When the GNS content exceeded 3 wt %, the solution viscosity of the PTT/GNS solution increased, which resulted in the higher flow retardation during electrospinning. This phenomenon resulted in an increase in the d_f of the PTT/GNS composite fiber when the GNS content was further increased. Therefore, the initial change in d_f was attributed to a greater increase in solution conductivity than in solution viscosity, whereas the final change in d_f was attributed to a greater increase in solution viscosity than in solution conductivity. This finding is in agreement with PET/CNT [46], PDLLA/CNC [23], and PVA/GNS [17] composite fibers.

The increased GNS content to 10 wt % in the 14 wt % PTT solution results in the viscosity of PTT/GNS solution to become too high to manipulate a stable cone-jet mode. Eventually, the PTT/GNS solution was dried at the nozzle end during electrospinning. Non-Newtonian flow behavior occurred for the 14 wt % PTT solution filled with 10 wt % GNS during the measurement of the shear rate, and η reached 11.52 Pa-S at 8.4 1/s (Figure S4). Therefore, the PTT solution with high GNS content must acquire a lower solution viscosity to improve the PTT/GNS solution fluidity for the stable manipulation of the electrified jet. When the PTT content was decreased to 11 wt % in the PTT/GNS/TFA solution, non-Newtonian flow behavior still occurred during measurement of the shear rate, and η became 1/3-fold lower at 8.4 1/s (Figure S4). The η value of the 11 wt % PTT/GNS 90/10 solution was in the same order as that of the 7 wt % PTT/GNS 93/7 solution. Thus, reducing the PTT content in a PTT/GNS solution was found to be a good route to manipulate the stable cone-jet mode when PTT/GNS composite fibers are electrospun for high GNS content. We selected an 11 wt % PTT solution filled with 10 and 12 wt % GNS, a 9 wt % PTT solution filled with 14 wt % GNS, and a 7 wt % PTT solution filled with 16 wt % GNS to ensure a completely stable cone-jet formation during electrospinning (Table 1).

Figure S5 shows the SEM and TEM images of the fiber collected from electrospinning the 11 wt % PTT solutions with 10 and 12 wt % GNS, the 9 wt % PTT solution filled with 14 wt % GNS, and the

7 wt % PTT solution filled with 16 wt % GNS. The irregular PTT composite fiber structure increased in size along the fiber. When the GNS content in the PTT solution exceeded 12 wt %, the GNS protrusion size on the electrospun fiber increased. GNS aggregation particles higher than 10 μm were observable for the PTT/GNS 86/14 and 84/16 fibers. Moreover, the nanowebs composed of nanofibrils could be observed from the electrospun PTT/GNS composite fibers, such as PTT/GNS 88/12 fibers (Figure S5c) (the dashed circles indicate the nanowebs). The TEM images (Figure S5f,h) show that the GNS aggregation particles are identifiable. When the GNS content in the PTT fiber was increased to 12 wt %, the GNS aggregation particles exceeded 2 μm and started darkening. Moreover, when the GNS content in the PTT fiber exceeded 12 wt %, the GNS aggregation particles darkened and increased in size. This observation indicates that the electrons could not pass through the GNS aggregation particles because the GNS aggregation particles were too thick. This result revealed that the GNS was layered in the PTT fiber. The GNS dispersed well in the PTT composite powders; however, the prepared PTT/GNS solution with extremely high GNS content still could not prevent GNS aggregation due to the GNS–GNS interaction at a high GNS content. The formation of PTT nanofibril is interesting. One of the hypotheses for nanofibril formation could be attributed to the fast phase separation of charged droplets generated during electrospinning [47]. However, this issue is beyond the scope of this study and deserves future work.

Based on the SEM and TEM images of the PTT/GNS composite fibers, four modes of arrangements for the GNSs in the PTT fiber were assumed upon increasing the GNSs (Figure 6). First, PTT fibers filled with 1–3 wt % GNS content were individually dispersed in the PTT fiber at intervals. Second, for the PTT fibers filled with 5–10 wt % GNS content, the GNSs were close together in the PTT fiber. Third, for PTT fibers filled with 12 wt % GNS content, some parts of the GNS overlapped in the PTT fibers. Fourth, for PTT fibers filled with 14–16 wt % GNS content, the GNS was layered in the PTT fibers. As the GNS content increased, the arrangement changed from the first arrangement to the fourth. The GNS–GNS network was apparently not formed along the PTT/GNS fiber. Therefore, the GNSs in the composite fibers were arranged differently from those in the composites, which was proposed by Zhao et al. [48].

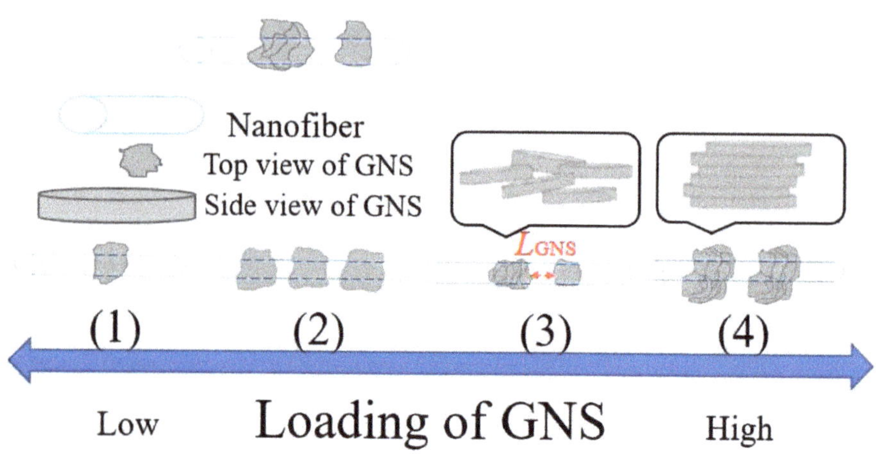

Figure 6. Schematic models of various dispersion types of GNSs in PTT fiber: (**1**) GNSs are individually dispersed in the PTT fiber at intervals. (**2**) GNSs are close together in the PTT fiber. (**3**) Some parts of the GNSs overlap with one another in the PTT fiber. L_{GNS} indicates the inter-GNS distance assemblies in the PTT composite fiber. (**4**) The GNS are layered in the PTT fiber. When the GNS content is increased, the resulting dispersion form gradually changes.

3.3. Cold Crystallization of Electrospun PTT/GNS Composite Fibers

Figure 7a shows the WAXD intensity profiles of the as-spun PTT/GNS composite fiber samples. Notably, the diffraction peak at 26°, which was associated with the d-spacing of the graphitic structure, was not observed in the as-spun PTT/GNS composite fibers. According to our previous studies [29], the WAXD intensity profiles of the amorphous PTT showed a halo at 2θ = 20.3° and full width at half-maximum=10.8 and the characteristic diffraction peaks of the PTT crystals should be observed at 2θ of 15.3°, 16.8°, 19.4°, 21.8°, 23.6°, 24.6°, and 27.3° for the PTT samples. However, the WAXD intensity profiles of the as-spun neat PTT fibers showed the diffraction humps at 15°–18° and 22°–27°. This outcome was in contrast to the behavior of the PTT amorphous characteristics. The order structure is associated with the mesomorphic phase, which ws reported for neat PTT and PTT/GNS composites cold crystallization [29,49]. Therefore, the mesomorphic phase was formed during PTT electrospinning. Furthermore, when the GNS content was increased to 3 wt %, the three diffraction peaks of 16.4°, 20.0°, and 24.0° were observed at 2θ. The mesomorphic phase was significant during the PTT electrospinning process after the addition of GNS. Thus, GNSs can help to induce the order structure formation of the PTT chains. Given such results, the changes in GNS-filled PTT chain conformation in fiber morphologies during the subsequent cold crystallization should be further analyzed.

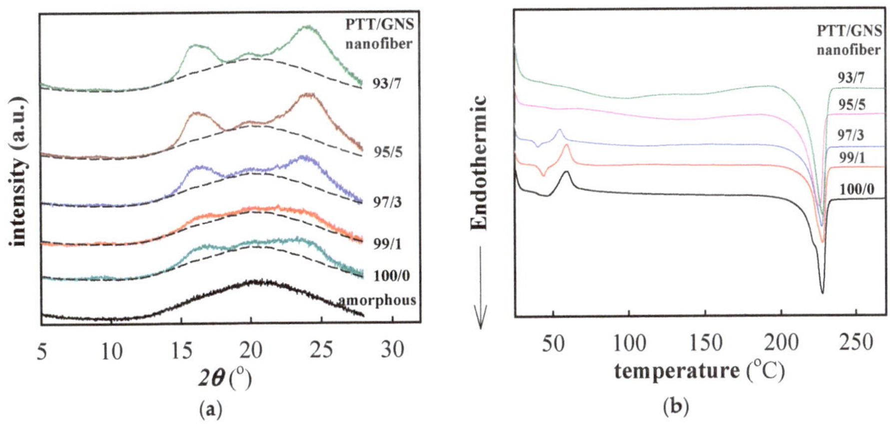

Figure 7. (**a**) Wide-angle X-ray diffraction (WAXD) intensity profiles of the as-spun PTT/GNS composite fibers, and (**b**) differential scanning calorimetry (DSC) heating traces of the as-spun PTT/GNS composite fibers.

Figure 7b shows the DSC heating traces of the as-spun PTT/GNS composite fibers. T_g was determined from the mid-point of the heat capacity jump, whereas the peak temperature of the cold crystallization was denoted as T_c. Crystallization enthalpy, which was determined from the integral area of the exotherm, was represented by ΔH_c. To represent the crystallizability of the PTT fiber in the presence of GNSs, ΔH_c was normalized with the PTT content. The data listed in Table 2 show that the T_g of PTT/GNS 100/0, 99/1, and 97/3 fibers are 40.4, 43.4, and 39.3 °C, respectively. T_g remained unchanged at approximately 41.0 °C with increased GNS content. T_c slightly shifted to a low temperature with increased GNS content. When the GNS content increased to 5 wt %, the exothermic peak of the PTT cold crystallization could not be clearly observed. The normalized ΔH_c decreased with increased GNS content because some mesomorphic phases already developed in the as-spun PTT/GNS composite fibers prior to heating. Moreover, the melting temperature of the cold-crystallized PTT/GNS composite fibers was relatively unchanged, as shown in Figure 7b.

Table 2. Thermal properties of the as-spun PTT/GNS composite fibers.

PTT/GNS fibers	T_g (°C)	T_c (°C)	ΔH_c (J/g)	T_m (°C)	ΔH_m (J/g)
100/0	40.4	59.4	10.1	227.4	66.21
99/1	43.4	59.7	9.9	227.9	41.20
97/3	39.3	55.3	5.6	227.4	46.95
95/5	-	-	-	225.8	48.51
93/7	-	-	-	225.9	43.24

Figure 8 shows the FTIR spectra of the neat PTT and PTT/GNS 99/1 composite fibers during stepwise heating to 225 °C. According to literature [50,51], the absorbance bands at 875 and 870 cm^{-1} are associated with the B_{3u}–CH out-of-plane bending of phenylene rings in the amorphous and crystalline phases, respectively. By contrast, the absorbance bands at 948 and 935 cm^{-1} are related to CH_2 rocking in a gauche conformation in the crystalline phase. In the current work, the absorbance band, resulting from the trans-conformation was located at 815 and 978 cm^{-1}. Prior to heating, the neat PTT and PTT/GNS 99/1 composite fibers were in the mesomorphic phase, as indicated by the intensity of the WAXD (Figure 7a). Upon heating, the neat PTT and PTT/GNS 99/1 composite fibers were gradually crystallized, as shown by the gradual absorbance increase in the 870, 935, and 948 cm^{-1} bands and the gradual absorbance decrease in the 815 and 978 cm^{-1} bands. These absorbance changes, which are consistent with previous findings on isothermal cold crystallization [49], revealed the trans-to-gauche transition behavior in PTT cold crystallization [52].

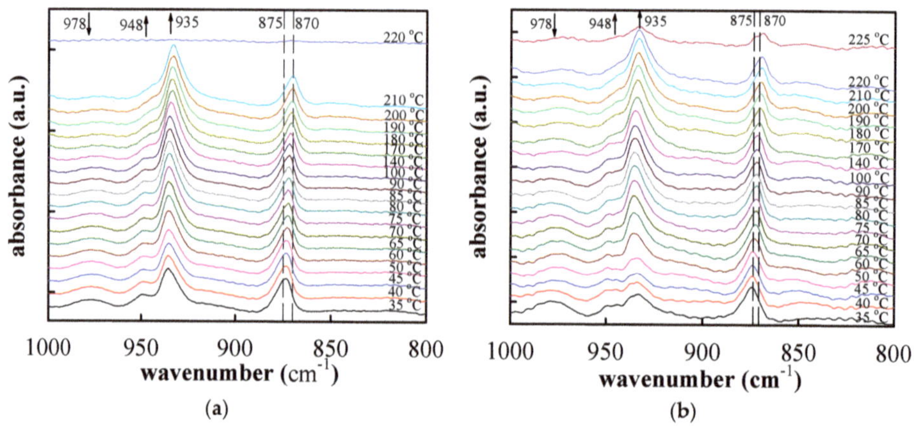

Figure 8. Fourier transform infrared spectroscopy (FTIR) spectra of (**a**) neat PTT and (**b**) PTT/GNS 99/1 composite fibers during stepwise heating to 225 °C.

Figure 9a shows the normalized area of the 948 and 935 cm^{-1} bands and the variation in the absorbance peak at 875–870 cm^{-1} as a function of temperature for the neat PTT and PTT/GNS 99/1 composite fibers. The normalized area of the bands at 935 and 948 cm^{-1} was used in the following equation [29]:

$$A_{935+948} = \frac{A_T - A_{35}}{A_{max} - A_{35}} \quad (2)$$

where $A_{935+948}$ is the relative crystallinity of PTT, and A_{max} is the maximum area of the 935 and 948 cm^{-1} bands. A_T and A_{35} are the areas of the 935 and 948 cm^{-1} bands at different temperatures and at 35 °C, respectively. The cold crystallization process was characterized by the increased $A_{935+948}$ (Figure 9a), which was accompanied by a gradual band shift from 874 to 870 cm^{-1} (Figure 9b). Furthermore, a band shift was observed during heating at ~870 cm^{-1} and prior to crystal melting at approximately 200 °C, at which spectral recording became infeasible owing to the sample flow. Notably, the $A_{935+948}$ value of the PTT/GNS 99/1 composite fibers at 45–50 °C was lower than that at 35 °C, and then significantly increased above 55 °C. This result indicates that the gauche conformation of the PTT/GNS composite fibers decreased at temperatures higher than T_g. The gauche conformation content of the PTT chains for PTT/GNS composite fibers could be higher than that for neat PTT fibers during the solution jet and bending instability in the electrostatic field due to the presence of GNSs, but the mild increase in regularity is not sufficient to cause the formation of crystalline structures. Thus, the chain structure of PTT was rearranged again after subsequent heating treatments. The $A_{935+948}$ of the PTT/GNS 99/1 composite fibers at 65 °C was higher than that of the neat PTT fiber at 65 °C. The saturated $A_{935+948}$ of the PTT/GNS 99/1 composite fibers at 80 °C was also higher than that of the neat PTT fiber at 90 °C. Meanwhile, the band between 875 and 870 cm^{-1} of the PTT/GNS 99/1 composite fibers was slightly lower than that of the neat PTT fiber at a high-temperature region (Figure 9b). These results indicate that the cold crystallization of the PTT composite fibers was enhanced by the GNSs. Consistent results were obtained between DSC and FTIR measurements with regard to the kinetics of cold crystallization in PTT fibers.

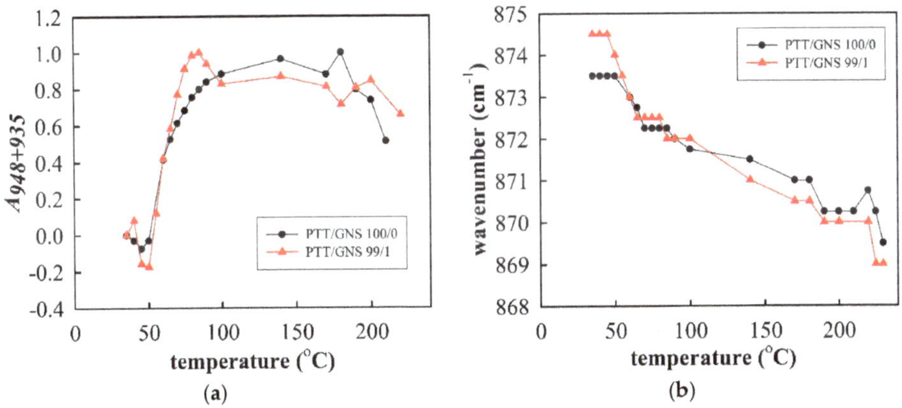

Figure 9. (**a**) Normalized area of the 948 and 935 cm^{-1} bands ($A_{935+948}$) and (**b**) variation in the absorbance peak at 875–870 cm^{-1} for the neat PTT and PTT/GNS 99/1 composite fibers during stepwise heating.

Figure 10a shows the WAXD of the neat PTT fibers during stepwise heating to 240 °C. The WAXD intensity profiles of as-spun neat PTT fibers showed the mesomorphic phase characteristics prior to heating. When the temperature was increased to 60 °C, the intensity of diffraction humps at 15.0°–18.0° and 22.5°–25.5° began to increase. When the temperature was increased to 120 °C, the characteristic diffraction peaks at 15.6° and 17.2° appeared. When the temperature was increased to 170 °C, the characteristic diffraction peaks of the PTT crystals were clearly observed and increased gradually at the 2θ of 15.6°, 17.2°, 19.4°, 21.4°, 23.2°, and 24.5°. Figure 10b shows the WAXD of the PTT/GNS 99/1 composites fibers during stepwise heating to 240 °C. A similar increase in the WAXD intensity profiles was observed in the PTT/GNS 99/1 composite fibers with the increased temperature. Therefore, neat PTT and PTT/GNS 99/1 composite fibers crystallizes from a mesomorphic phase into a triclinic structure at 170 °C during cold crystallization. The neat PTT and PTT/GNS 99/1 composite fibers were still mesomorphic phases during $A_{935+948}$ increase in the FTIR spectra and exothermic peak in DSC, which is associated with the PTT cold crystallization kinetics.

Based on our previous studies [29], the PTT lamellae could be epitaxially grown on top of the GNS. Furthermore, the mesomorphic phase was formed during PTT electrospinning, and GNSs could induce the PTT mesomorphic phase significantly during PTT/GNS electrospinning in this study. During cold crystallization, the PTT mesomorphic phase could be the PTT nuclei, which were randomly developed on the GNS surface, to form the lamellae in PTT/GNS composite fiber (Figure 11). The PTT cold crystallization rate in the PTT/GNS composite fiber increased with the gradual addition of GNSs. The enhanced cold crystallization kinetics was attributed to the high nucleation ability of GNSs, which results from PTT mesomorphic phase on their surfaces, while the overgrown lamellae from the GNS surface of the PTT/GNS composite fibers may result in the retardation alignment of PTT crystalline lamellae in the PTT/GNS composite fibers under quiescent heating. Therefore, the orientation of PTT lamellae in the PTT fibers is higher than that in the PTT/GNS composite fibers.

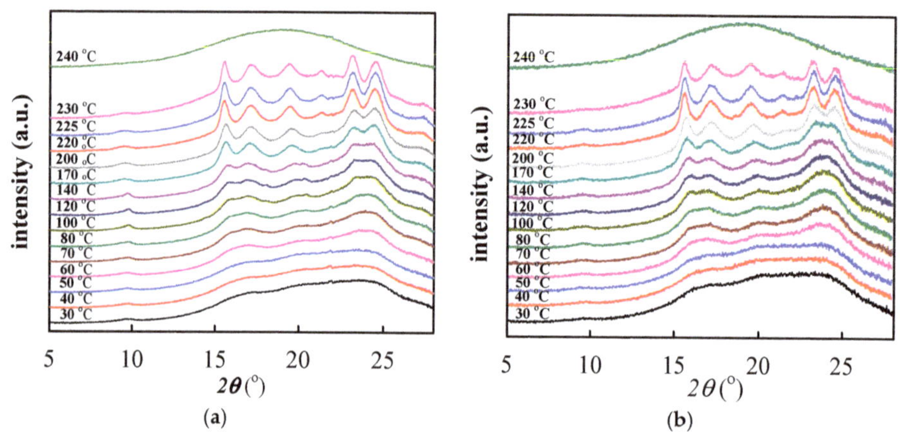

Figure 10. WAXD intensity profiles of (**a**) neat PTT and (**b**) PTT/GNS 99/1 composite fibers during stepwise heating to 240 °C.

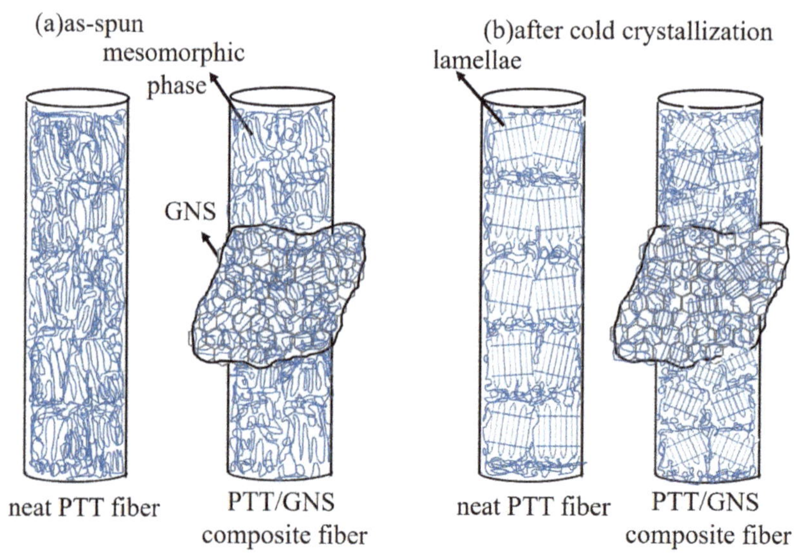

Figure 11. Schematic of the microstructure change of the PTT/GNS composite fibers during cold crystallization.

3.4. Mechanical Properties of Electrospun PTT/GNS Composite Fibers

Based on cold crystallization results, the PTT/GNS composite fibers crystallizes into a triclinic structure at 170 °C. Therefore, 170 °C was selected for annealing. Figure 12 shows the stress-strain curves of PTT/GNS CFMs with and without annealing at 170 °C. The determined Young's modulus (*E*), tensile strength (σ_{max}), elongation at break (ε_{max}), porosity, and d_f are displayed in Table 3. The as-spun neat PTT fiber mats prior to 170 °C annealing possessed higher E, larger σ_{max}, and longer ε_{max} than after 170 °C annealing. Similar results were obtained in the PTT/GNS CFMs. Annealing could lead to an increase in the porosity and d_f. Moreover, the E and σ_{max} of the PTT/GNS CFMs were decreased, and the ε_{max} of the PTT/GNS CFMs were increased with increased GNS content, regardless of annealing. Based on the cold crystallization and solution viscosity results of the as-spun PTT/GNS CFMs, the present mechanical reduction is attributed to the alignment of the PTT chains in the PTT/GNS composite fibers during electrospinning. When the viscosities of the PTT/GNS solution were increased with the increase in GNS content, the d_f of the electrospun PTT/GNS composite fibers was decreased first and then increased with increased GNS content. This phenomenon resulted in retarded alignment of the PTT chains in the PTT/GNS composite fibers during electrospinning. Therefore, the PTT/GNS CFMs became ductile with the addition of GNSs. In addition, the PTT chains in the PTT/GNS composite fibers relaxed via thermal expansion during thermal heating when the temperature exceed the T_g of PTT. Thus, the d_f values of the PTT/GNS composite fibers at 170 °C annealing temperature was slightly larger than those of the as-spun PTT/GNS composite fibers. The alignment of the PTT crystalline lamellae in the PTT/GNS composite fibers was inhibited by the PTT nuclei on the GNS surface after annealing at this temperature (Figure 11).

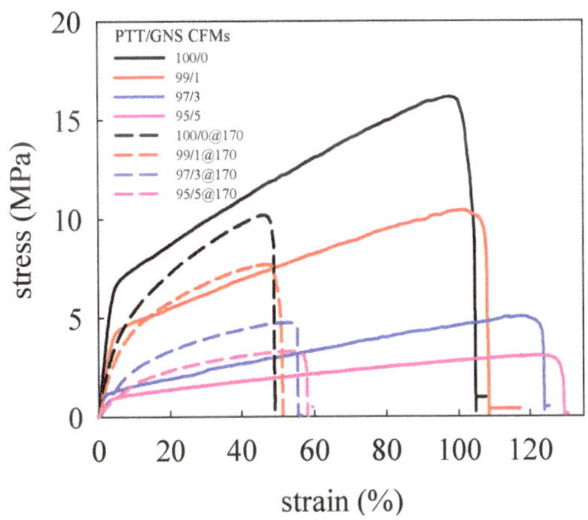

Figure 12. Stress-strain curves of PTT/GNS CFMs with and without annealing at 170 °C for 30 min. The solid lines indicate the as-spun PTT/GNS CFMs, and the dashed lines indicate PTT/GNS CFMs with annealing at 170 °C.

Chen et al. [24] found that the PTT filaments have a high instantaneous elastic recovery even at a high elongation of 20%, whereas the clothing materials, such as filaments, are usually used at elongations below 20%. Therefore, we selected 20% elongation for the instantaneous elastic recovery measurement. Figure S6 show the stress-strain curves of PTT/GNS CFMs filled with various amounts of GNS in five stretch cycles by the 20% elongation limit with and without annealing at 170 °C. The elastic recovery results are listed in Table 3. When the crosshead returned to its original position, no PTT/GNS CFMs could return to their original position, and the remnant elongation slightly increased with increasing cycles. The elastic recoveries of the PTT/GNS CFMs with 170 °C annealing are better than those of the as-spun PTT/GNS CFMs. The elastic recoveries of the PTT/GNS CFMs were slightly reduced with increased GNS content. The WAXD and FTIR results indicate that the as-spun PTT/GNS CFMs are of the mesomorphic phase, and the PTT/GNS CFMs with 170 °C annealing have a triclinic crystal structure with numerous gauche conformations. This results indicates that the elastic recoveries of the PTT/GNS CFMs are mainly caused by the gauche conformations of the PTT crystals in the fiber.

Table 3. Mechanical properties of PTT/GNS CFMs with and without annealing at 170 °C for 30 min.

PTT/GNS CFMs	E (MPa)	σ_{max} (MPa)	ε_{max} (%)	d_f (nm)	Porosity (%)	Fifth Elastic Recovery (%)
100/0	141.6 ±30.7	15.1 ± 2.0	102.5 ±7.7	256 ± 92	46	85.8
99/1	89.4 ± 5.0	10.0 ± 0.5	106.9 ±2.8	141 ± 28	57	84.4
97/3	52.9 ± 11.2	4.4 ± 0.5	109.8 ± 8.7	310 ± 116	43	83.4
95/5	22.6 ± 7.2	3.0 ± 0.1	123.8 ± 3.6	359 ± 133	39	83.9
100/0@170	102.0 ± 6.5	9.9 ± 0.4	48.6 ± 3.5	290 ± 87	47	91.8
99/1@170	77.4 ± 5.4	7.4 ± 0.2	49.6 ± 1.9	237 ± 81	59	91.7
97/3@170	37.0 ± 3.0	4.9 ± 0.2	53.7 ±1.8	347 ± 128	43	90.9
95/5@170	35.0 ± 3.9	3.5 ± 0.1	56.9 ± 3.3	366 ± 134	43	90.4

3.5. Electrical Properties of Electrospun PTT Composite Fiber Mats

Figure 13a shows the σ of PTT composites and PTT fibers filled with GNS. When the GNS contents were increased, the GNS gradually formed a conductive path. The percolation scaling laws are considered suitable in revealing the σ of the PTT/GNS CFMs. Thus, in this section, the addition of GNS to PTT/GNS CFMs have to be presented in the volume percentage on the basis of the percolation equation. For the PTT/GNS composite fibers, the PTT/GNS CFMs exhibited an improvement in conductivity, and a gradual transition was observed from 1.77 to 9.98 vol % (3–16 wt %). The σ of the PTT/GNS CFMs became 3.18×10^{-5} S/cm following a GNS content of 9.98 vol %. Compared with that of the neat PTT fibers, a 9-order increase in σ was observed for the PTT/GNS CFMs, with 7.35 vol % GNS content. However, compared with our previous study [21], wherein GNSs were used in PTT/GNS composites using the same GNSs as this study, the PTT/GNS CFMs in the present study exhibited significantly lower conductivity than that of PTT/GNS composites with the same GNS content. PTT/GNS composites could form a conductive network more easily than PTT/GNS CFMs. This discrepancy could be attributed to the highly porous electrospun fiber mats with voids filled with insulating air. Chien et al. [23] demonstrated that the conductivity of electrospun PDLLA/CNC CFMs may be affected by the porosity of CFMs. Therefore, we prepared a PTT/GNS composite film via hot pressing PTT/GNS CFMs (HP-PTT/GNS CFMs) at 280 °C. The σ of the HP-PTT/GNS CFMs was similar to that of the PTT/GNS composites that were prepared from hot-pressed dried powders. The porous structure of the PTT/GNS CFMs is one of the major reasons causing the difference in conductive behavior between the PTT/GNS composites and the PTT/GNS CFMs. The hot-pressed fiber mats eliminated the insulating voids.

To determine quantitatively the minimum volume fraction of GNS required for developing the network for electron transportation, the percolation scaling law was applied to describe the relationship between σ and $\phi - \phi_c$. Corresponding plots are provided in Figure 13b. ϕ_c and β values were determined based on the regression analysis. The experimental ϕ_c values of the PTT/GNS CFMs is 6.10 vol %. The derived ϕ_c of the PTT/GNS CFMs is higher than that of the PTT/GNS composites (0.84 vol %) [21]. The derived β values of the PTT/GNS CFMs is 6.52, which is substantially different from the universal value of 2.0 for a three-dimensional lattice. In addition, several studies analyzed similar polymer/carbon nanofiller CFM samples. For polymer/carbon nanofiller CFMs, the β of the PDLLA/CNC CFMs is 3.20 [23], while one of the PS/CNT CFMs is 0.80 [22]. The β of the PTT/GNS CFMs is significantly higher than those of the PDLLA/CNC and PS/CNT CFMs. These discrepancies could be attributed to the route of dispersion state of the fillers, their geometrical structures, the intrinsic differences in the polymer/GNS pair, or the porosity of the electrospun fiber mats. Nevertheless, the β values of the PTT/GNS CFMs are significantly higher than those of the PTT/GNS composites (3.91) [21]. However, this result is different from the percolation exponent result for the PDLLA/CNC composite fiber mats [23]. The β of the PDLLA/CNC CFMs is similar to that of the PDLLA/CNC casting composite film. This phenomenon implies that the conductivity of the PTT/GNS fiber mats is not merely affected by the porosity of the electrospun fiber mats.

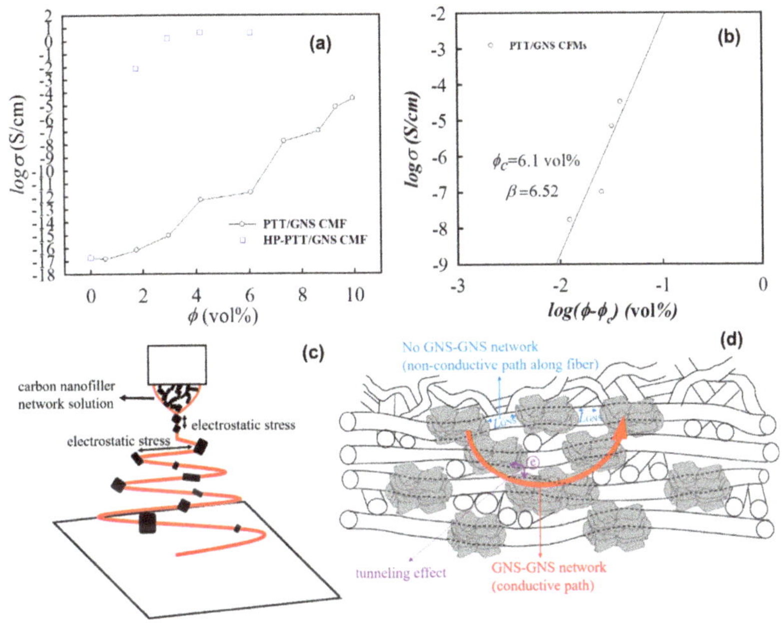

Figure 13. (**a**) Conductivity versus GNS volume contents of PTT/GNS CFMs and HP-PTT/GNS CFM composites. (**b**) Percolation scaling law between σ and $\phi - \phi_c$ for PTT/GNS CFMs. (**c**) Schematic representation of electrospun PTT/GNS fiber. (**d**) Schematic representation of GNS-GNS conductive path in the PTT/GNS CFMs.

L_{GNS} in the PTT/GNS/TFA solution was significantly decreased with 5 wt % GNS content due to the significant increase of G' and solution conductivity (Figures 3a and S3), whereas the conductivity of the electrospun PTT/GNS 95/5 CFMs (8.90×10^{-16} S/cm) is significantly low. Moreover, the value of the GNS–GNS network formation in the PTT/GNS CFMs was larger than those in the PTT/GNS composites and the PTT/GNS/TFA solution. These results imply that the GNS–GNS network in the solution did not form a GNS–GNS conductive network along the PTT/GNS fiber during electrospinning. However, a GNS conductive fiber could be successfully made using another spinning method. Jalili et al. [53] showed that GO fibers could be produced at a low GO content (0.075 wt %, the GO liquid crystalline network) via wet-spinning, and that a GO–GO network would be formed in the GO fibers. The difference of the filler network formation between GO fibers and PTT/GNS fibers is caused by the processing method. The wet-spinning process could put GO in order along the GO fiber because the jet stretch ratio (stretching velocity/injection velocity of the flow) could be low (0.5–1.5) during wet-spinning [54]. The electrospun PTT/GNS fiber process is illustrated in Figure 13c. Wang et al. [32] found that the drawdown ratio could be estimated at approximately 27,000 from the Taylor cone apex up to the straight jet end (spinneret diameter/jet-end diameter)^2. The drawdown ratio was approximately 20–80 at the whipping region (jet-end diameter/fiber diameter)^2. When the GNS–GNS network in the composite solution flowed from the Taylor cone apex into a straight jet, L_{GNS} was stretched in the straight jet and was subsequently stretched again at the whipping region under an electric field. Therefore, L_{GNS} could be increased; the GNS–GNS network would not be formed along the PTT/GNS composite fibers despite the electrospun PTT/GNS composite fibers from the GNS–GNS network solution.

When L_{GNS} was increased via electrostatic stress, the electrons could not be transported along the PTT/GNS composite fiber. However, the lateral dimension of GNS was larger than the diameter of the electrospun PTT fibers. Hence, GNS aggregated particles protruded from the smooth PTT fiber.

The protruding parts of GNS in the PTT/GNS composite fibers and in the neighboring PTT/GNS fibers helps in the formation of the GNS–GNS network. The electrons were transported in the GNS-GNS network that was formed via GNS aggregations in the neighboring PTT/GNS fibers with a non-woven structure. (Figure 13d) The percolated GNS–GNS networks in the PTT/GNS CFMs are different from those in the PTT/GNS composite. Several studies [21,55–57] have indicated that when conduction-insulation composites experience a tunneling effect between conducting particles, the percolation exponent will be in the range of β_o and β_{o+10}, where β_o is the universal conductivity exponent. The percolation exponent value of the PTT/GNS CFMs was higher than that of the PTT/GNS composites because the porous structures and GNS–GNS network of the PTT/GNS CFMs were increased in L_{GNS}, thereby promoting a tunneling effect [58] (Figure 13d).

4. Conclusions

PTT composite powders filled with well-dispersed GNSs were prepared via coagulation, and the composite powders were re-dissolved in a TFA solvent to obtain well-dispersed PTT/GNS solutions. The effects of increased GNS concentration on the electrospinning solution and process, as well as the morphologies and property variations of PTT/GNS composite fibers, were investigated using several analytical techniques, including rheometers, conductivity meters, SEM, and TEM. The addition of GNS increased the PTT solution viscosity and conductivity. After electrospinning the 14 wt % PTT solution with various GNS contents, the d_f of the PTT/GNS fibers initially decreased and then increased as the GNS was added into the PTT solution. The decreased d_f and the increased d_f of the PTT/GNS fiber were initially dominated by the solution conductivity, and subsequently by the solution viscosity, respectively. The morphologies of the GNS dispersion in the PTT composite fiber changed as the GNS content increased. Moreover, GNSs were embedded and protruded from the fibers. Regardless of GNS content, a PTT mesomorphic phase was formed during electrospinning, whereas GNSs could induce PTT mesomorphic phase significantly during PTT/GNS electrospinning. The PTT cold crystallization rate of PTT/GNS composite fibers increased with the GNS content. During cold crystallization, the PTT mesomorphic phase could be the PTT nuclei, which were randomly developed on the GNS surface, to form the lamellae in the PTT/GNS composite fiber. The increased viscosities of PTT/GNS solution retarded the alignment of PTT chains in the PTT/GNS composite fibers during the electrospinning process, whereas the alignment of PTT crystalline lamellae in the PTT/GNS composite fibers was inhibited by the PTT nuclei on the GNS surface after annealing at 170 °C. Therefore, the mechanical properties of PTT/GNS CFMs became ductile with the addition of GNSs. The elastic recoveries of the PTT/GNS CFMs with 170 °C annealing are better than those of the as-spun PTT/GNS CFMs. The elastic recoveries of the PTT/GNS CFMs were slightly reduced with increased GNS content. The PTT/GNS composites could form a conductive network more easily than PTT/GNS CFMs. The conductivity thresholds of the PTT/GNS CFMs are higher than that of the PTT/GNS composite films because the electrospun fiber mats were highly porous and contained voids filled with insulating air, and the GNS–GNS network could not be formed along the composite fibers. The porous structure and GNS dispersion of the elicited PTT/GNS CFMs are important to the CFMs' performance, especially for σ. The σ of the HP-PTT/GNS CFMs were similar to those of the PTT/GNS composites that were prepared from hot-pressed dried powders.

Supplementary Materials: The following are available online at http://www.mdpi.com/2073-4360/11/1/164/s1, Figure S1: Thermo-gravimetric analysis curves of the PTT/GNS CFMs; Figure S2: SEM images of electrospun fibers from (a) 7 wt % PTT/TFA solution, (b) 8 wt % PTT/TFA solution, (c) 9 wt % PTT/TFA solution, (d) 11 wt % PTT/TFA solution, (e) 12 wt % PTT/TFA solution, and (f) 14 wt % PTT/TFA solution; Figure S3: Effects of GNS concentration on PTT solution conductivity at 25 °C; Figure S4: Effects of PTT/TFA concentration with 10 wt % GNS content on solution viscosity; Figure S5: SEM and TEM images of electrospun PTT fibers filled with: (a), (b) 10 wt %, (c), (d) 12 wt %, (e), (f) 14 wt %, and (g), (h) 16 wt % GNS; Figure S6: Elastic recovery of the (a) neat PTT CFMs, (b) neat PTT CFMs with annealing at 170 °C, (c) PTT/GNS 99/1 CFMs, (d) PTT/GNS 99/1 CFMs with annealing at 170 °C, (e) PTT/GNS 97/3 CFMs, (f) PTT/GNS 97/3 CFMs with annealing at 170 °C, (g) PTT/GNS 95/5 CFMs, and (b) PTT/GNS 95/5 CFMs with annealing at 170 °C during five cycles of loading–unloading with an elongation of 20%.

Author Contributions: C.-L.H. conceived and designed the experiments; H.-H.W., Y.-C.J. and W.-Z.L. performed the experiments; C.-L.H. and H.-H.W. analyzed the data; C.-L.H. wrote the paper; and all authors approved the final paper.

Funding: This research was funded by the National Science Council of Taiwan (ROC) and the Ministry of Science and Technology of Taiwan for the research grant (NSC 102-2221-E-035-086-) and (MOST 107-2632-E-035-001-).

Acknowledgments: The authors would like to thank the National Science Council of Taiwan (ROC) and the Ministry of Science and Technology of Taiwan for the research grant (NSC 102-2221-E-035-086-) and (MOST 105-2632-E-035-001-), respectively, which supported this work. The authors appreciate the Precision Instrument Support Center of Feng Chia University and Chi Wang in NCKU for providing the fabrication and measurement facilities.

Conflicts of Interest: The authors declare no conflict of interest.

References

1. Novoselov, K.S.; Geim, A.K.; Morozov, S.; Jiang, D.; Zhang, Y.; Dubonos, S.; Grigorieva, I.; Firsov, A. Electric field effect in atomically thin carbon films. *Science* **2004**, *306*, 666–669. [CrossRef] [PubMed]
2. Grossiord, N.; Loos, J.; van Laake, L.; Maugey, M.; Zakri, C.; Koning, C.E.; Hart, A.J. High-conductivity polymer nanocomposites obtained by tailoring the characteristics of carbon nanotube fillers. *Adv. Funct. Mater.* **2008**, *18*, 3226–3234. [CrossRef]
3. Thomassin, J.-M.; Jérôme, C.; Pardoen, T.; Bailly, C.; Huynen, I.; Detrembleur, C. Polymer/carbon based composites as electromagnetic interference (EMI) shielding materials. *Mater. Sci. Eng. R* **2013**, *74*, 211–232. [CrossRef]
4. Zhang, H.B.; Yan, Q.; Zheng, W.G.; He, Z.; Yu, Z.Z. Tough graphene-polymer microcellular foams for electromagnetic interference shielding. *ACS Appl. Mater. Interfaces* **2011**, *3*, 918–924. [CrossRef] [PubMed]
5. Sreeprasad, T.S.; Rodriguez, A.A.; Colston, J.; Graham, A.; Shishkin, E.; Pallem, V.; Berry, V. Electron-tunneling modulation in percolating network of graphene quantum dots: Fabrication, phenomenological understanding, and humidity/pressure sensing applications. *Nano Lett.* **2013**, *13*, 1757–1763. [CrossRef] [PubMed]
6. Potts, J.R.; Dreyer, D.R.; Bielawski, C.W.; Ruoff, R.S. Graphene-based polymer nanocomposites. *Polymer* **2011**, *52*, 5–25. [CrossRef]
7. Bisri, S.Z.; Gao, J.; Derenskyi, V.; Gomulya, W.; Iezhokin, I.; Gordiichuk, P.; Herrmann, A.; Loi, M.A. High performance ambipolar field-effect transistor of random network carbon nanotubes. *Adv. Mater.* **2012**, *24*, 6147–6152. [CrossRef]
8. Derenskyi, V.; Gomulya, W.; Rios, J.M.; Fritsch, M.; Frohlich, N.; Jung, S.; Allard, S.; Bisri, S.Z.; Gordiichuk, P.; Herrmann, A.; et al. Carbon nanotube network ambipolar field-effect transistors with 10(8) on/off ratio. *Adv. Mater.* **2014**, *26*, 5969–5975. [CrossRef]
9. Kim, K.S.; Zhao, Y.; Jang, H.; Lee, S.Y.; Kim, J.M.; Kim, K.S.; Ahn, J.-H.; Kim, P.; Choi, J.-Y.; Hong, B.H. Large-scale pattern growth of graphene films for stretchable transparent electrodes. *Nature* **2009**, *457*, 706–710. [CrossRef]
10. Wang, X.; Ding, B.; Yu, J.; Wang, M. Engineering biomimetic superhydrophobic surfaces of electrospun nanomaterials. *Nano Today* **2011**, *6*, 510–530. [CrossRef]
11. Su, Z.; Li, J.; Li, Q.; Ni, T.; Wei, G. Chain conformation, crystallization behavior, electrical and mechanical properties of electrospun polymer-carbon nanotube hybrid nanofibers with different orientations. *Carbon* **2012**, *50*, 5605–5617. [CrossRef]
12. Kim, B.-H.; Yang, K.S. Structure and electrochemical properties of electrospun carbon fiber composites containing graphene. *J. Ind. Eng. Chem.* **2014**, *20*, 3474–3479. [CrossRef]
13. Li, Y.; Zhang, P.; Ouyang, Z.; Zhang, M.; Lin, Z.; Li, J.; Su, Z.; Wei, G. Nanoscale graphene doped with highly dispersed silver nanoparticles: Quick synthesis, facile fabrication of 3D membrane-modified electrode, and super performance for electrochemical sensing. *Adv. Funct. Mater.* **2016**, *26*, 2122–2134. [CrossRef]
14. Jin, X.; Ni, Q.-Q.; Fu, Y.; Zhang, L.; Natsuki, T. Electrospun nanocomposite polyacrylonitrile fibers containing carbon nanotubes and cobalt ferrite. *Polym. Compos.* **2012**, *33*, 317–323. [CrossRef]
15. Li, B.; Yuan, H.; Zhang, Y. Transparent PMMA-based nanocomposite using electrospun graphene-incorporated PA-6 nanofibers as the reinforcement. *Compos. Sci. Technol.* **2013**, *89*, 134–141. [CrossRef]

16. Ramazani, S.; Karimi, M. Electrospinning of poly(ε-caprolactone) solutions containing graphene oxide: Effects of graphene oxide content and oxidation level. *Polym. Compos.* **2016**, *37*, 131–140. [CrossRef]
17. Huang, C.-L.; Peng, S.-Y.; Wang, Y.-J.; Chen, W.-C.; Lin, J.-H. Microstructure and characterization of electrospun poly(vinyl alcohol) nanofiber scaffolds filled with graphene nanosheets. *J. Appl. Polym. Sci.* **2015**, *132*, 41891. [CrossRef]
18. Gao, J.; Hu, M.; Dong, Y.; Li, R.K. Graphite-nanoplatelet-decorated polymer nanofiber with improved thermal, electrical, and mechanical properties. *ACS Appl. Mater. Interfaces* **2013**, *5*, 7758–7764. [CrossRef] [PubMed]
19. Garboczi, E.J.; Snyder, K.A.; Douglas, J.F.; Thorpe, M.F. Geometrical percolation threshold of overlapping ellipsoids. *Phys. Rev. E* **1995**, *52*, 819–828. [CrossRef]
20. Heaney, M.B. Measurement and interpretation of nonuniversal critical exponents in disordered conductor–insulator composites. *Phys. Rev. B* **1995**, *52*, 12477–12480. [CrossRef]
21. Huang, C.-L.; Wang, Y.-J.; Fan, Y.-C.; Hung, C.-L.; Liu, Y.-C. The effect of geometric factor of carbon nanofillers on the electrical conductivity and electromagnetic interference shielding properties of poly(trimethylene terephthalate) composites: A comparative study. *J. Mater. Sci.* **2016**, *52*, 2560–2580. [CrossRef]
22. Mazinani, S.; Ajji, A.; Dubois, C. Morphology, structure and properties of conductive PS/CNT nanocomposite electrospun mat. *Polymer* **2009**, *50*, 3329–3342. [CrossRef]
23. Chien, H.-S.; Wang, C. Morphology, microstructure, and electrical properties of poly(D,L-lactic acid)/carbon nanocapsule composite nanofibers. *J. Appl. Polym. Sci.* **2013**, *128*, 958–969. [CrossRef]
24. Chen, K.; Tang, X. Instantaneous elastic recovery of poly(trimethylene terephthalate) filament. *J. Appl. Polym. Sci.* **2004**, *91*, 1967–1975. [CrossRef]
25. Zhang, J. Study of poly(trimethylene terephthalate) as an engineering thermoplastics material. *J. Appl. Polym. Sci.* **2004**, *91*, 1657–1666. [CrossRef]
26. Chen, T.-K.; Tien, Y.-I.; Wei, K.-H. Synthesis and characterization of novel segmented polyurethane/clay nanocomposites. *Polymer* **2000**, *41*, 1345–1353. [CrossRef]
27. Launay, A.; Thominette, F.; Verdu, J. Water sorption in amorphous poly(ethylene terephthalate). *J. Appl. Polym. Sci.* **1999**, *73*, 1131–1137. [CrossRef]
28. Gupta, A.; Choudhary, V. Electromagnetic interference shielding behavior of poly(trimethylene terephthalate)/multi-walled carbon nanotube composites. *Compos. Sci. Technol.* **2011**, *71*, 1563–1568. [CrossRef]
29. Huang, C.-L.; Wang, Y.-J.; Fan, Y.-C. Morphological features and crystallization behavior of the conductive composites of poly(trimethylene terephthalate)/graphene nanosheets. *J. Appl. Polym. Sci.* **2016**, *133*, 43419. [CrossRef]
30. Wu, D.; Shi, T.; Yang, T.; Sun, Y.; Zhai, L.; Zhou, W.; Zhang, M.; Zhang, J. Electrospinning of poly(trimethylene terephthalate)/carbon nanotube composites. *Eur. Polym. J.* **2011**, *47*, 284–293. [CrossRef]
31. Khil, M.S.; Kim, H.Y.; Kim, M.S.; Park, S.Y.; Lee, D.-R. Nanofibrous mats of poly(trimethylene terephthalate) via electrospinning. *Polymer* **2004**, *45*, 295–301. [CrossRef]
32. Wang, C.; Fang, C.-Y.; Wang, C.-Y. Electrospun poly(butylene terephthalate) fibers: Entanglement density effect on fiber diameter and fiber nucleating ability towards isotactic polypropylene. *Polymer* **2015**, *72*, 21–29. [CrossRef]
33. Wang, C.; Lee, M.-F.; Wu, Y.-J. Solution-electrospun poly(ethylene terephthalate) fibers: Processing and characterization. *Macromolecules* **2012**, *45*, 7939–7947. [CrossRef]
34. Wang, C.; Lee, M.-F.; Jao, C.-H. Phase transition of poly(ethylene terephthalate) in nanofibers electrospun from phenol-based solution. *Eur. Polym. J.* **2014**, *52*, 127–136. [CrossRef]
35. Zhang, H.-B.; Zheng, W.-G.; Yan, Q.; Yang, Y.; Wang, J.-W.; Lu, Z.-H.; Ji, G.-Y.; Yu, Z.-Z. Electrically conductive polyethylene terephthalate/graphene nanocomposites prepared by melt compounding. *Polymer* **2010**, *51*, 1191–1196. [CrossRef]
36. Zhang, H.-B.; Zheng, W.-G.; Yan, Q.; Jiang, Z.-G.; Yu, Z.-Z. The effect of surface chemistry of graphene on rheological and electrical properties of polymethylmethacrylate composites. *Carbon* **2012**, *50*, 5117–5125. [CrossRef]
37. Stankovich, S.; Dikin, D.A.; Dommett, G.H.; Kohlhaas, K.M.; Zimney, E.J.; Stach, E.A.; Piner, R.D.; Nguyen, S.T.; Ruoff, R.S. Graphene-based composite materials. *Nature* **2006**, *442*, 282–286. [CrossRef]

38. McKee, M.G.; Wilkes, G.L.; Colby, R.H.; Long, T.E. Correlations of solution rheology with electrospun fiber formation of linear and branched polyesters. *Macromolecules* **2004**, *37*, 1760–1767. [CrossRef]
39. Shenoy, S.L.; Bates, W.D.; Frisch, H.L.; Wnek, G.E. Role of chain entanglements on fiber formation during electrospinning of polymer solutions: Good solvent, non-specific polymer–polymer interaction limit. *Polymer* **2005**, *46*, 3372–3384. [CrossRef]
40. Tsou, S.-Y.; Lin, H.-S.; Cheng, P.-J.; Huang, C.-L.; Wu, J.-Y.; Wang, C. Rheological aspect on electrospinning of polyamide 6 solutions. *Eur. Polym. J.* **2013**, *49*, 3619–3629. [CrossRef]
41. Huang, C.-L.; Wang, C. Rheological and conductive percolation laws for syndiotactic polystyrene composites filled with carbon nanocapsules and carbon nanotubes. *Carbon* **2011**, *49*, 2334–2344. [CrossRef]
42. Cloupeau, M.; Prunet-Foch, B. Electrostatic spraying of liquids in cone-jet mode. *J. Electrost.* **1989**, *22*, 135–159. [CrossRef]
43. Chen, D.-R.; Pui, D.Y.; Kaufman, S.L. Electrospraying of conducting liquids for monodisperse aerosol generation in the 4 nm to 1.8 μm diameter range. *J. Aerosol. Sci.* **1995**, *26*, 963–977. [CrossRef]
44. Bhattarai, S.R.; Bhattarai, N.; Yi, H.K.; Hwang, P.H.; Cha, D.I.; Kim, H.Y. Novel biodegradable electrospun membrane: Scaffold for tissue engineering. *Biomaterials* **2004**, *25*, 2595–2602. [CrossRef] [PubMed]
45. Wang, C.; Hsu, C.-H.; Lin, J.-H. Scaling laws in electrospinning of polystyrene solutions. *Macromolecules* **2006**, *39*, 7662–7672. [CrossRef]
46. Ahn, B.W.; Chi, Y.S.; Kang, T.J. Preparation and characterization of multi-walled carbon nanotube/poly(ethylene terephthalate) nanoweb. *J. Appl. Polym. Sci.* **2008**, *110*, 4055–4063. [CrossRef]
47. Ding, B.; Li, C.; Miyauchi, Y.; Kuwaki, O.; Shiratori, S. Formation of novel 2D polymer nanowebs via electrospinning. *Nanotechnology* **2006**, *17*, 3685–3691. [CrossRef]
48. Zhao, X.; Zhang, Q.; Chen, D.; Lu, P. Enhanced mechanical properties of graphene-based poly(vinyl alcohol) composites. *Macromolecules* **2010**, *43*, 2357–2363. [CrossRef]
49. Chuang, W.-T.; Su, W.-B.; Jeng, U.S.; Hong, P.-D.; Su, C.-J.; Su, C.-H.; Huang, Y.-C.; Laio, K.-F.; Su, A.-C. Formation of mesomorphic domains and subsequent structural evolution during cold crystallization of poly(trimethylene terephthalate). *Macromolecules* **2011**, *44*, 1140–1148. [CrossRef]
50. Ward, I.; Wilding, M. Infra-red and Raman spectra of poly(m-methylene terephthalate) polymers. *Polymer* **1977**, *18*, 327–335. [CrossRef]
51. Kim, K.; Bae, J.; Kim, Y. Infrared spectroscopic analysis of poly(trimethylene terephthalate). *Polymer* **2001**, *42*, 1023–1033. [CrossRef]
52. Poulin-Dandurand, S.; Pérez, S.; Revol, J.-F.; Brisse, F. The crystal structure of poly(trimethylene terephthalate) by X-ray and electron diffraction. *Polymer* **1979**, *20*, 419–426. [CrossRef]
53. Jalili, R.; Aboutalebi, S.H.; Esrafilzadeh, D.; Shepherd, R.L.; Chen, J.; Aminorroaya-Yamini, S.; Konstantinov, K.; Minett, A.I.; Razal, J.M.; Wallace, G.G. Scalable one-step wet-spinning of graphene fibers and yarns from liquid crystalline dispersions of graphene oxide: Towards multifunctional textiles. *Adv. Funct. Mater.* **2013**, *23*, 5345–5354. [CrossRef]
54. Chen, S.; Ma, W.; Cheng, Y.; Weng, Z.; Sun, B.; Wang, L.; Chen, W.; Li, F.; Zhu, M.; Cheng, H.-M. Scalable non-liquid-crystal spinning of locally aligned graphene fibers for high-performance wearable supercapacitors. *Nano Energy* **2015**, *15*, 642–653. [CrossRef]
55. Balberg, I.I. Tunneling and nonuniversal conductivity in composite materials. *Phys. Rev. Lett.* **1987**, *59*, 1305–1308. [CrossRef] [PubMed]
56. Vionnet-Menot, S.; Grimaldi, C.; Maeder, T.; Strässler, S.; Ryser, P. Tunneling-percolation origin of nonuniversality: Theory and experiments. *Phys. Rev. B* **2005**, *71*, 064201–064204. [CrossRef]
57. Grimaldi, C.; Balberg, I. Tunneling and nonuniversality in continuum percolation systems. *Phys. Rev. Lett.* **2006**, *96*, 066602. [CrossRef] [PubMed]
58. Hu, N.; Karube, Y.; Yan, C.; Masuda, Z.; Fukunaga, H. Tunneling effect in a polymer/carbon nanotube nanocomposite strain sensor. *Acta Mater.* **2008**, *56*, 2929–2936. [CrossRef]

polymers

MDPI

Article

Evaluation of Exfoliated Graphite to Graphene in Polyamide 66 Using Novel High Shear Elongational Flow

Justin W. Hendrix [1,*], Ryan Szeto [2], Thomas Nosker [2,*], Jennifer Lynch-Branzoi [2] and Thomas J. Emge [3]

1 Department of Chemical and Biochemical Engineering, Rutgers University, 607 Taylor Road, Piscataway, NJ 08854, USA
2 Department of Materials Science and Engineering, Rutgers University, 607 Taylor Road, Piscataway, NJ 08854, USA; ryanxszeto@yahoo.com (R.S.); jklynch@soe.rutgers.edu (J.L.-B)
3 Department of Chemistry and Chemical Biology, Rutgers University, 610 Taylor Road, Piscataway, NJ 08854, USA; emge@chem.rutgers.edu
* Correspondence: justin.hendrix.w@gmail.com (J.W.H.); tjnosker@soe.rutgers.edu (T.N.); Tel.: +1-202-321-1791 (J.W.H.); +1-848-445-3631 (T.N.)

Received: 29 November 2018; Accepted: 11 December 2018; Published: 17 December 2018

Abstract: Graphene has been publicized as the game changing material of this millennium. To this day, scalable production leading to exceptional material properties has been difficult to attain. Most methods require harsh chemicals, which result in destroying the graphene surface. A method was developed, exploiting high speed elongational flow in a novel designed batch mixer; creating a distribution of pristine few to many layer graphene flakes. The method focuses on exfoliating in a molten polyamide 66 (PA66) matrix, creating a graphene reinforced polymer matrix composite (G-PMC). The process revealed that high speed elongational flow was able to create few layer graphene. Graphite exfoliation was found driven in part by diffusion, leading to intercalation of PA66 in graphite. The intercalated structure lead to increases in the hydrogen bonding domain, creating anisotropic crystal domains. The thermal stability of the G-PMC was found to be dependent to the degree of exfoliation, PA66 crystal structure and composite morphology. The aim of this research is to characterize uniquely produced graphene containing polymer matrix composites using a newly created elongational flow field. Using elongational flow, graphite will be directly exfoliated into graphene within a molten polymer.

Keywords: graphene; graphene polymer matrix composite; polyamide 66; elongational flow; hydrogen bond

1. Introduction

Graphene has been publicized as the leading edge material of this millennia. In the past two decades, the 2-D material has been considered for use in applications, such as thermal management [1], environmental remediation and filtration [2,3], biotechnology [4,5] and lightweight, structural materials [6]; due to its exceptional mechanical properties [7]. Many exfoliation methods have been developed to separate graphene from graphite and produce un-oxidized graphene flakes, including electrochemical [8,9], chemical oxidation and expansion [10,11] and liquid phase [12,13]. However, these approaches are costly, multistep methods that use or produce harmful byproducts. Since the graphene is often extracted from a sacrificial medium, defects are often introduced on the graphene surfaces and edges, which may reduce some properties [14,15]. The challenge is finding a viable, inexpensive, less complex process that permits property retention.

An alternative approach would be to cultivate an exfoliation method, for which the matrix is not a sacrificial phase but rather becomes part of the end use application, for example, a graphene contained reinforced polymer nanocomposite. In order to do that, the matrix must impart significant shear forces to exfoliate flake graphite to graphene nanoflakes. Thus, it is important to study scalable polymer processing techniques that are optimized for efficient graphite exfoliation.

Direct exfoliation has been known to providing an economical, top-down approach in converting the graphite to graphene [3]. The processes that provide the most advantage involve traditional plastic processing utilizing extensional flow. Using elongational flow to exfoliate graphite has been attempted but inefficient exfoliation was achieved [16]. Traditionally, current mixing technologies are not independently able to efficiently create graphene without the addition of another step that causes exfoliation. Utilizing elongational flow with the addition of an exfoliation aid, has been the direction for research in this field. Ellingham et al. [17] has shown that using twin screw extrusion and the addition of supercritical CO_2, exfoliation was further improved through bubble formation. The bubbles were able to generate an additional extension mechanism, upon expansion of CO_2 to further exfoliate the graphite. Those that are specifically designed for elongational flow start with pre-expanded graphite, attempting to reduce the size of graphite with inherent defects [18]. There are mixers that exploit extensional flow without the addition of an additive exfoliating component. Work conducted by Oxfall et al. [16] uses dual piston driven operation injected through a capillary to cause elongational flow mixing. A drawback is the creation of hydrostatic pressure, causing intermittent stages of no mixing. Oxall's work disproportionally hinders efficient exfoliation. Supported in our work, we're able to create a succession of shear strain events without any chemical aid or pretreatment. A succession of shearing events in Nosker et al. [19] produces continuous mixing, making our process an economical advantage.

Some of the earliest experimental and theoretical work in fluid flow of two immiscible liquids was pioneered by Taylor in his experiments with droplet suspension breakup in a dilute medium of dissimilar viscosity [20]. Taylor designed two methods to observe droplet deformation contributing to a difference in mechanism of deformation. The first consisted of two counter moving surfaces, creating planar shear flow. From this first experimental setup at maximum roller speed, the droplet deformed to a spheroid that distorted without breaking. The second experiment conducted was of a four roll apparatus producing planar elongational flow under the same matrix conditions. What was observed, was droplet extension occurred parallel to the direction of deformation. The extension lead to droplet rupture and breakage.

Taylor found the mechanism for breakage of droplets to be related to apparent differences in kinematic viscosity ratio defined by p to a droplets critical deformability; where μ_D is the viscosity of the dispersed phase and μ_C, the viscosity of the continuous phase (1).

$$p = \frac{\mu_D}{\mu_C} \quad (1)$$

The critical deformability was found to work but only up to a maximum viscous ratio of 4. Similar results were confirmed by Rumscheidt and Mason of this critical viscosity ratio to deformation [21]; suggesting that beyond this critical viscosity exist no deformation and breakage of droplets. An expansion of work by Taylor & Rumscheidt from Grace produced a slightly different view on the viscous ratio. For droplet deformation leading to breakage, Grace found that a specific deformation rate above a critical capillary value Ca_{crit} was needed; where $Ca_{crit} < Ca$ [22]. The capillary number Ca represented the ratio of the viscous force of the matrix to the restorative force of the dispersed phase to deformation or shear; where μ is the kinematic viscosity, γ is the shear rate, σ interfacial tension and R_d the droplet radius (2).

$$Ca = \frac{\mu R_d \gamma}{\sigma} \quad (2)$$

Associating elongational flow to simple shear flow figure, Grace found the critical capillary number to be smaller for the case of elongational flow. Leaving for simple shear, the critical capillary number for breakage approaching a near infinite value as the viscous ratio exceeds $p > 1$ in Figure 1 [23]. This forms the bases of 2D elongational flow as being essential for particulate breakup and deformation, based on the idea of it being seemingly independent of viscosity. This viscous independence suggests that any material system, that undergoes deformation by the application of stress, has the potential to create new surfaces. These systems include liquid-liquid in Graces work and solid-liquid systems. Particularly for solids that deform by shear stress under an inherent slip plane, like graphite.

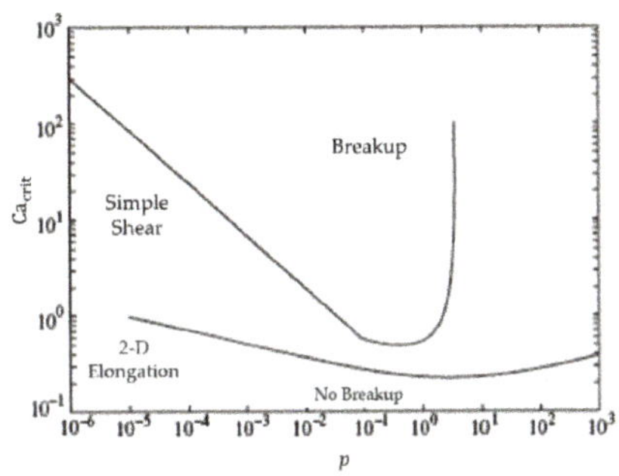

Figure 1. Capillary number Ca vs. viscous ratio p [23].

The aim of this research is to characterize uniquely produced graphene containing polymer matrix composites using a newly created elongational flow field. Using elongational flow, the drive is to exfoliate graphite into graphene directly within a molten polymer. This work is an expansion of a method to create graphene reinforced polymer matrix composite by Nosker et al. [19]. In using a continuous polymer phase and a dispersive graphite phase, the previous art exploits fluid mechanics of droplet breakup by exceeding Ca crit across large viscous regimes created by continuous exfoliation. By exfoliating in-situ the polymer phase seeks to exceed the resistive force to exfoliation and detail a scalable method of creating a polymer matrix composites, where graphene is used as a reinforcing phase. The process may show that this method has subsequent improved particle-matrix interaction between newly formed nanographite and graphene surfaces; which is not found in traditional plastic-graphene composites.

In this work, we use elongational flow to exfoliate 35 wt % mined graphite within polyamide 66 (PA66) to create graphene containing polyamide 66 nanocomposites (G-PA66). The components are subject to 10, 20, 30 and 45 min of elongational flow mixing. The mixed components are then characterized for their changes in structure, morphology and thermal properties. Using this single step approach to convert graphite directly to nanographite and graphene, it is suggested that a molten elongational phase can apply unique hydrodynamics on the graphite flake to produce exfoliated few, many layer graphene and nanographite at shear strain rates exceeding 10^3 s^{-1}.

2. Materials and Methods

Natural flake graphite (Asbury Carbons, mills grade 3627 with 99.2% purity, diameter = 250 μm, ρ = 2.26 g/cm^3, Asbury Carbons, Asbury, NJ, USA) was used as the exfoliating species. Polyamide 66 (PA66, Zytel 101 NC010, T_m = 262 °C, T_g = 60 °C, ρ = 1.14 g/cm^3, DuPont, Wilmington, DE, USA) was used as the high temperature polymer in this study. Prior to processing, both components were dried in an oven to eliminate water. PA66 was dried at 85 °C under vacuum and graphite was dried at 350 °C for 4 h.

PA66, 20 wt % and 35 wt % flake mineral graphite were mixed using a high shear batch mixer imparting elongational flow. The blends were mixed at 276 °C and a shear strain rate of 2876 s^{-1}, under an Ar gas atmosphere to reduce polymer degradation. 35 wt % graphite and in PA66 and PA66 were mixed for 10, 20, 30 and 45 min to produce G-PA66 samples and PA66 control samples at each mixing time Table 1, PA66 and G-PA66 extrudate (approximate diameter of 2.94 mm) was prepared for subsequent characterization.

Table 1. Processing Matrix.

Mixing time (min)	Control materials	Composite materials
10	PA66	G-PA66
20	PA66	G-PA66
30	PA66	G-PA66
45	PA66	G-PA66

Wide Angle X-ray Diffraction (WAXD) was performed using a Philips X-Pert Powder X-ray Diffractometer (PANalytical, Almelo, Netherlands) with a Cu (λ = 1.54 Å) K-alpha source at 45 kV/40 mA and scanned from 4° to 70° 2θ. The samples were scanned with a step size of 0.02° 2θ to detect graphitic and nylon characteristic peaks. Extrudates were notched and cryogenically fractured to produce a pristine surface.

Thermal properties of PA66 and G-PA66 samples were determined using a Q1000 Differential Scanning Calorimeter (TA instruments, New Castle, DE, USA). PA66 and G-PA66 samples weighing 10 mg, were sectioned from the extrudate and subject to a heat, cool, reheat method at 10 °C/min from 0–300 °C under a nitrogen atmosphere. The effect of processing on the melting (T_m), crystallization (T_c) and glass transition (T_g) temperatures of the PA66 component was determined, to include enthalpy of crystallization ΔH_c, enthalpy of melting ΔH_m and % crystallinity in PA66.

Morphology of G-PA66 samples was investigated using a Sigma Field Emission Scanning Electron Microscope (SEM) (Zeiss, Oberkochen, Germany) with Oxford EDS.

In order to better investigate the nanoflakes produced during this in situ exfoliation method, G-PA66 nanocomposites were granulated using a Spex 6700 Freezer Mill (SPEX Sample Prep, Metuchen, NJ, USA) to cryogenically impact samples and form a fine powder for viewing using the SEM. The fine powder was ultra-sonicated in isopropyl alcohol (IPA) for 5 min, then the suspension was drop coated on to a lacey carbon grid for transmission and characterization in a JEOL 2010F Transmission Electron Microscope (JEOL, Tokyo, Japan) operating at 200 kV for High Resolution Transmission Electron Microscopy (HR-TEM).

3. Results and Discussion

3.1. Wide Angle X-ray Diffraction

In our assessment of the crystalline characteristics of the X-ray diffraction results, the Scherrer Equation (3) was used to calculate the crystalline domain size of the phases in PA66, G-PA66 and Graphite. K = 0.9 represents the shape factor, λ is the incident radiation wavelength, β is the line broadening at half the maximum intensity of the detectable peak and θ is the Bragg angle in degrees.

$$L = \frac{K\lambda}{\beta cos} \tag{3}$$

The starting graphite is analyzed to having a (002) lattice spacing of 0.334 nm, indicative of the stable hexagonal phase [24]. After exfoliated in the presence of elongational flow, the existing hexagonal phase shows an increase in (002) d-spacing, in Table 2. The results from the Figure 2 and Table 2 show a progressive reduction in the average domain size, leading to a 76% reduction in domain size for the graphite portion compared to its native structure. The increase in (002) lattice spacing suggests that the starting hexagonal phase is converting to a turbostratic phase of exfoliated graphite containing composite. The turbostratic form of exfoliate has been known to contain stacks of graphene planes, which have undergone rotations and distortions [25]. Comparing samples G-PA66 30 min and

G-PA66 45 min, there is a large difference in both the reduction in (002) domain size and the increase in lattice spacing. This suggests that after 30 min of mixing, a transition in the conversion of the exfoliated graphite and the resulting morphology has occurred.

Table 2. Diffraction data for Graphite, PA66 and G-PA66.

Materials	Mixing time (min)	PA66 d (100) (nm)	PA66 (100) Domain (nm)	PA66 d (010)/(110) (nm)	PA66 (010)/(110) Domain (nm)	Graphite d (002) (nm)	Graphite (002) Domain (nm)
Graphite	-	-	-	-	-	0.3344	62.7 (4.1)
PA66	-	0.4407	4.4 (0.2)	0.3878	4.4 (0.7)	-	-
PA66	10	0.4427	8.6 (0.4)	0.3753	5.6 (0.2)	-	-
PA66	20	0.4453	9.7 (0.6)	0.3768	5.3 (0.2)	-	-
PA66	30	0.4414	8.3 (0.4)	0.3758	6.1 (0.2)	-	-
PA66	45	0.4435	2.9 (0.9)	0.3796	2.9 (0.5)	-	-
G-PA66	10	0.4404	8.4 (0.4)	0.3719	7.0 (0.2)	0.3360	21.4 (0.3)
G-PA66	20	0.4405	9.3 (0.5)	0.3720	6.8 (0.1)	0.3358	21.1 (0.2)
G-PA66	30	0.4393	11.6 (0.8)	0.3735	6.0 (0.1)	0.3363	19.1 (0.2)
G-PA66	45	0.4423	8.9 (1.7)	0.3772	7.0 (0.3)	0.3386	14.8 (0.1)

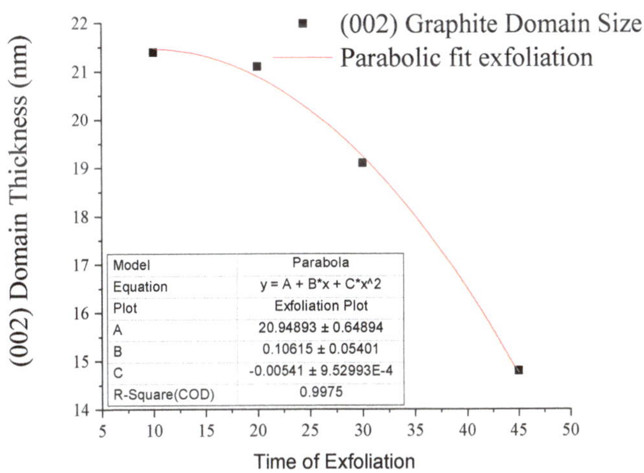

Figure 2. Graphite (002) domain size curve fit.

In Figure 2, a curve fit was constructed to establish a model for exfoliation by the reduction in (002) domain size in graphite. The exfoliation of graphite in PA66 is shown to fit a parabolic model, described as the mechanism for exfoliation. Parabolic models are closely tied to diffusion reaction systems [26,27]. In the diffusion reaction system, the movement of species are modeled as a projecting wave front. In the creation of our graphene reinforced composite, the projecting wave behaves as the diffusing PA66 species between the graphite layers.

The results of the curve fit for the diffusion model shows that the linear equation is second-order with respect to mixing time. Second-order approximations accounts for a linear driving force (LDF) for diffusion a graphite intercalant, from Budzianowski [28]. In his work a second-order model a cyclic adsorption/desorption step change for fluid concentration at the surface was assumed. He found that the half-life for adsorption or distortion was too long and provided no instance for full saturation, while diffusing. In our case, the PA66 diffuses, adsorbs and then desorbs continuously. This process is successive until mixing stops, making it diffusion limited.

$$(002)\ \text{Domain (nm)} = 20.949 + 0.106\,t - 0.005\,t^2 \quad (4)$$

Diffraction patterns of PA66 under extensional mixing are shown in Figure 3. In PA66, the first crystalline peak at 20.7° 2θ, corresponding to the inter-chain hydrogen bonded (100) plane of the amide group for semi-crystalline PA66 [29]. The second crystalline peak for PA66 at 23.6° 2θ, for the overlapping diffraction peaks for the (010) and (110) planes, namely the key intra-chain and intersheet periodicities perpendicular to the (100) direction [29]. As mixing time increased for PA66 samples to 30 min, the results do not indicate progressive changes in diffracted domain size or lattice, from Table 2. After 45 min of mixing, (100) domain is drastically reduced by 5.4 nm and (010) by 3.2 nm. The additional mixing time also results in a lattice increase of +0.0038 nm, for the (010) plane. The diffraction data of PA66 suggest that extensive mixing times lead to a reduction in crystal domain and a shift in the (010) plane. The extensive mixing is likely to cause partial degradation which would ultimately results in these crystal morphological changes.

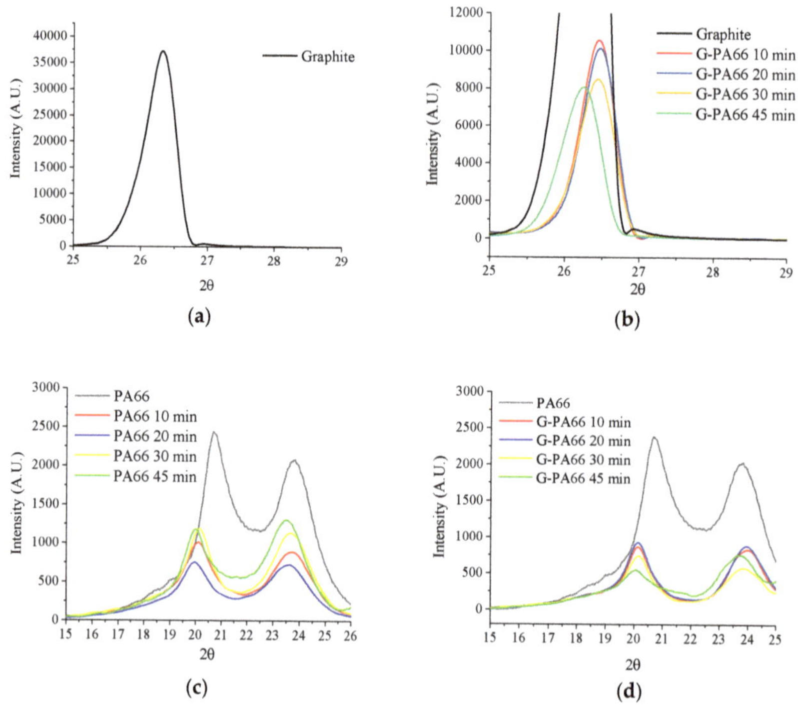

Figure 3. XRD diffraction patterns (**a**) Graphite (**b**) Graphite, G-PA66 10 min, G-PA66 20 min and G-PA66 30 min at Graphite's (002) c-axis peak-scaled for composite exfoliation (**c**) PA66, PA66 10 min, PA66 20 min and PA66 30 min at the (100) & (010) lattice peaks for PA66 (**d**) PA66, G-PA66 10 min, G-PA66 20 min and G-PA66 30 at the (100) & (010) lattice peaks for PA66.

XRD patterns of PA66, relative to the extensional mixing with graphite, appear in Figure 3d. Relative to Figure 3c, the PA66 diffraction peaks are found to contain close similarities to the graphite exfoliating PA66. After analyzing the diffraction peaks by fitting the resulting lattice and domains, crystal structure information is obtained, Table 2. During graphite exfoliation to 30 min, the crystalline domain of PA66 increases in the hydrogen bonded direction (100) and decreases in the van der Waals direction (010)/(110). The change in domain size suggests a preferential crystallization direction in PA66, likely due to the extensional flow process. In Figure 4, anisotropy of the diffraction peak profiles in pure PA66 versus graphene reinforced PA66 reveals a non-linear dependence on exfoliation time. By the increasing mixing time to 30 min for G-PA66, the inter-sheet (010)/(110) d-spacing increase,

Table 2. The non-uniform changes in crystal domain size may allow for preferred crystallization along the surface of the nanoflake and in the hydrogen bonded plane. This suggests preferred direction for PA66 in the (010)/(110) directions and limited growth in the (100) direction.

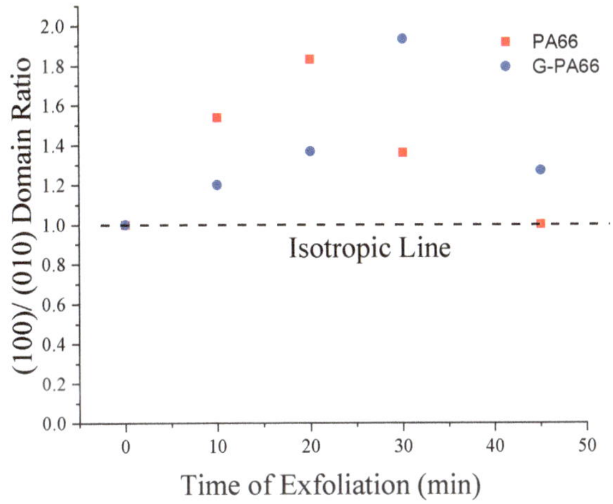

Figure 4. PA66 (100)/(010) domain size isotropy parameter vs. graphite concentration; (010) is (010)/(110).

Extending the time of mixing from 30 min to 45 min with graphite, produces a distinctive change in the G-PA66 lattice structure. G-PA66 shows a lattice increase for both the (100) and (010)/(110) by +0.0030 nm and +0.0037 nm, respectively. This material also changes with a reduction in the hydrogen bonded (100) domain and an increase in the intersheet (010)/(110) domain. The change in the van der Walls (010)/(110) direction trends parallel to the lattice changes in graphite (002). The reduction in (100) d-spacing is likely due to disruptions in the hydrogen bonded network due to graphene and nanographite lattice mismatches, that may increase available surfaces for binding [30].

It is likely that the lattice parameters in the hydrogen bond and the van der Waals direction follows not with the degree at which the graphene is reduced in size but with the (002) lattice spacing between graphene layers beyond 30 min of mixing. The stacking faults in exfoliated graphite invoke changes to the lattice parameters in PA66, which lead to parameter increase. These faults are the results of created turbostratic graphite [31], consisting of an increased and shifted lattice spacing.

3.2. *Differential Scanning Calorimetry*

Thermal properties for PA66 and G-PA66 after 10, 20, 30 and 45 min of elongational mixing appear in Table 3, with corresponding thermograms in Figure 5. Though difficult to notice in Figure 5a,c, the glass transition temperatures for PA66 and G-PA66 were calculated using a baseline tangential method. A line representing the baseline was drawn from 0 °C in the direction of increasing temperature. A deviation from the baseline, at an inflection point in the thermogram, was found to be the glass transition temperature (T_g). Transition temperatures for 10, 20, 30 and 45 min of exfoliation mixing of PA66 & 10, 20, 30, 45 min G-PA66 were found. The glass transitions (T_g) were found to occur at 60, 56, 57, 53, 56, 57, 51, 50 °C respectively. For all the values listed in Table 2. The bulk crystallinity of PA66 was calculated according to (3), for which enthalpy of melting for 100% crystalline PA66 is ΔH_m = 197 J/g [32]. The values for ΔH_m were adjusted for percent PA66, since G-PA66 samples only have 65 wt % PA66.

$$\text{Weight \% Crystallinity} = \frac{\Delta H_m}{\Delta H_{m100}} \times 100 \tag{5}$$

Table 3. Differential scanning calorimetry results for PA66 and G-PA66.

Material	Mixing time (min)	T_g (°C)	Onset melting (°C)	T_m (°C)	ΔH_m (J/g)	Onset crystallization (°C)	T_c (°C)	ΔH_c (J/g)	Crystallinity %
PA66	10	60	246	261	72	235	233	70	~36
PA66	20	56	244	260	69	233	231	68	~35
PA66	30	57	243	259	71	233	231	66	~36
PA66	45	53	245	256	73	238	234	59	~37
G-PA66	10	56	241	252	97	243	239	85	~49
G-PA66	20	57	243	254	103	245	241	94	~52
G-PA66	30	51	226	239	69	229	222	74	~35
G-PA66	45	50	218	236	48	227	217	38	~24

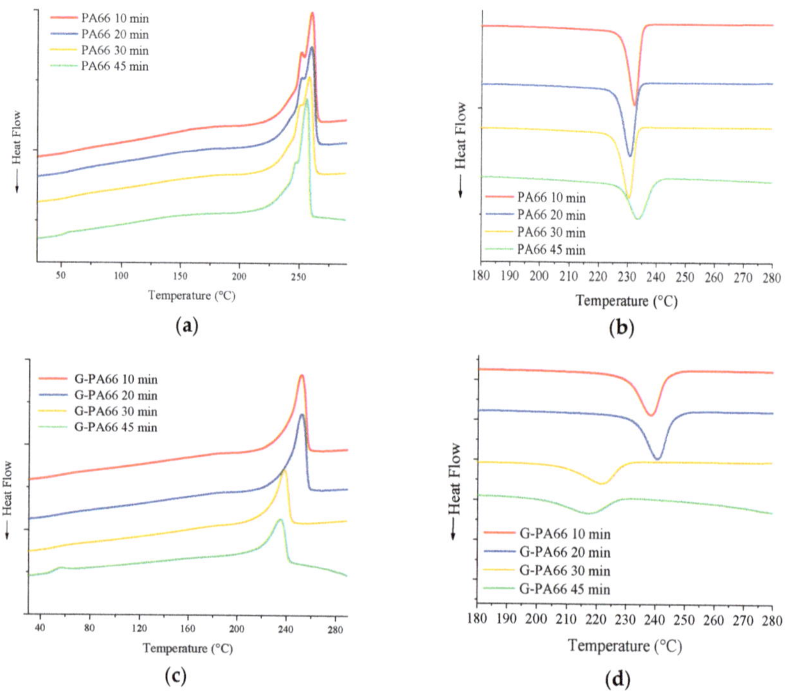

Figure 5. DSC thermograms for PA66 10 min, PA66 20 min and PA66 30 min (**a**) melting and (**b**) crystallization and for G-PA66 10 min; G-PA66 20 min and G-PA66 30 min (**c**) melting (**d**) crystallization.

In Figure 5a,c, the reheat cycle is shown for PA66 samples, we see the formation of two distinct melting peaks in Figure 5a. Two melting peaks are due to a bimodal distribution of crystallites formed by an intermediate heating and cooling rate of 10 °C/min [33]. Since the rate chosen is constant, the thermal characteristics of the crystallites are dependent on the time of mixing and crystal structure formed. As the mixing time increases, the first melting peak reduces in intensity and the secondary melting peak broadens. This broadening of the overall melting in PA66 is indicated by the reduction in onset melting temperature relative to T_m. As the mixing time increases, T_g, T_m, T_c and ΔH_c all decrease. However, there is no consistent trend with ΔH_m results in Table 3 which produces a small increase in % crystallinity as mixing continues beyond 20 min.

In Figure 5c, the reheat scans of the G-PA66 samples show a single melt peak; suggesting a narrow distribution of crystallites formed, compared to the bimodal curve in neat PA66. G-PA66 produces an increased crystallization temperature (T_c) and percent crystallinity for 10 and 20 min of mixing; symptomatic to heterogeneous nucleation of PA66 in graphene/nanographite presence. As the exfoliation time continues beyond 30 min, there a reduction in T_g, T_m and T_c. These properties are much lower than all the other materials, suggesting a drastic change occurs at 30 min and beyond, in its structure.

Compared to pristine PA66 at 30 min, G-PA66 at the same exfoliating time, has nearly equivalent values of percent crystallinity but has a much lower T_m and T_c. The reduced crystallization previously mentioned, suggests the suppression of crystallization to occur. The exfoliated graphitic nanoflake in the matrix is likely to cause a subdued rate of crystallization [34,35]. This suppression lead to a reduced crystallization temperature T_c of 9 °C. The same results appear for the T_g, with 45 min of exfoliation resulting to a 6 °C reduction than the lowest T_g. The crystallization suppression is shown sensitive to the percent crystallinity calculated, which suggest lower T_c depression leads to lower crystallinity.

Identifying how the crystal structure relates to the thermal stability of G-PA66 composite, we see that the hydrogen bonded lattice parameters play an important role. This is not represented in the neat PA66, since the differences between 20 and 30 min are nominal. G-PA66 mixed at 20 min; having the least modified hydrogen bonded lattice, has relatively higher thermal properties and percent crystallinity. The thermal property variations occurring through reduction of the hydrogen bonded lattice, suggesting there is a relation to the structure in the PA66 of the composite. From the increased crystallization temperature Tc and percent crystallinity, we know this is evidence of surface crystallization occurring in the G-PMC (10 & 20 min). The cause of crystallization temperature depression in G-PA66 mixed at 30 min might be explained by the exfoliated graphene acting as unstable crystals growth surfaces. These surfaces form disrupted hydrogen bonded sites; imparting themselves between the formed crystals, suppressing their crystallinity and thermal stability [36].

3.3. Scanning Electron Microscopy

In all stages of the exfoliation process, it is important to note that there exist both nanographite and many layer graphene flakes, even though the average graphite domain is calculated to be 14.8 nm. This is suggested, due to the fact that few to many layer graphene crystallites in the (002) direction are weakly diffracted. The weakly diffracted domains make the crystals unresolved from the background. That being said, the first thing to notice is the lack of large flakes of graphite present on the fractured surface. The lack of visibly large flakes suggest that nearly all the graphite were reduced in thickness but retained relatively large surface area in Figure 6. The reduction in graphite flake thickness provides improved dispersion of nanographite and many layer graphene sheets, making it difficult in locating the surfaces of the PA66 matrix. When cryogenically fracturing the extrudate in Figure 6a, voids are left behind on the surface of the samples. These voids are holes left by the sliding of graphitic sheets not interacting with the matrix, as they are being pulled out from their center. The nature of the pullout structure left behind suggests some flakes are 'locked in.' It is also interesting to note that pulled out sections leave behind corrugated steps at the edges of exfoliate, seen in Figure 6.

When looking at Figure 6b, a more detailed image is shown of the step morphology left by the process. Following along the steps in Figure 6b we notice what appears to be a rolled edge, with the direction of rolling applied to the left in the micrograph. The development of this rolled edge structure seen in the image, is presumably caused by frictional forces imposed by the matrix. The peeling leaves behind a nanosheet, which is shown to be ripped at the edge. If you follow the edge where the ripping occurs, you can see PA66 attached or adhered to its surface. This rolled structure with the edged attached PA66 is an indication of matrix interactions with the nanoflake graphitic structure. The morphology provides clues to the mechanism of exfoliation in this process.

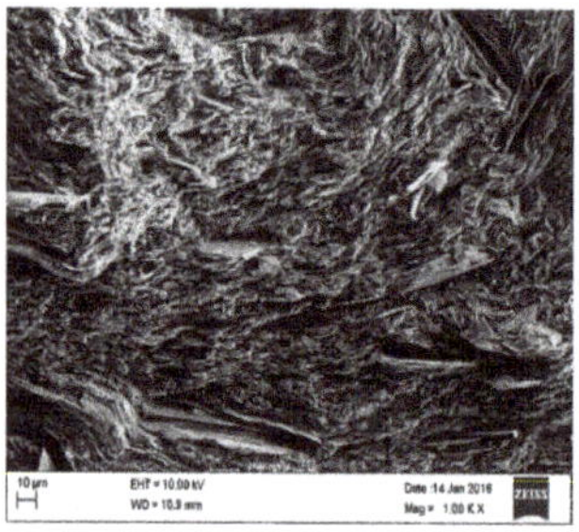

(a)

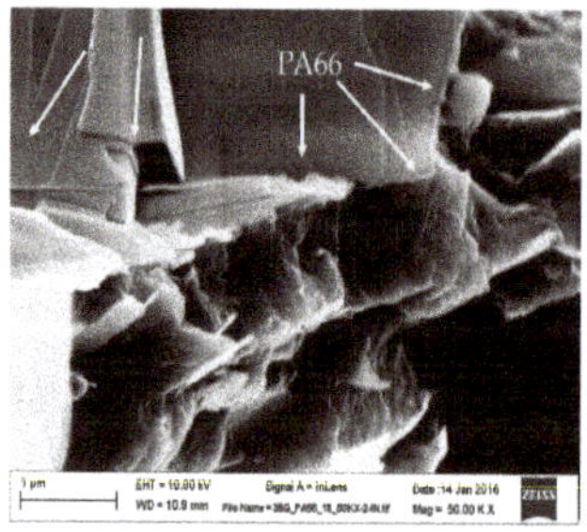

(b)

Figure 6. (**a**) Surface of cryogenically fractured G-PA66 20 min composite at low magnification; (**b**) Graphitic edge overhanging polymer in G-PA66 20 min composite.

To observe the created graphene in the absence of the extrudate matrix, cryogenic milling was performed on the 30 min G-PA66. In Figure 7, large area nanographite and many layer graphene is shown having what looks like an elongated shape. Focusing on the edges of the newly created flakes, a jagged and then smooth structure is noticed. By following those edges on the right side of the micrograph, the rolled section appears smooth. Comparing this discovery to the previous Figures, this suggests that the other smooth sections are folded over sheets of many layer graphene, created by the frictional force of the molten polymer.

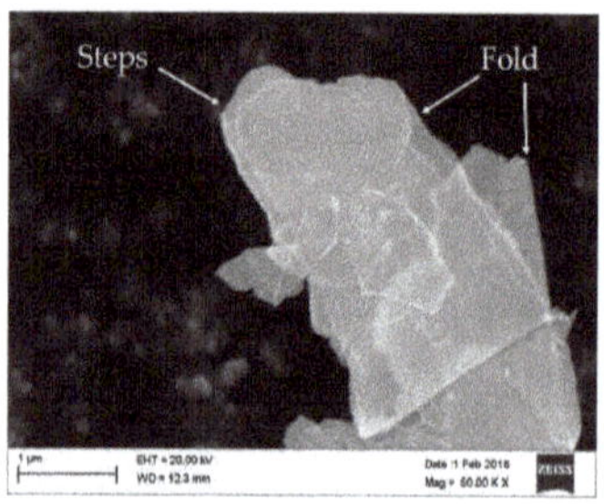

Figure 7. Cryogenically milled G-PA66 20 min showing transparency of exfoliated graphene flake, at high magnification.

What is interesting in Figure 7, is the transparency of the large area flakes to the electron microscope. Folded sections, as previously mentioned, shows the presence of additional wrinkles and folding structures under the layered graphene sheets. The transparency suggests that graphene is being created, otherwise transmission would be extremely difficult for nanographite flakes.

What is difficult to find, is a distinct indication of the PA66 with the graphene in Figure 6b. What we do find is that the transparent surfaces of the flake are littered with bright surface and edge features. What these features represent is still not fully known. We do know that given the analysis technique, Insulators/nonconductors in SEM cause charge buildup in their surface. In that this charge buildup presents itself as bright spots in produced micrographs [37]. PA66 is known to have such low conductivity, that it is an electrical insulator. The bright spots are likely to be edges containing PA66, PA66 alone, or a combination of both. Observing the micrographs, we see the pristine surfaces, which stress the point that our novel exfoliation preserves the in-plane surface structure of graphene. A traditional chemical method of graphene creates surface defects, wrinkles and holes; which never simply retain such pristine surfaces [38,39].

3.4. Transmission Electron Microscopy

The first thing noticed from the HR-TEM analysis is that from the low magnification image, the process produces graphene and nanographite composite flakes of varying sizes Figure 8a. Theses sizes consist of a combination of few layer graphene (2–5 layers), many layer graphene (5–10 layers) and the rest containing nanographite (<100 nm in thickness) [40]. In Figure 8b, 5 individual layers of graphene sheets are identified, while having the on the edge structure in focus. Following along the same figure, the sheets are noted to folding upward; representative of a feature inherent to graphene, due to stresses at the edge [41]. Looking closer in Figure 8c, a continuous structure is noticed on the steps of the nanographite (the number of counted steps exceeds 10 layers), containing a smaller periodicity on the order of ~0.246 nm in size. 0.246 nm is known to be the lattice parameter of graphite, representing the distance between carbon atoms along a single direction in the surface of a graphene flake. Given the size of the smaller periodicity, the surface structure is believed to be a 1 to 2 layer graphene sheet draped along the edge. This is an indication from the previous statement that nanographite to graphene with a (002) smaller than 14.8 nm exist within the G-PA66 composite. This provides evidence that a multiscale layers of graphene and nanographite exists within this composite.

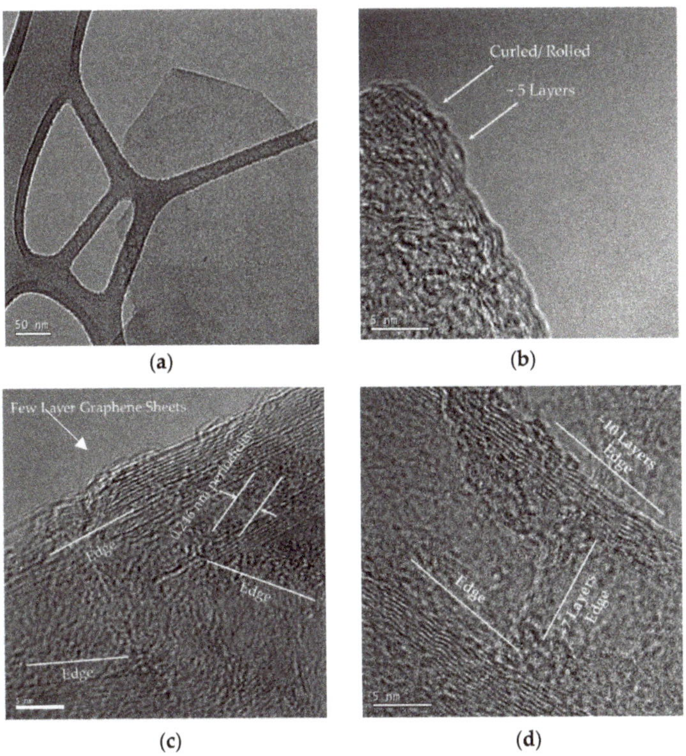

Figure 8. (**a**) Low Magnification micrograph of isolated 20 min G-PA66 flake on lacey carbon surface; (**b**) High magnification micrograph of rolled edge of many layer graphene; (**c**) High magnification micrograph of edge containing nanographite, many layer and few layer graphene; (**d**) High magnification micrograph containing many layer graphene.

Taking a step back and looking at the arrangement of the flakes, many layer graphene and nanographite are shown to be orientated in multiple directions in Figure 8c,d. These structural disruptions are indicative of turbostratic graphite; where the c-axis stacking sequence is either rotated or completely disrupted. These turbostratic layers are confirmed by the diffraction results; where they exhibit an increase in the c-axis spacing of the graphene layers. In this case, G-PA66 that was characterized, was mixed for 45 min; which suggests this to have a spacing of 0.3386 nm; having a 0.0042 nm lattice increase.

4. Conclusions

In our novel process, elongational exfoliation in the molten phase led to the formation of pristine, large area nanographite and multilayer graphene flakes. These results are confirmed from HR-TEM and morphology in electron microscopy. From HR-TEM and XRD, the resulting process identified that the graphite was exfoliated, leading to the disruption of the graphene layers in the composite and a 76% reduction in the initial (002) domain size. In-situ, a misalignment is described by a rotation of the graphene layers and conversion to a turbostratic phase. The model of graphite exfoliation revealed that the process is diffusion driven, with intercalation of the PA66 within the graphite galleries. In addition to diffusion, the process is found to tear and fold the outer layer sheets in graphite creating instance for the formation of dangling bonds. The results of which, preclude to a solvent-less method to functionalization graphene. The morphology created changes in the hydrogen bonded structure, given by changes in the lattice and domain in the (100) and (010) crystalline plane. Changes lead to preferential crystal growth in the hydrogen bonded direction, resulting in changes to the T_m and T_g. Due to these changes, the graphene exfoliate is found suppresses the crystallization in PA66, leading to a modified microstructure.

The G-PMC created by the process, shows graphene to be created due high shear. The results show a single step approach to create as well as modify graphene. Further investigation into this work may reveal a better description of the interfacial properties of these two species and the type of bonding present. This may provide ways of modifying the interface functionality in a scalable process and further exploiting the capabilities of elongational flow. This in-turn drives down cost, which is an important variable in adaptation and scalability.

Author Contributions: Conceptualization, J.H., T.N. and J.L.-B.; Methodology, J.H., R.S. and T.E.; Resources, T.N. and J.L.-B.; Data Curation, J.W.H.; Writing—Original Draft Preparation, J.W.H. and R.S.; Writing—Review & Editing, J.W.H., J.L.-B., T.N. and T.E.

Funding: This research received no external funding.

Acknowledgments: This work was supported by AMIPP Advanced Polymer Center at Rutgers University.

Conflicts of Interest: The authors declare no conflict of interest.

References

1. Yan, Z.; Nika, D.L.; Balandin, A.A. Thermal properties of graphene and few-layer graphene: Applications in electronics. *IET Circuits Devices Syst.* **2015**, *9*, 4–12. [CrossRef]
2. Miculescu, M.; Thakur, V.K.; Miculescu, F.; Voicu, S.I. Graphene-based polymer nanocomposite membranes: A review. *Polym. Adv. Technol.* **2016**, *27*, 844–859. [CrossRef]
3. Kim, H.; Abdala, A.A.; Macosko, C.W. Graphene/Polymer Nanocomposites. *Macromolecules* **2010**, *43*, 6515–6530. [CrossRef]
4. Perreault, F.; de Faria, A.F.; Elimelech, M. Environmental applications of graphene-based nanomaterials. *Chem. Soc. Rev.* **2015**, *44*, 5861–5896. [CrossRef] [PubMed]
5. Talat, M.; Srivastava, O. Deployment of New Carbon Nanostructure: Graphene for Drug Delivery and Biomedical Applications. In *Advances in Nanomaterials*; Springer: New Delhi, India, 2016; pp. 383–395.
6. Hao, F.; Fang, D.; Xu, Z. Mechanical and thermal transport properties of graphene with defects. *Appl. Phys. Lett.* **2011**, *99*, 041901. [CrossRef]
7. Chen, Y.; Tan, C.; Zhang, H.; Wang, L. Two-dimensional graphene analogues for biomedical applications. *Chem. Soc. Rev.* **2015**, *44*, 2681–2701. [CrossRef] [PubMed]
8. Frank, I.W.; Tanenbaum, D.M.; van der Zande, A.M.; McEuen, P.L. Mechanical properties of suspended graphene sheets. *J. Vacuum Sci. Technol. B Microelectron. Nanometer Struct.* **2007**, *25*, 2558–2561. [CrossRef]

9. Yang, S.; Brüller, S.; Wu, Z.-S.; Liu, Z.; Parvez, K.; Dong, R.; Richard, F.; Samorì, P.; Feng, X.; Müllen, K. Organic radical-assisted electrochemical exfoliation for the scalable production of high-quality graphene. *J. Am. Chem. Soc.* **2015**, *137*, 13927–13932. [CrossRef]
10. Bélanger, D. One-Pot Electrochemical Exfoliation and Functionalization of Graphene Sheets. In Proceedings of the 229th ECS Meeting, San Diego, CA, USA, 29 May–2 June 2016.
11. Wassei, J.K.; Kaner, R.B. Oh, the Places You'll Go with Graphene. *Acc. Chem. Res.* **2013**, *46*, 2244–2253. [CrossRef]
12. Chua, C.K.; Pumera, M. Chemical reduction of graphene oxide: A synthetic chemistry viewpoint. *Chem. Soc. Rev.* **2014**, *43*, 291–312. [CrossRef]
13. Manna, K.; Hsieh, C.-Y.; Lo, S.-C.; Li, Y.-S.; Huang, H.-N.; Chiang, W.-H. Graphene and graphene-analogue nanosheets produced by efficient water-assisted liquid exfoliation of layered materials. *Carbon* **2016**, *105*, 551–555. [CrossRef]
14. Haar, S.; Ciesielski, A.; Clough, J.; Yang, H.; Mazzaro, R.; Richard, F.; Conti, S.; Merstorf, N.; Cecchini, M.; Morandi, V. Graphene: A Supramolecular Strategy to Leverage the Liquid-Phase Exfoliation of Graphene in the Presence of Surfactants: Unraveling the Role of the Length of Fatty Acids (Small 14/2015). *Small* **2015**, *11*, 1736. [CrossRef]
15. Vicarelli, L.; Heerema, S.J.; Dekker, C.; Zandbergen, H.W. Controlling Defects in Graphene for Optimizing the Electrical Properties of Graphene Nanodevices. *ACS Nano* **2015**, *9*, 3428–3435. [CrossRef] [PubMed]
16. Oxfall, H.; Rondin, J.; Bouquey, M.; Muller, R.; Rigdahl, M.; Rychwalski, R.W. Elongational flow mixing for manufacturing of graphite nanoplatelet/polystyrene composites. *J. Appl. Polym. Sci.* **2013**, *128*, 2679–2686. [CrossRef]
17. Ellingham, T.; Duddleston, L.; Turng, L.-S. Sub-critical gas-assisted processing using CO_2 foaming to enhance the exfoliation of graphene in polypropylene + graphene nanocomposites. *Polymer* **2017**, *117*, 132–139. [CrossRef]
18. Ibarra-Gómez, R.; Muller, R.; Bouquey, M.; Rondin, J.; Serra, C.A.; Hassouna, F.; Mouedden, Y.E.; Toniazzo, V.; Ruch, D. Processing of nanocomposites PLA/graphite using a novel elongational mixing device. *Polym. Eng. Sci.* **2015**, *55*, 214–222. [CrossRef]
19. Nosker, T.; Lynch, J.; Hendrix, J.; Kear, B.; Chiu, G.; Tse, S.; Rutgers State University of New Jersey. In Situ Exfoliation Method to Fabricate a Graphene-Reinforced Polymer Matrix Composite (G-PMC). U.S. Patent 9,896,565, 20 February 2018.
20. Taylor, G.I. The Formation of Emulsions in Definable Fields of Flow. *Proc. R. Soc. Lond.* **1934**, *146*, 501–523. [CrossRef]
21. Rumscheidt, F.D.; Mason, S.G. Particle motions in sheared suspensions XII. Deformation and burst of fluid drops in shear and hyperbolic flow. *J. Colloid Sci.* **1961**, *16*, 238–261. [CrossRef]
22. Aguilera, J.M.; Simpson, R.; Welti-Chanes, J.; Aguirre, D.B.; Barbosa-Cánovas, G.V. *Food Engineering Interfaces*; Springer: New York, NY, USA, 2010.
23. Grace, H.P. Dispersion phenomena in high viscosity immiscible fluid systems and application of static mixers as dispersion devices in such systems. *Chem. Eng. Commun.* **1982**, *14*, 225–277. [CrossRef]
24. Delhaes, P. *Graphite and Precursors*; Taylor & Francis: London, UK, 2000.
25. Varin, R.A.; Czujko, T.; Wronski, Z.S. *Nanomaterials for Solid State Hydrogen Storage*; Springer: New York, NY, USA, 2009.
26. Goto, M.; Smith, J.M.; McCoy, B.J. Parabolic profile approximation (linear driving-force model) for chemical reactions. *Chem. Eng. Sci.* **1990**, *45*, 443–448. [CrossRef]
27. Christofides, P.D.; Daoutidis, P. Nonlinear control of diffusion-convection-reaction processes. *Comput. Chem. Eng.* **1996**, *20*, S1071–S1076. [CrossRef]
28. Buzanowski, M.A.; Yang, R.T. Approximations for intraparticle diffusion rates in cyclic adsorption and desorption. *Chem. Eng. Sci.* **1991**, *46*, 2589–2598. [CrossRef]
29. Jones, N.A.; Atkins, E.D.T.; Hill, M.J. Investigation of solution-grown, chain-folded lamellar crystals of the even-even nylons: 6 6, 8 6, 8 8, 10 6, 10 8, 10 10, 12 6, 12 8, 12 10, and 12 12. *J. Polym. Sci. Part B Polym. Phys.* **2000**, *38*, 1209–1221. [CrossRef]
30. Starr, F.W.; Schrøder, T.B.; Glotzer, S.C. Molecular Dynamics Simulation of a Polymer Melt with a Nanoscopic Particle. *Macromolecules* **2002**, *35*, 4481–4492. [CrossRef]

31. Lopes dos Santos, J.M.B.; Peres, N.M.R.; Castro Neto, A.H. Graphene Bilayer with a Twist: Electronic Structure. *Phys. Rev. Lett.* **2007**, *99*, 256802. [CrossRef]
32. Haberkorn, H.; Illers, K.H.; Simak, P. Molekülordnung und Kristallinität in Polyhexamethylenadipamid. *Colloid Polym. Sci.* **1979**, *257*, 820–840. [CrossRef]
33. Khanna, Y.P. Overview of transition phenomenon in nylon 6. *Macromolecules* **1992**, *25*, 3298–3300. [CrossRef]
34. Zhang, F.; Wang, B.; Man, R.; Peng, Z. Isothermal crystallization kinetics of in situ Nylon 6/graphene composites by differential scanning calorimetry. *Polym. Eng. Sci.* **2014**, *54*, 2610–2616. [CrossRef]
35. Aoyama, S.; Park, Y.T.; Ougizawa, T.; Macosko, C.W. Melt crystallization of poly(ethylene terephthalate): Comparing addition of graphene vs. carbon nanotubes. *Polymer* **2014**, *55*, 2077–2085. [CrossRef]
36. Kim, S.; Poostforush, M.; Kim, J.; Lee, S. Thermal diffusivity of in-situ exfoliated graphite intercalated compound/polyamide and graphite/polyamide composites. *Express Polym. Lett.* **2012**, *6*, 476–484. [CrossRef]
37. Sawyer, L.C. *Polymer Microscopy*; Springer: Dordrecht, The Netherlands, 2012.
38. Stankovich, S.; Dikin, D.A.; Piner, R.D.; Kohlhaas, K.A.; Kleinhammes, A.; Jia, Y.; Wu, Y.; Nguyen, S.T.; Ruoff, R.S. Synthesis of graphene-based nanosheets via chemical reduction of exfoliated graphite oxide. *Carbon* **2007**, *45*, 1558–1565. [CrossRef]
39. Eda, G.; Fanchini, G.; Chhowalla, M. Large-area ultrathin films of reduced graphene oxide as a transparent and flexible electronic material. *Nat. Nanotechnol.* **2008**, *3*, 270. [CrossRef] [PubMed]
40. Bianco, A.; Cheng, H.-M.; Enoki, T.; Gogotsi, Y.; Hurt, R.H.; Koratkar, N.; Kyotani, T.; Monthioux, M.; Park, C.R.; Tascon, J.M.D.; et al. All in the graphene family—A recommended nomenclature for two-dimensional carbon materials. *Carbon* **2013**, *65*, 1–6. [CrossRef]
41. Shenoy, V.B.; Reddy, C.D.; Zhang, Y.-W. Spontaneous Curling of Graphene Sheets with Reconstructed Edges. *ACS Nano* **2010**, *4*, 4840–4844. [CrossRef] [PubMed]

polymers

MDPI

Article

Effect of Compatibilizer on the Interface Bonding of Graphene Oxide/Polypropylene Composite Fibers

Miao Miao, Chunyan Wei *, Ying Wang and Yongfang Qian

Department of Textile and Material Engineering, Dalian Polytechnic University, Dalian 116034, China; m1a0_1993@163.com (M.M.); wangying@dlpu.edu.cn (Y.W.); qianyf@dlpu.edu.cn (Y.Q.)
* Correspondence: weicy@dlpu.edu.cn; Tel.: +86-137-9514-6016

Received: 17 October 2018; Accepted: 15 November 2018; Published: 18 November 2018

Abstract: To improve the interfacial bonding and thermal stability of graphene oxide (GO)/polypropylene (PP) composite fibers, a composite fiber with PP as the matrix, GO as reinforcement and maleic anhydride-grafted PP (PP-g-MAH) as a compatibilizer was prepared by a simple and efficient melt-blending method. The GO content was 0.0–5.0 wt %. According to the Fourier Transform Infrared (FT-IR) spectroscopy results, the interfacial bonding in the PP/MAH/GO composite fibers was improved. The Dynamic Mechanical Analysis (DMA) results show that the addition of GO resulted in better interfacial adhesion and higher storage modulus (E'). The loss modulus (E'') of the PP/MAH/GO-x composite fibers increased with increasing amount of added GO, whereas the loss factor (tan δ) decreased. GO and PP-g-MAH were analyzed by Thermogravimetric Analysis (TGA). The thermal stability of the composite fibers was improved compared to PP. Differential Scanning Calorimetry (DSC) analysis showed that the addition of PP-g-MAH to the composite fiber improved the interfacial bonding of GO in the PP matrix. Thus, compatibility between the two components was obtained. Based on the Scanning Electron Microscopy (SEM) results, the PP fibers exhibited relative orientation due to the strong crystalline morphology. The rough section, PP/GO blend fiber exhibits a very clear phase separation morphology due to the incompatibility between the two and the compatibility of GO and PP in PP/MAH/GO-3 composite fiber is improved, resulting in the interface between the two has improved.

Keywords: graphene oxide; polymer composite fiber; interfacial bonding; polypropylene; thermal stability

1. Introduction

Polypropylene (PP), a thermoplastic polyolefin, has excellent characteristics such as low cost, high strength, good dimensional stability, friction resistance and chemical resistance. PP is widely used in textile fibers, packaging and labelling and automobiles parts, industries, agriculture and other fields have been widely concerned in recent years [1]. However, the non-polar chain structure of PP makes it difficult to combine with reinforcing materials [2]. Nowadays, this is a problem which should be solved. Therefore, considerable research has been done on the development of materials with modified PP. The modification of PP can be done using the ultraviolet light irradiation grafting method [3,4], plasma graft modification method [5,6] and melt-blending method [7]. The melt blending of polymers is a convenient and effective method for designing new high-performance polymers [8]. Blends of two or more polymers can balance performance, highlighting the respective advantages of the blends to mask the disadvantages. However, due to their different chemical structures and polarities, most polymer blends are thermodynamically immiscible and exhibit poor mechanical properties, poor interfacial adhesion and unstable morphology [9]. Therefore, we used maleic anhydride-grafted PP (PP-g-MAH) as a compatibilizer. The MAH group on the PP-g-MAH segment helps to improve the compatibility between graphene oxide (GO) and PP [10,11], thereby improving the interfacial bonding between GO and PP.

In recent years, inorganic nanomaterials have provided new modification methods for incompatible blends, such as carbon black [12], nanoclay [13], carbon nanotubes [14], graphene and its derivatives [15] and glass fibers [16]. Among inorganic nanomaterials, GO has a typical quasi-two-dimensional structure [17] and its layer contains many oxygen-containing groups. Because of its high specific surface energy, good hydrophilicity and good mechanical properties [18], GO is considered as an ideal polymer nano-based composite fiber additive for polymers. Tiwari et al. investigated the effects of different GO loadings on the compatibility, thermomechanical and morphological properties of incompatible PP/polycarbonate polymer blends [19]. Botlhoko et al. comparatively investigated the effects of graphite (G) and GO dispersions on the thermal, mechanical and rheological properties of biodegradable polylactide (PLA)/poly(07-caprolactone) (PCL) blends [20]. Wang et al. employed GO as a two-dimensional nanofiller and nucleating agent to improve the properties of immiscible PLA/PCL blends with weight ratios of 70/30, 50/50 and 30/70 [21]. Chen et al. prepared polyamide-6/graphene-graphene oxide composites with super-high thermal conductivity through in situ polymerization and the interface adhesion enhanced by adding small amounts of graphene-GO [22]. Meng et al. prepared polyamide-6/graphite nanoflakes by situ intercalation polymerization approach for enhanced thermal conductivity [23].

In this study, PP was used as the matrix, GO was used as a reinforcement and PP-g-MAH was used as a compatibilizer. Composites containing GO contents of 0.0–5.0 wt % were prepared using a melt-blending method. The interfacial bonding properties, thermal properties, mechanical properties and morphological characteristics of the composite fibers were studied with GO content as the independent variable.

2. Experiment

2.1. Materials

Polypropylene (melt flow rate 8.0 g/10 min) was purchased from Sinopec Hainan petrochemical Co., Ltd. (Yangpu Economic Development Zone, Hainan, China); Maleic anhydride grafted polypropylene (MA content of 1.3%, melt flow rate of 150 g/10 min at 230 °C and 2.16 kg) was supplied from Hebei Xintianqi Plastic Co., Ltd., (Hebei, China); Multilayer graphene oxide (thickness 3.4–7.0 nm, sheet diameter 10–50 μm)was purchased from Suzhou Tanfeng Technology Co., Ltd., Suzhou, China.

2.2. Preparation of Composite Fiber

PP and PP-g-MAH were first mixed in proportion and the mixture was melt-compounded and extruded with different contents of GO using a Huck single-screw extruder (HAAKE Polylab OS, Karlsruhe, Germany). This experiment used the single hole spinneret with a diameter of 1.5 mm. To effectively mix the mixture, the temperature of the single-screw extruder from the barrel to the head was set to 150, 180, 200 and 200 °C. The screw speed was fixed at 20 rpm. A schematic diagram of the melt spinning device is shown in Figure 1. The extruded mixture was cooled by cold water and dried in an oven for 24 h for subsequent testing and analysis. The GO content in PP/PP-g-MAH (3/1) ranged from 0.0 to 5.0 wt %. The composite fibers were named PP/MAH/GO-x, where x represents the GO weight percentage.

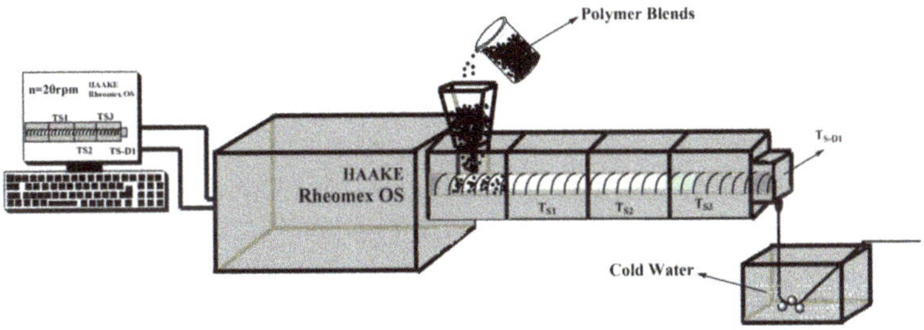

Figure 1. Schematic diagram of the melt spinning device.

2.3. Characterization

The molecular structure of the composite fiber was analyzed by Fourier Transform Infrared spectroscopy (FT-IR, Waltham, MA, USA). The sample to be tested and the potassium bromide powder were thoroughly ground and mixed in an agate mortar at a mass ratio of 1:100. Press into the mold and press into a sheet on the press. The spectral range is 4000–500 cm^{-1}.

The dynamic storage modulus and loss tangent angle of the composite fiber were measured by a Dynamic Mechanical Analyzer (DMA-Q800, TA, New Castle, DE, USA). The sample clamping length was 10.5 ± 0.5 mm and the diameter was 1.3 ± 0.1 mm. In the tension mode, the temperature range was −50 °C–150 °C. The heating rate was 5 °C/min and the frequency was 1 Hz.

The thermal stability of the composite fibers was tested using a Thermogravimetric Analyzer (TGA, SDT-Q600, TA, New Castle, DE, USA). About 10 mg of the sample was placed in the SDT-Q600 at a heating rate of 10 °C/min, rising from room temperature to 550 °C.

The Differential Scanning Calorimeter (DSC-Q2000, TA, New Castle, DE, USA) was used to identify the thermal transition behavior of the composite fiber. About 10mg of the sample was placed in the DSC-Q2000 at a heating rate of 10 °C/min, rising from −50 °C to 200 °C, naturally cooling to 30 °C, eliminating the heat history and then, the glass transition temperature (Tg) of the sample was tested under a nitrogen atmosphere at a temperature increase rate of 10 °C/min from −50 °C to 200 °C.

The cross-sectional morphology of the composite fibers was characterized by Scanning Electron Microscopy (SEM, JSM-7800F, JEOL, Tokyo, Japan). The fibers were quenched at low temperature in liquid nitrogen and their cross sections were observed.

3. Results and Analysis

3.1. FT-IR Analysis of Composite Fibers

Figure 2 shows the infrared spectra of pure PP, GO, PP/GO blend fibers and PP/MAH/GO-3 composite fibers. The absorption band of PP (Figure 2a) has four main characteristic peaks and the characteristic absorption peaks at 2951 and 2872 cm^{-1} correspond to the asymmetric and symmetric stretching of the methyl group. The characteristic absorption peaks at 2920 and 2850 cm^{-1} are attributed to the asymmetric and symmetric stretching of the methylene groups. The characteristic absorption peak in the spectrum of GO (Figure 2b) at 1727 cm^{-1} corresponds to the stretching of C=O. The characteristic absorption peaks at 1621 and 1384 cm^{-1} are attributed to the stretching of C=C and –OH, respectively. The hydroxyl group in GO acts as a reactive group that chemically interacts with the MAH group to improve the interfacial bonding between GO and PP. The infrared spectrum of the PP/MAH/GO composite fiber shows new absorption peaks at 2923 and 2854 cm^{-1}, corresponding to the stretching of –CH_2 on the PP-g-MAH molecular chain. The characteristic peak at 2951 cm^{-1} is attributed to –CH_3 stretching. When two polymer materials are blended, if simple physical blending alone does not produce intermolecular forces, the FT-IR spectrum of the blend will simply be a superposition of the spectra of the two components, as for the PP/GO fibers (Figure 2c). If intermolecular interactions (e.g., hydrogen bonds) occur between the two components of the blend, a new spectrum that is not simply the superposition of peaks will be generated, as for the PP/MAH/GO composite fibers. The generation of hydrogen bonds causes some absorption peaks to shift in position and shape. The anhydride group of MAH can be esterified with the macromolecular alcohol hydroxyl group under melt-blending conditions and the intermolecular reaction improves the interfacial compatibility between the blend components to some extent. After adding MAH, the PP/MAH/GO-3 (Figure 2d) composite fibers and pure PP fibers showed similar peaks. However, it belongs to the vibration absorption peak of the ester-based band (at 1127 cm^{-1}) and the stretching vibration peak of C=C (at 1621 cm^{-1}) and the intensity increases remarkably. This indicates that in the experimental range, in addition to the physical state mixing of GO and PP-g-MAH molecular chains in the PP/MAH/GO composite fibers, a certain degree of esterification occurred. The reaction mechanism is shown in Figure 3. The esterification reaction promotes the compatibility between the molecular chains, thereby improving interfacial bonding in the PP/MAH/GO composite fibers.

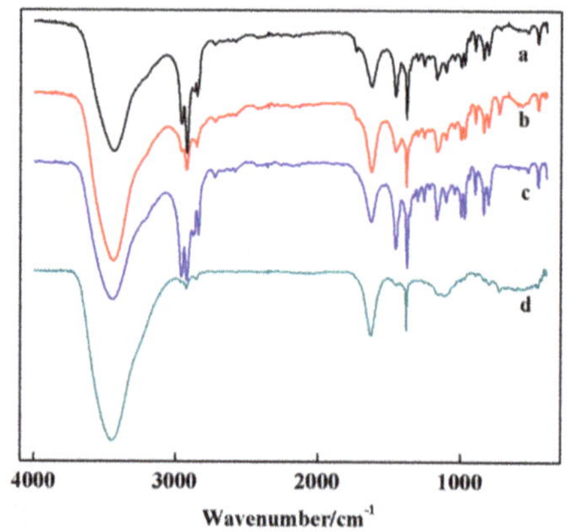

Figure 2. FT-IR analysis chart (**a**) PP; (**b**) GO; (**c**) PP/GO blend fiber; (**d**) PP/MAH/GO-3 composite fiber).

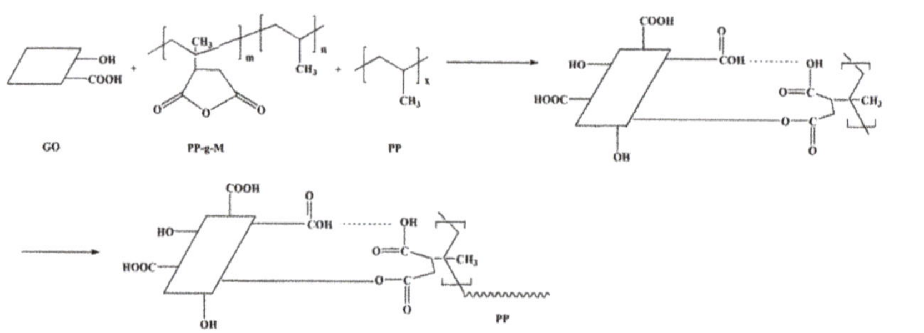

Figure 3. Composite fiber reaction mechanism diagram.

3.2. *Dynamic Mechanical Properties of Composite Fibers*

DMA is an effective method for evaluating the interfacial interactions in reinforced composite fibers. To characterize the mechanical properties of composite fibers with different GO weight fractions as a function of temperature, the dynamic mechanical data, storage modulus (E'), loss modulus (E'') and loss factor (tan δ) were evaluated (Figure 4).

Figure 4a shows the E′ values of pure PP fibers, PP/GO blended fibers and PP/MAH/GO-x composite fibers as functions of temperature. E' is similar to the bending modulus (E), which describes the stiffness of the material. Compared to pure PP, the storage modulus temperature of PP/GO blend fibers and the change are similar to those of pure PP fibers. However, the E' values of the blended fibers are higher than that of pure PP because of the presence of GO. The high E' value of the PP/MAH/GO-x composite fibers indicates good interfacial adhesion.

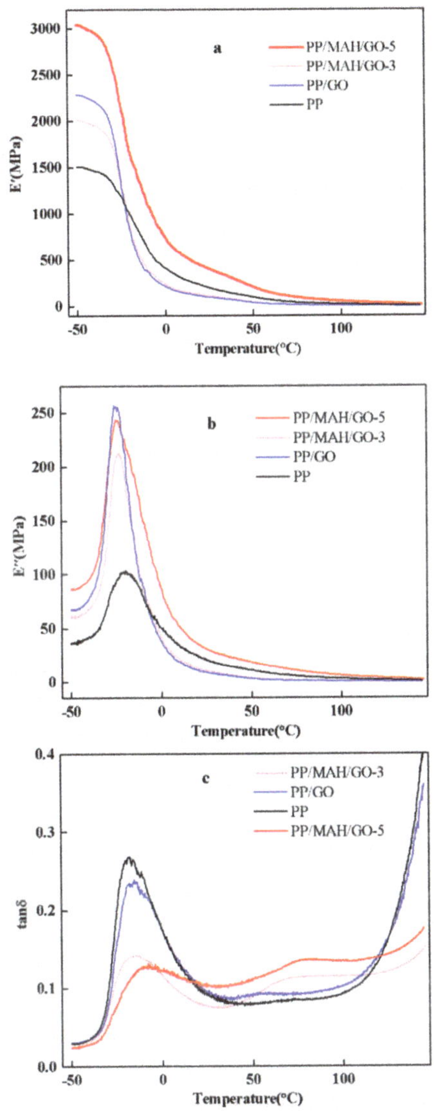

Figure 4. DMA image of pure PP and composite fibers: (**a**) E'; (**b**) E''; (**c**) tan δ.

Figure 4b shows E'', which generally indicates the viscosity of the material, as a function of time for the pure PP fibers, PP/GO blended fibers and PP/MAH/GO-x composite fibers. Similar to E', E'' also increased upon the addition of GO. This phenomenon is attributed to the interactions between the fibers and matrix, which limit the movement of PP molecules, resulting in a higher viscosity and increased E''. On the other hand, if the fibers interact very strongly with the matrix, E'' will decrease because of the absence of fiber slip and energy dissipation. This is why E'' of PP/MAH/GO-5 is lower than E'' of PP/MAH/GO-3. This conclusion is consistent with Wei et al. [2].

In general, the peak temperature of the loss factor corresponds to the glass transition temperature (Tg) of a material. In a binary blend without a compatibilizer, the loss factor has two peaks in the curve over time. These peaks correspond to the Tg values of both materials. After the addition of compatibilizer, the dispersibility of the dispersed phase is improved, the interaction between interfaces is enhanced, the degree of bonding between particles is improved and the compatibility of the system is remarkably improved. However, a homogeneous blended system is equivalent to a random copolymer. Thus, there is only one Tg value and only one peak in the loss factor curve. As the compatibility of the blend improves, the Tg values of the two phases in the blend will interact with each other. Close, there is a tendency to form a Tg [2]. Figure 4c shows plots of tan δ for pure PP, PP/GO blended fibers and PP/MAH/GO composite fibers. The loss factor of the PP/MAH/GO composite fibers showed only one peak, which corresponded to the Tg of PP. This indicates that the compatibility between PP and GO increased after adding PP-g-MAH, which is consistent with the FT-IR results. Moreover, the tan δ values of all composite fibers are lower than those of pure PP fibers, which is attributed to the strengthening of GO and the limitation of polymer molecular motion. As shown in Table 1, the maximum loss factor decreased as the amount of added GO increased.

Table 1. The DMA analysis results of pure PP and composite fibers.

Sample Name	E'(MPa)	E''(MPa)	$\tan\delta_{max}$	Tg (°C)
PP	2913	253	0.296	−17.91
PP/GO	2188	241	0.235	−16.56
PP/MAH/GO-0.3	1923	212	0.14	−16.15
PP/MAH/GO-0.5	1439	99.8	0.12	−10.61

3.3. TGA of Composite Fibers

To analyze the degradation temperature and thermal stability of the composite fibers, TGA was conducted. Figure 5 shows weight losses of pure PP and composite fibers with temperature. The thermal degradation of pure PP mainly occurred in the range of 410 °C to 470 °C. With the addition of PP-g-MAH, the initial degradation temperature of the blend shifted into the high-temperature zone, indicating that the addition of PP-g-MAH increased the thermal stability of the composite fibers. This phenomenon is explained as follows. On one hand, PP-g-MAH itself has a certain degree of crystallinity and heat resistance and is capable of absorbing a certain amount of heat and storing it in the form of heat energy. On the other hand, the interaction between GO and PP-g-MAH causes the concentration of the entanglement point to increase, which increases the outward diffusion of thermal energy to a certain extent and hinders alkyl oxygen cleavage. Thus, the thermal stability of the composite fibers is improved with respect to PP. These results are consistent with the FT-IR spectra.

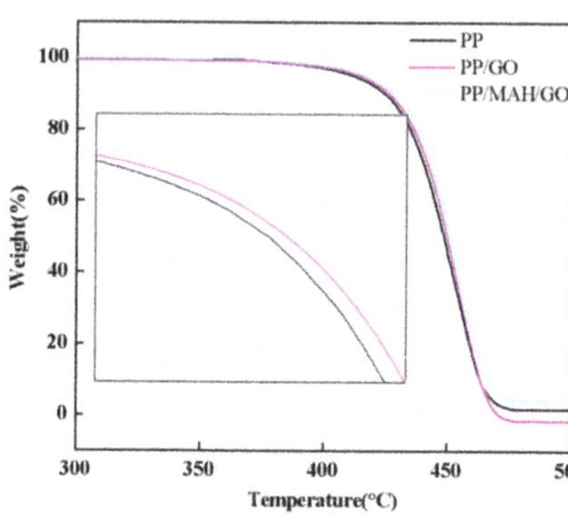

Figure 5. TGA image of composite fibers.

3.4. Thermal Transition Behavior Analysis of Composite Fibers

The crystallization temperature (*Tc*) of the PP fibers, PP/GO fibers and PP/PP-g-MAH/GO composite fibers were determined by DSC (Figure 6). The *Tc* peak is usually used to indicate the crystal structure. The *Tc* of the PP fibers was 115.8 °C and a semi-crystalline state formed as a result of the rapid crystallization rate of PP. No glass transition pattern or cold crystallization temperature was detected in the PP fibers. After adding GO, the *Tc* of the composite fibers increased, which is attributed to an increase in GO nucleation ability. Adding PP-g-MAH to the composite fiber improved the interfacial bonding of GO in the PP matrix along with the compatibility between the reinforcement and matrix and it is difficult to crystallize. Therefore, the *Tc* of the PP/PP-g-MAH/GO composite fibers shifted towards lower temperature.

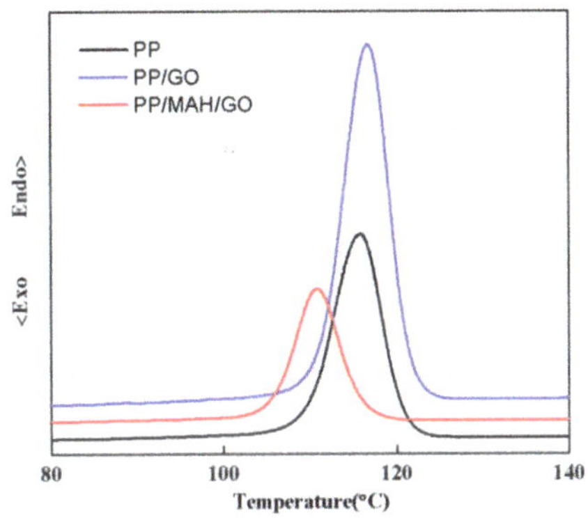

Figure 6. Crystallization temperature curve of PP, PP/GO and PP/PP-g-MAH/GO composite fibers.

3.5. SEM of Composite Fibers

To detect the effect of interfacial bonding between GO and the PP matrix, the images of PP fibers, PP/GO blended fibers and PP/MAH/GO composite fibers at low temperature fracture interface were obtained by SEM (Figure 7). Figure 7a shows a cross-sectional view of pure PP fibers under liquid nitrogen, revealing a relatively rough cross section attributed to the strong crystalline morphology of PP. Figure 7b shows a cross-sectional SEM image of a PP/GO blended fiber. As a result of the incompatibility between PP and GO, the interface in the PP/GO blended fiber is clear and a distinct phase separation is apparent. Figure 7c shows a cross-sectional SEM image of a PP/MAH/GO-3 composite fiber. The compatibility between GO and PP in the composite fiber was improved with respect to the PP/GO fiber, in agreement with the FT-IR and DMA results.

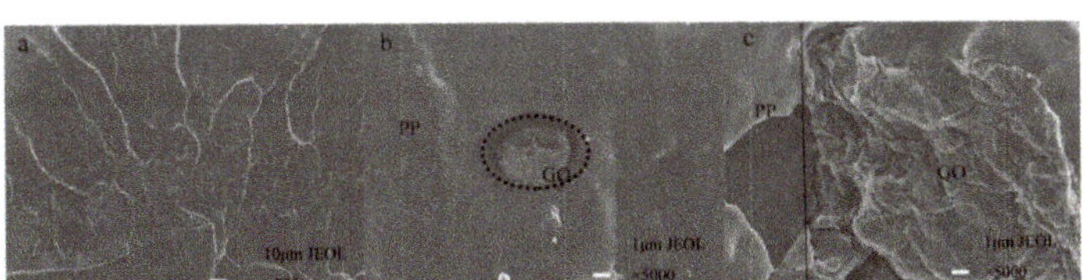

Figure 7. SEM image of low temperature fracture interface of PP, PP/GO blended fiber and PP/PP/GO composite fiber (**a**) pure PP fiber; (**b**) PP/GO blend fiber; (**c**) PP/MAH/GO-3 composite fiber).

4. Conclusions and Prospects

In this study, PP was used as the matrix, GO was the reinforcement and PP-g-MAH was the compatibilizer. Composite fibers with GO contents of 0.0–5.0 wt % were prepared by a simple and efficient melt-blending method. FT-IR, DMA and DSC showed that the compatibility between GO and PP along with the interfacial compatibility in the PP/MAH/GO composite fibers were improved compared to in the PP/GO blended fibers. TGA demonstrated that the thermal stability of the composite fibers was higher than that of PP. SEM analysis verified that the PP fibers showed relatively rough cross sections resulting from the strong crystalline morphology of PP. The PP/GO blended fibers showed clear phase separation resulting from the incompatibility between the two phases. Compared to the PP/GO blended fibers, the compatibility between GO and PP was much improved in the PP/MAH/GO-3 composite fibers, resulting in enhanced interfacial bonding between GO and PP. The current study is only the first step in studying the performance of PP nonwoven materials. In the future, composite fibers will be used for the preparation of PP nonwoven fabrics and their antistatic properties, aging resistance, electrical conductivity and comfort performance will be evaluated.

Author Contributions: Methodology, M.M.; Validation, M.M.; Formal Analysis, M.M.; Resources, C.W.; Data Curation, M.M.; Writing-Original Draft Preparation, M.M.; Writing-Review & Editing, M.M. and C.W.; Supervision, Y.W.; Project Administration, Y.Q.

Funding: This research received no external funding.

Conflicts of Interest: The authors declare no conflict of interest.

References

1. Lee, T.W.; Jeong, Y.G. Enhanced electrical conductivity, mechanical modulus, and thermal stability of immiscible polylactide/polypropylene blends by the selective localization of multi-walled carbon nanotubes. *Compos. Sci. Technol.* **2014**, *103*, 78–84. [CrossRef]
2. Luo, W.; Zhang, B.; Zou, H.; Liang, M.; Chen, Y. Enhanced interfacial adhesion between polypropylene and carbon fiber by graphene oxide/polyethyleneimine coating. *J. Ind. Eng. Chem.* **2017**, *51*, 129–139. [CrossRef]
3. Stragliotto, M.F.; Strumia, M.C.; Gomez, C.G.; Romero, M. Optimization of an UV-Induced Graft Polymerization of Acrylic Acid on Polypropylene Films Using CdS as a Light Sensor. *Ind. Eng. Chem. Res.* **2018**, *57*, 1188–1196. [CrossRef]
4. Chagas, G.R.; Weibel, D.E. UV-induced switchable wettability between superhydrophobic and superhydrophilic polypropylene surfaces with an improvement of adhesion properties. *Polym. Bull.* **2017**, *74*, 1965–1978. [CrossRef]
5. Sanbhal, N.; Mao, Y.; Sun, G.; Xu, R.F.; Zhang, Q.; Wang, L. Surface modification of polypropylene mesh devices with cyclodextrin via cold plasma for hernia repair: Characterization and antibacterial properties. *Appl. Surf. Sci.* **2018**, *439*, 749–759. [CrossRef]
6. Honarvar, Z.; Farhoodi, M.; Khani, M.R.; Mohammadi, A.; Shokri, B.; Ferdowsi, R.; Shojaee-Aliabadi, S. Application of cold plasma to develop carboxymethyl cellulose-coated polypropylene films containing essential oil. *Carbohydr. Polym.* **2017**, *176*, 1–10. [CrossRef] [PubMed]
7. Mao, Q.; Wyatt, T.P.; Chien, A.; Chen, J.; Yao, D. Melt spinning of high-strength fiber from low-molecular-weight polypropylene. *Polym. Eng. Sci.* **2016**, *56*, 233–239. [CrossRef]
8. Vogel, R.; Gedan-Smolka, M.; Häussler, L. Evaluation of the crystallization of polypropylene at melt spinning conditions using the green chemical orotic acid as nucleating agent. *Adv. Res. Text. Eng.* **2018**, *3*, 1023.
9. Banerjee, S.S.; Janke, A.; Gohs, U.; Heinrich, G. Electron-induced reactive processing of polyamide 6/polypropylene blends: Morphology and properties. *Eur. Polym. J.* **2017**, *98*, 295–301. [CrossRef]
10. Ezat, G.S.; Kelly, A.L.; Mitchell, S.C.; Youseffi, M.; Coates, P.D. Effect of maleic anhydride grafted polypropylene compatibilizer on the morphology and properties of polypropylene/multiwalled carbon nanotube composite. *Polym. Compos.* **2012**, *33*, 1376–1386. [CrossRef]
11. Szentes, A.; Varga, C.; Horvath, G.; Bartha, L.; Kónya, Z.; Haspel, H.; Szél, J.; Kukovecz, Á. Electrical resistivity and thermal properties of compatibilized multi-walled carbon nanotube/polypropylene composites. *Express Polym. Lett.* **2012**, *6*, 494–502. [CrossRef]
12. Li, P.L.; Ma, S.S.; Li, J.G.; Zhang, D.Y. Analysis of Dynamic Response Master Curve of Carbon Black Modified Asphalt Mixture. *J. Zhengzhou Univ.* **2018**, *39*, 12–17.

13. Zhou, C.; Liu, T.; Liu, J.; Lu, X.; Shi, Y.; Zhou, S.; Xin, Z. Polybenzoxazine/organoclay composite coatings with intercalated structure: Relationship between solubility parameters and corrosion protection performance. *Prog. Org. Coat.* **2018**, *115*, 188–194. [CrossRef]
14. Yao, X.; Gao, X.; Jiang, J.; Xu, C.; Deng, C.; Wang, J. Comparison of carbon nanotubes and graphene oxide coated carbon fiber for improving the interfacial properties of carbon fiber/epoxy composites. *Compos. Part B* **2018**, *132*, 170–177. [CrossRef]
15. Zhan, Y.; Wan, X.; He, S.; Yang, Q.; He, Y. Design of durable and efficient poly(arylene ether nitrile)/bioinspired polydopamine coated graphene oxide nanofibrous composite membrane for anionic dyes separation. *Chem. Eng. J.* **2018**, *333*, 132–145. [CrossRef]
16. Teng, X.; Wen, L.; Lv, Y.; Tang, W.; Zhao, X.; Chen, C. Effects of potassium titanate whisker and glass fiber on tribological and mechanical properties of PTFE/PEEK blend. *High Perform. Polym.* **2017**, *30*, 752–764. [CrossRef]
17. Su, W.; Kumar, N.; Krayev, A.; Chaigneau, M. In situ topographical chemical and electrical imaging of carboxyl graphene oxide at the nanoscale. *Nat. Commun.* **2018**, *9*, 2891. [CrossRef] [PubMed]
18. Zhang, Q.; Zhang, H.; Cheng, X.-L. Highly stable two-dimensional graphene oxide: Electronic properties of its periodic structure and optical properties of its nanostructures. *Chin. Phys. B* **2018**, *27*, 027301. [CrossRef]
19. Tiwari, S.K.; Oraon, R.; De Adhikari, A.; Nayak, G.C. A thermomechanical study on selective dispersion and different loading of graphene oxide in polypropylene/polycarbonate blends. *J. Appl. Polym. Sci.* **2017**, *134*, 45062. [CrossRef]
20. Joseph, B.O.; James, R.; Sinha, R.S. Thermal, mechanical, and rheological properties of graphite- and graphene oxide-filled biodegradable polylactide/poly(ε-caprolactone) blend composites. *J. Appl. Polym. Sci.* **2017**, *134*, 45373.
21. Wang, X.; Gao, Y.; Li, X.; Xu, Y.; Jiang, J.; Hou, J.; Li, Q.; Turng, L.-S. Selective localization of graphene oxide in electrospun polylactic acid/poly(ε-caprolactone) blended nanofibers. *Polym. Test.* **2017**, *59*, 396–403. [CrossRef]
22. Chen, J.; Chen, X.; Meng, F.; Li, D.; Tian, X.; Wang, Z.; Zhou, Z. Super-high thermal conductivity of polyamide-6/graphene-graphene oxide composites through in situ polymerization. *High Perform. Polym.* **2017**, *29*, 585–594. [CrossRef]
23. Meng, F.; Huang, F.; Guo, Y.; Chen, J.; Chen, X.; Hui, D.; He, P.; Zhou, X.; Zhou, Z. In situ, intercalation polymerization approach to polyamide-6/graphite nanoflakes for enhanced thermal conductivity. *Compos. Part B* **2017**, *117*, 165–173. [CrossRef]

polymers

MDPI

Article

Effect of Conducting Polyaniline/Graphene Nanosheet Content on the Corrosion Behavior of Zinc-Rich Epoxy Primers in 3.5

Yanhua Lei [1,2,*], Zhichao Qiu [1], Jiurong Liu [1], Dongdong Li [1], Ning Tan [1], Tao Liu [1], Yuliang Zhang [1], Xueting Chang [1], Yanhong Gu [2,*] and Yansheng Yin [1]

1. Institute of Marine Materials Science and Engineering, College of Ocean Science and Engineering, Shanghai Maritime University, Shanghai 201306, China; qzc4321@163.com (Z.Q.); liujingrong@126.com (J.L.); 201730410071@stu.shmtu.edu.cn (D.L.); tanning986@163.com (N.T.); liutao@shmtu.edu.cn (T.L.); ylzhang@shmtu.edu.cn (Y.Z.); xtchang@shmtu.edu.cn (X.C.); ysyin@shmtu.edu.cn (Y.Y.)
2. Beijing Key Laboratory of Pipeline Critical Technology and Equipment for Deepwater Oil & Gas Development, Beijing Institute of Petrochemical Technology, Beijing 102617, China

* Correspondence: yhlei@shmtu.edu.cn (Y.L.); guyanhong@bipt.edu.cn (Y.G.); Tel.: +86-021-3828-4824 (Y.L.)

Received: 11 April 2019; Accepted: 6 May 2019; Published: 10 May 2019

Abstract: The corrosion behavior of zinc-rich epoxy primers or paints (ZRPs) with different conducting polyaniline-grafted graphene (PANI/Gr) contents was investigated. Conductivity of the formed PANI/Gr nanosheets was significantly improved by employing the Gr as the inner template to synthesize the PANI. The protective properties and electrochemical behavior of coatings with artificial defects were investigated by monitoring the free corrosion potential versus time and by using localized electrochemical impedance spectroscopy (LEIS). A synergetic enhancement of the physical barrier role of the coating and the zinc sacrificial cathodic protection was achieved in the case of ZRP including PANI/Gr nanosheets. In addition, the ZRP mixed with the PANI/Gr at a content of 0.6% exhibited the best anticorrosion performance across the range of investigated PANI/Gr contents.

Keywords: conducting polymer; PANI; LEIS; graphene; corrosion

1. Introduction

Organic coatings are applied to protect steel from corrosion due to their inhibition behavior, barrier function, and cathodic protection [1]. Zinc-rich epoxy primers or paints (ZRPs) are widely used as cathodic protection in a variety of severe environments via two protective mechanisms. First, sufficient electrical contact between coating and substrate is established because of the high zinc content. After the electrolyte diffuses into the coating, the zinc particles provide cathodic protection via the sacrificial anodic dissolution, thereby promoting the electrochemical process of a Zn–Fe microcell. Second, the zinc corrosion products (i.e., zinc oxide, zinc hydroxide, and hydrozincite) fill the coat pores and may also act as an electrical insulator under normal conditions [2]. As the organic and inorganic binders which are commonly used in ZRPs are naturally non-conductive, the electrical contact between the spherical zinc particles and steel substrates could not be maintained for a long time. Therefore, high zinc content (normally higher than 65 wt %) is necessary to ensure electrical conductivity [3]. Nevertheless, the utilization ratio of the zinc particles is very low [4], which is a waste of resources and results in environmental pollution.

Numerous efforts have been made to prolong the ZRP's lifetime through two routes. One way is to improve the barrier effect of the ZRP with graphene oxide (GO) [5], nano clay [6], TiO_2 [7], and so on. The addition of these fillers could extend the diffusion path of the electrolyte, which can reduce the rate of zinc particle consumption. However, the utilization ratio is still low. The other way is to improve the conductive property of the coatings by introducing different types of electrical conducting additives, for example, grephene (Gr) [8], polypyrrole (PPy) [4], polyaniline (PANI) [9,10], carbon nanotube (CNT) [11], carbon black [12], Al [13–15], and Co [16]. The improved conductive property could make it easier to link zinc particles and the steel substrate. In this way, the utilization ratio of zinc particles is increased while good conductivity often results in a decreased barrier effect due to the easy charge transfer.

Hayatdavoudi and Rahsepar [8] applied Gr to modify a ZRP coating, and pointed out that the presence of Gr nanosheets in the ZRP coating could provide not only an improvement in electrical contact but also an enhanced barrier against aggressive species. The impedance value was very low, almost under 1 kΩ·cm^2, which may be attributed to the superior conductivity of the coating by adding the good conductive Gr. Interestingly, Ramezanzadeh et al. [5] employed highly crystalline and conductive PANI/GO composites as an additional filler into a ZRE coating. They revealed that GO acted as a barrier and PANI enhanced the conductive behavior, which resulted in a higher cathodic protection duration. We compared the effects of Gr, PANI, and polyaniline-grafted graphene (PANI/Gr) on the corrosion performance of ZRPs [17], and considered that the combined PANI/Gr was a more effective strategy to enhance the protection performance of ZRP coatings compared to Gr. However, the quantity of added PANI/Gr nanosheets must be optimized to enhance the protection performance of ZRP coatings. Thus, in this work, the corrosion behavior of ZRP samples was studied with different PANI/Gr contents in the range of 0–0.8 wt %. The influence of PANI/Gr content on the corrosion protection behavior was evaluated in 3.5% NaCl solution by electrochemical impedance spectroscopy (EIS) and by monitoring the free corrosion potential versus time. Localized electrochemical impedance spectroscopy (LEIS), scanning electron microscopy (SEM) coupled with energy-dispersive X-ray spectroscopy (EDS), and X-ray photoelectron spectroscopy (XPS) were also utilized to support the electrochemical mechanism findings.

2. Materials and Methods

Gr nanosheets with thicknesses of 1–5 nm and widths of several microns were obtained from XianFeng Nano Co. (Suzhou, China). The conductivity of the purchased Gr nanosheets was in the range of 1200–1500 S cm^{-1}. Aniline monomer (Ani) with purity of 99% purchased from Aladdin Co. (Shanghai, China) was used without any pretreatment. All the other chemicals were of analytical grade and were used without further purification. The solutions were prepared using double-distilled and subsequently deionized Millipore water. DH32 steel obtained from Baowu Steel Co. (Shanghai, China) was used as the substrate. Specimens for the corrosion measurements were cut from the DH32 steel to a size of 10 × 10 mm. Prior to the experiment, the specimen surface was ground to an 800-grit SiC paper, then degreased ultrasonically with acetone and ethane, and washed with triply distilled water. Following this pretreatment, the steel as work electrode was immediately transferred to the electrochemical cell.

Synthesis of PANI/Gr: PANI/Gr composites were prepared by in situ chemical polymerization of Ani on Gr. In a typical synthesis, a 100 mg portion of Gr was first dispersed in 200 mL 0.1 M sodium dodecyl benzene sulfonate (SDBS) by ultrasonic treatment for 2 h under ambient conditions to give a dark dispersion. Then, the suspension was cooled to 2 °C in an ice-water bath. Ani monomer with a volume of 1 mL was dissolved in 20 mL 0.2 M HCl and the mixture was injected into the previous suspension with an injection springe, then further dispersed for 1 h with strong stirring. Afterwards, an equal volume of a 0.25 M ammonium persulfate (APS) solution was slowly added dropwise into the above mixture and kept at 0–4 °C for 24 h with strong stirring. Finally, the PANI/Gr composites were washed with distilled water and ethanol, dried in a vacuum oven at 60 °C for 24 h. The weight ratio was evaluated by comparing the product weight with the presence and absence of Gr during the PANI synthesis procedure under the same conditions. The synthesis procedures are schematically presented in Figure 1.

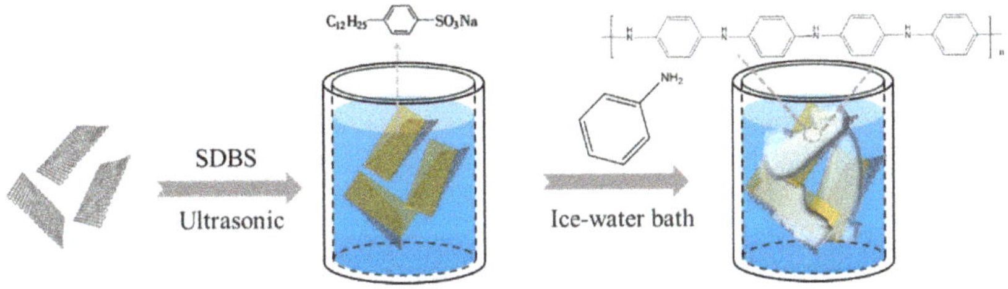

Figure 1. The synthesis procedures of polyaniline-grafted graphene (PANI/Gr).

2.1. Characterization of the Functional Nanoparticle or Composites

The synthesized PANI powder and PANI/Gr were characterized by SEM, TEM, Fourier-transform infrared spectroscopy (FT-IR), and Raman spectroscopy (RS). The surface morphologies and compositions were characterized using a SEM-EDS, JSM-6510LA (JEOL, Tokyo, Japan). The TEM observation were carried out on a JEM-2100F (JEOL, Tokyo, Japan). The formed PANI layer was analyzed by RS and FT-IR. For RS, a Raman polychromator spectrometer (Bunko-Keiki M30-TP-M, Tokyo, Japan) was used with excitation by YVO_4 solid-state laser at 532.0 nm wavelength. The laser power for the excitation was controlled below 5 mW to avoid damage to the PANI films. FT-IR spectra were measured in the wavenumber range of 4000–400 cm^{-1} at a resolution of 4.0 cm^{-1} by using Vertex-70 (Bruker, Ettlingen, Germany). XPS measurements were performed on a JPS-9200 (JEOL, Tokyo, Japan) with a Mg Ka (1253.6 eV) radiation. The calibrations of the binding energies were referenced to the Au 4f7/2 electron peak at 84.00A eV. The DC electrical conductivity of the powder was measured using a multifunction digital four-probe electrical conductivity measuring instrument (ST4742B, Suzhou Jingge Electronic Co. Ltd., Jiangsu, China). The AC dielectric properties of the ZRP coatings treated with different pigments were determined using an impedance analyzer Agilent E4980AL (Agilent, Santa Clara, CA, USA).

2.2. Preparations of Coated Specimens

A commercially available ZRP (provided by Sanmu Group Co., Jiangsu, China) was used to prepare various coating compositions. The ZRP containing 80 wt % zinc particles was composed of an epoxy resin (E44) and a polyamide hardener (T31) with the weight ratio of 4:1. To prepare the PANI/Gr-modified ZRP (PANI/Gr-ZRP) coating, different weight percentages were dissolved in dimethylbenzene by ultrasonic stir for 30 min and then added into the epoxy resin containing 80 wt % zinc. After stirring at room temperature until volatilization of dimethylbenzene finished, the resin was coated onto the DH32 steel. The samples were exposed to air for one week to ensure the absolute solidification of the resin. The ZRP formulation containing PANI/Gr is numbered and given in Table 1. Thickness of the coatings was determined using a PosiTector6000FNS2 probe (DeFelsko Co. Ltd., Ogdensburg, NY, USA), a coating thickness measuring instrument.

Table 1. Composition of the investigated zinc-rich paint (ZRP) coatings.

Sample	Zinc (wt %)	PANI/Gr (wt %)	Thickness (μm)
PANI/Gr_0-ZRPs (control)	80	0	60 ± 3.4
PANI/$Gr_{0.4}$-ZRPs	80	0.4	60 ± 2.7
PANI/$Gr_{0.6}$-ZRPs	80	0.6	60 ± 5.3
PANI/$Gr_{0.8}$-ZRPs	80	0.8	60 ± 4.6

2.3. Corrosion Test of the Modified ZRPs

The protection properties of the PANI/Gr on the DH32 steel were studied in a solution containing Cl^- at 25.0 ± 0.3 °C under atmospheric conditions, in which the open circuit potential (OCP) and EIS were measured. In order to study the effect of the PANI/Gr nanocomposite segments on the cathodic sacrificial behavior of the ZRPs, the coated samples were first immersed into a 3.5 wt % NaCl solution for 24 h. Then, an artificial defect with a size of 10 × 0.2 mm (which means the defect penetrated the coating) was induced for each coated sample. The samples with artificial defect were then immersed into the NaCl solution, and the OCP was monitored during the whole period of immersion.

Electrochemical impedance spectroscopy (Autolab, Metrohm, Zurich, Switzerland) was used to evaluate the protection capacity of the steel sheets coated with ZRP and PANI/Gr-ZRP. A conventional three-electrode electrochemical cell, composed of platinum as counter electrode, a saturated calomel electrode (SCE) as reference, and the coated steel panel as working electrode, was used to carry out the electrochemical tests in 3.5 wt % NaCl solution. The test was done at OCP in the frequency range and amplitude sinusoidal voltage of 10^{-2} Hz to 10^5 Hz (peak to zero) and 20 mV, respectively. The software program ZSimpwin (E Chem Software, Ann Arbor, MI, USA) were used to fit the impedance data.

Localized electrochemical impedance spectroscopy (LEIS) of the coatings in the presence of an artificial defect of 2 × 0.2 mm was performed on an M470 scanning electrochemical workstation (Bio-Logic, Paris, France). A three-electrode cell was used to perform the LEIS measurement. An SCE and a carbon rod were used as the reference electrode and the counter electrode, respectively. The coated sample was studied as the working electrode. During LEIS testing, a distance of 100 μm was adjusted from the electrode surface to a microprobe Pt tip with a tip diameter of 5 μm. A current signal with amplitude of 10 μA at a single frequency of 10 Hz was generated by the electrochemical workstation to finish the LEIS measurements. All the potentials in this paper were plotted with respect to the Ag/AgCl/sat. KCl (0.197 V vs. Standard hydrogen electrode(SHE)). The sacrificial protection performance of the samples with X-cut scribes was investigated in a 5 wt % NaCl solution at 45 °C by salt spray test in a salt spray cabin KLT-90 (China Electronics Technology Group Co., Beijing, China).

3. Results

3.1. Characterization of the PANI/Gr Composites

Since the conductivity of additions would significantly influence the corrosion performance of the ZRP primer, the conductivity of the addition was therefore first measured. Figure 2a shows the conductivities of the synthesized PANI particles and PANI/Gr nanosheets. The conductivity of the PANI sample was evaluated at 5 S cm^{-1}, which was comparable to the results of Stejskal and Gilbert [18] and Meroufel et al. [19], indicating the formation of the emeraldine salt (PANI-ES) state. However, the conductivity after introducing the Gr during the in situ synthesis process increased to 150 S cm^{-1}, which is more than one order of magnitude. It must be noted that although the modification of Gr with SDBS decreased the conductivity of the Gr, the conductivity of the pretreated Gr was still much higher than that of the synthesized PANI/Gr composites. The results of the DC conductivity measurement indicated that the PANI was successfully synthesized on the surface of Gr, and the formation of PANI on the surface of Gr reduced the conductivity of the Gr. During the synthesis procedure, when Ani monomers were added into the Gr suspension, the Ani monomers (being the electron donor) could immediately absorb onto the surfaces of Gr (being the electron acceptor) and form a type of weak charge transfer complex due to the electrostatic attraction [20]. The weight ratio of the PANI to Gr in the synthesized PANI-Gr composites was analyzed by comparing the product weight with the presence and absence of Gr during the PANI synthesis procedure under the same conditions, the results of which indicated a ratio of nearly 2:1. Then, the AC conductivity of the ZRP coatings treated with the above pigments was evaluated. As shown in Figure 2b, compared to the ZRP coating, the AC conductivity of the ZRP coatings was improved to different degrees after the dispersion of the PANI, PANI/Gr, and Gr conductive pigments. Therefore, considering the excellent barrier property of PANI, it was assumed herein that the inclusion of PANI/Gr nanocomposites with a reasonable content in ZRP would obtain a better synergistic protection of the barrier and cathodic protection of the ZRP coating.

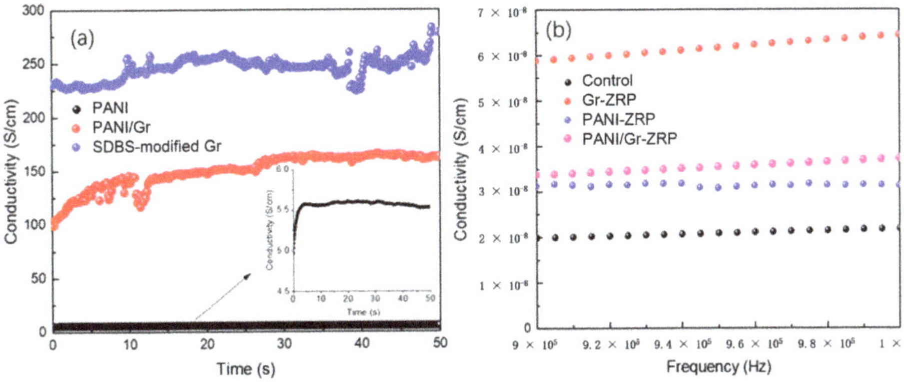

Figure 2. (**a**) The DC conductivity results of the synthesized PANI, the PANI/Gr nanocomposites, and the sodium dodecyl benzene sulfonate (SDBS)-modified Gr; (**b**) the AC conductivity results of the ZRP coatings with the addition of different conductive pigments.

The characterization of the synthesized PANI/Gr composites was carried out by SEM, TEM, FT-IR, and Raman, as shown in Figure 3. The SEM images reveal that the synthesized PANI/Gr nanocomposites consisted of nanosheets with a thickness of nearly 10–20 nm, which is thicker than the Gr nanosheets. A high-resolution SEM image is given in Figure 3b, in which PANI nanosheets consisting of small nanodots are observed. In the synthesis process, the aniline monomer was initially adsorbed onto the surface of the graphene nanosheets due to electrostatic attractions, and then in situ polymerization took place. From the TEM observation, the typical wrinkled structure of Gr remained after PANI grafting, as demonstrated in Figure 3c,d. The formation of PANI on Gr was characterized with FT-IR and Raman spectroscopy. Figure 3e compares the FT-IR spectroscopy of pristine Gr with the PANI/Gr nanocomposites. The stretching modes of C–C and C=C in the Gr occur at the absorptions of 1402 and 1580 cm^{-1}, respectively. The presence of the surface functional groups on graphene after PANI grafting was further confirmed by FT-IR. The O–H and N–H stretching modes give rise to the strong broad band at 3429 cm^{-1} and weak band at 3227 cm^{-1}, respectively. The C–H stretching band of the aromatic ring occurs at 2922 cm^{-1}, and the absorption peak intensity increased obviously due to the PANI formation to the surface of Gr. Other characteristic peaks related to the PANI structure can also be seen at 1580 cm^{-1}, 1491 cm^{-1}, and 1299 cm^{-1}, which are assigned to C–C=C conjugative backbone stretching, C=N stretching, and C–N stretching vibration with aromatic conjugation, respectively [21–24]. The peak at 1121 cm^{-1} corresponds to the in-plane bending vibration of C–H [25], whereas the peak at 804 cm^{-1} corresponds to the out-of-plane bending vibration of C–H. All the above peaks can be distinguished from the spectrum of PANI/Gr composites, which indicates the successful formation of PANI on the surface of graphene. Figure 3f shows the Raman spectra for the synthesized PANI/Gr nanocomposite and graphene. The G and D peaks of Gr are recorded at 1580 and 1350 cm^{-1}, arising from the stretching of the C–C bond and the disorder in the sp2-hydridized Gr structure, respectively [26]. For the neat Gr, it is observed that the I_D/I_G is quite high, which indicates that the Gr used is rich in defects. After the polymerization of PANI on Gr, it is evident that the I_D intensity was enhanced due to the sp2 C in the aromatic ring. Thus, the I_D/I_G of the PANI-Gr decreased compared to that of the neat Gr. In addition, the PANI/Gr composite presents a weak shift of the D peak towards low wavenumbers resulting from the π–π* electron interaction between Gr and the Ani monomer [20]. Besides the wavenumber shift of the G and D peaks, the characteristic vibrational peaks of PANI are also evident in the PANI-Gr nanocomposite spectrum. The 1562 cm^{-1}, 1372 cm^{-1}, and 1189 cm^{-1} peaks can be attributed to the bipolaronic N–H bending vibration, the C–N stretching vibration of the cation radical species, and the C–H bending of the quinoid ring [27], respectively. All of the above results definitely indicate the formation of PANI on the surface of Gr.

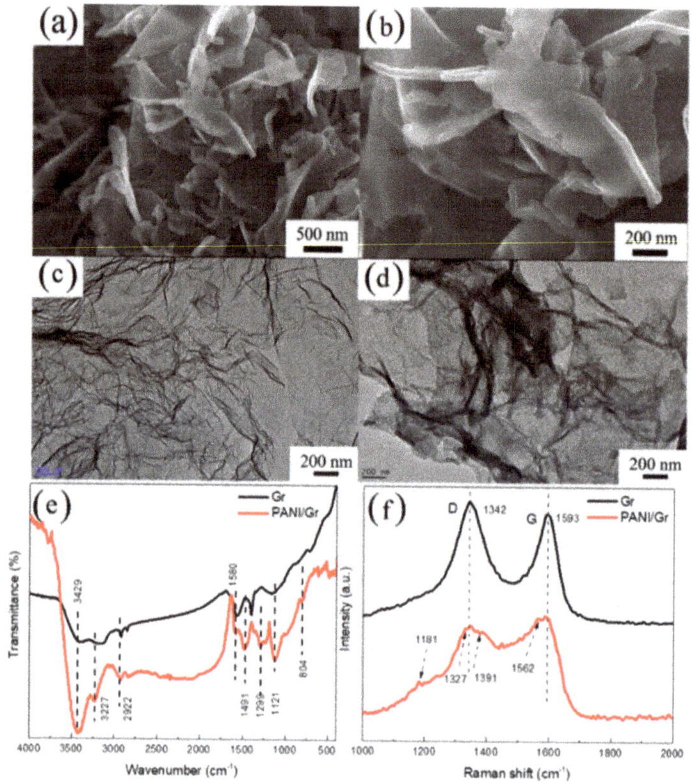

Figure 3. (**a**) SEM image of the synthesized PANI/Gr; (**b**) High magnification of the SEM image; (**c**) TEM image of graphene; (**d**) TEM image of the synthesized PANI/Gr; (**e**) Comparison FT-IR spectra; (**f**) Raman spectra of the modified Gr and PANI/Gr.

3.2. OCP Measurements

It is generally accepted that the OCP criterion for cathodic protection is lower than −0.76 V vs. SCE [28]. Thus, it is considered that the cathodic protection ability of substrate will be maintained if the corrosion potential of the coated substrate is kept below −0.8 V vs. SCE. OCP was monitored continuously during 1000 h of immersion. The OCP changes in the ZRP with different contents of PANI/Gr during immersion for up to 1000 h are plotted in line a in Figure 4. It was observed that the OCP of the control ZRP coating continuously increased with the extension of immersion time, and reached −0.76 V after 1000 h. All potentials of the ZRP coatings containing PANI/Gr remained lower than −0.8 V for up to 1000 h of immersion, which indicates that the substrates were protected cathodically. For example, the OCP of the PANI/$Gr_{0.4}$-ZRP was raised to −0.84 V after 1000 h of immersion, although the OCP slowly shifted anodically with prolonged immersion, indicating the degradation of cathodic protection. The potential of both the PANI/$Gr_{0.6}$-ZRP and PANI/$Gr_{0.8}$-ZRP coatings maintained a low value during the whole period of immersion, although the potential of PANI/$Gr_{0.8}$-ZRP fluctuated. Oxidation of the Zn to ZnO or other corrosion products due to the reduction of oxygen or PANI caused the fluctuation in the potential and the increase in the potential. Compared to the control samples, the doping of PANI/Gr extended the ZRP coating cathodic protection duration. The OCP of the PANI/$Gr_{0.6}$-ZRP coating exhibited a more stable and negative potential during the long immersion test, which indicated that the ZRP coating with PANI/$Gr_{0.6}$ could provide a better effective cathodic protection than the others.

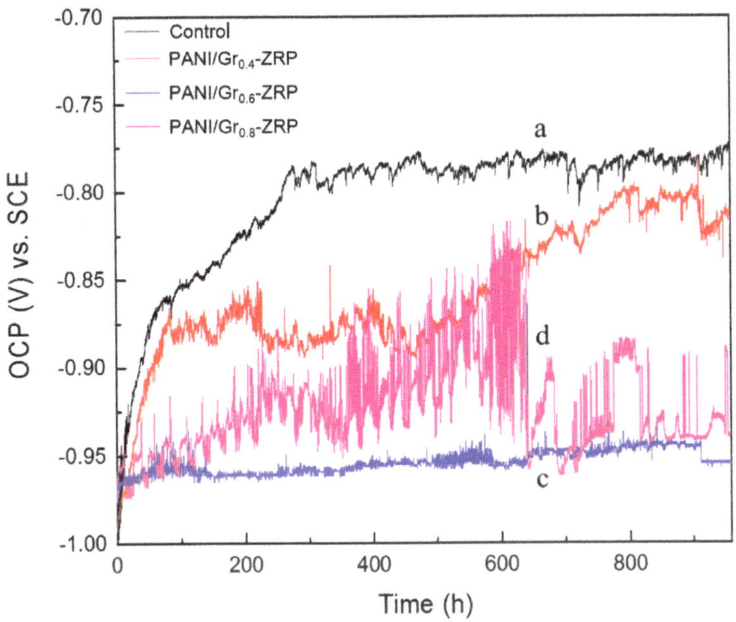

Figure 4. Open circuit potential (OCP) in 3.5% NaCl solution of DH32 steel coated with ZRP (control) (**line a**), PANI/$Gr_{0.4}$-ZRP (**line b**), PANI/$Gr_{0.6}$-ZRP (**line c**), and PANI/$Gr_{0.8}$-ZRP (**line d**).

In order to clearly study the content effects of PANI/Gr in the ZRPs on the cathodic sacrificial behavior, the changes in potential of the samples with artificial defects were also monitored as a function of immersion time. The ZRP coating with an artificial defect of 10 × 0.2 mm was immersed into the 3.5 wt % NaCl solution, and then the OCP was monitored throughout the whole process. Figure 5 presents the evolution of OCP for ZRPs with different contents of PANI/Gr and with immersion time. As can be seen in Figure 5, all potentials of the different coatings are initially in the range of −0.9 to −1.0 V, indicating the protection due to the cathodic scarification. For the control sample, the potential, initially at −1.0 V, shifts toward the positive direction with immersion, and reaches −0.6 V after 120 h of immersion. A decrease in the electroactive zinc area resulted in an increase in the potential, predicting a decrease in cathodic protection intensity. The introduction of defects to the coating accelerated the degradation of ZRP coatings' cathodic protection, and the quick shift of potential can be attributed to the oxidation of Zn. However, the addition of electrically enhanced PANI/Gr segments led to the slight changes in potential. The OCP of the ZRP containing 0.4 wt % PANI/Gr rapidly increased with the extension of immersion, whereas for the 0.6 wt % PANI/Gr coating, the potential first fluctuated in the range of −0.8 to −1.0 V, then slowly increased to −0.76 V after 200 h of immersion. When the content of PANI/Gr increased to 0.8 wt %, the OCP values shifted to the positive region rapidly after 120 h and stabilized at −0.7 V during further immersion. It should be noted that the positive potential shift for the PANI/Gr-added coatings results mainly from the reaction couples of zinc oxidation and PANI reduction. Furthermore, the corrosion potential values of the PANI/Gr-ZRP at different contents are always lower than that of the control sample, as shown in Figure 5. This further indicates that some of the zinc particles were sacrificially active, thus contributing to the cathodic protection of the substrate with the artificial defects.

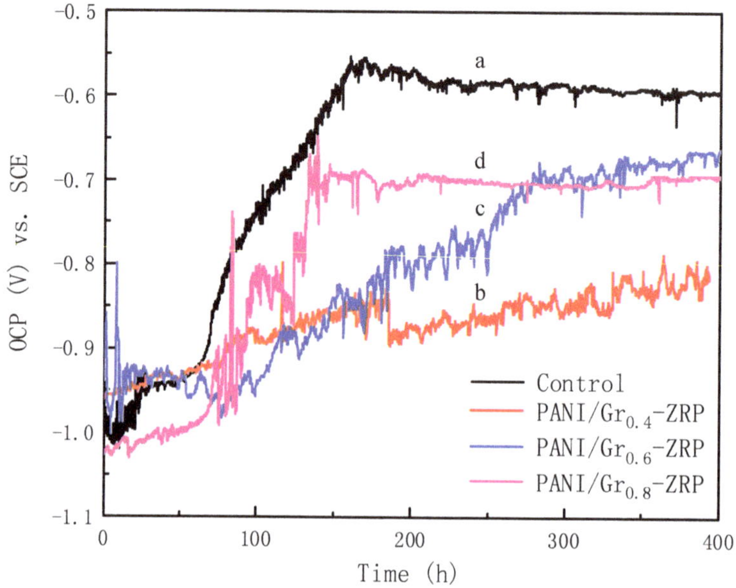

Figure 5. Open circuit potential (OCP) changes in 3.5% NaCl solution of samples with artificial defects: (**line a**) PANI/Gr_0-ZRPs (control), (**line b**) PANI/$Gr_{0.4}$-ZRP, (**line c**) PANI/$Gr_{0.6}$-ZRP, and (**line d**) PANI/$Gr_{0.8}$-ZRP.

3.3. LEIS Measurement of the PANI/Gr-ZRP Coatings with Artificial Defect

To further study the effect of PANI/Gr addition on the coating cathodic protection, localized electrochemical impedance spectroscopy measurements were performed on a scratch-coated steel specimen with artificial damage which was exposed to a 0.005 M NaCl solution. Figure 6(a1–a3) shows the LEIS spectra for PANI/Gr_0-ZRPs at different periods of immersion time, in which the local impedance in the scratched area remained at a low value during the exposure of 48 h. Additionally, it can be seen that the section of the scratched area with the low impedance value was enlarged after 48 h. The local impedance value in the scratched area of the PANI/$Gr_{0.4}$-ZRPs was slightly higher than the PANI/Gr_0-ZRPs at the beginning of immersion, approaching Figure 6b, and with the immersion time extension, the local impedance values initially increased lightly and decreased at 48 h, whereas the lowest impedance value barely changed during immersion. With the content of PANI/Gr reaching 0.6 wt %, as shown in Figure 6c, the local impedance value significantly improved with the extension of immersion. It is important to note that the scratched area with relatively low local impedance in the PANI/$Gr_{0.6}$-ZRP decreased in size during the immersion time extension. For the PANI/$Gr_{0.8}$-ZRP coating, the local impedance increased explicitly after 24 h of immersion, and no shrink in size of the scratched area with relatively low local impedance was observed during immersion.

Figure 7 reveals the values of the average and minimum local impedance of the whole scan area. It can be observed that both the minimum values and the average values of the marked area for the PANI/$Gr_{0.6}$-ZRP were the highest, compared to the other samples. An average impedance of 8000 $\Omega \cdot cm^2$ was obtained after 48 h for the PANI/$Gr_{0.6}$-ZRP sample, which indicated the excellent corrosion protection. In addition, the lowest impedance value of the PANI/$Gr_{0.6}$-ZRP also sharply increased, indicating the decrease in dissolution of the exposed substrate.

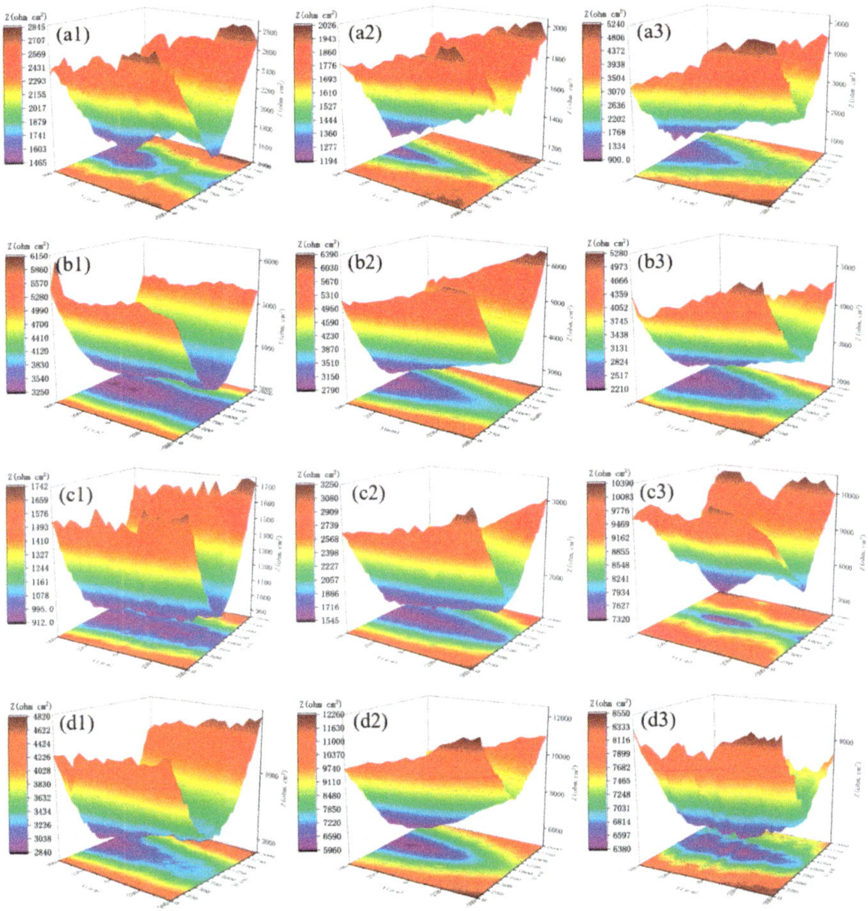

Figure 6. Localized electrochemical impedance spectroscopy (LEIS) results of the defect region for (**a**) PANI/Gr_0-ZRPs, (**b**) PANI/$Gr_{0.4}$-ZRPs, (**c**) PANI/$Gr_{0.6}$-ZRPs, and (**d**) PANI/$Gr_{0.8}$-ZRPs with artificial scratches immersed in 0.005 M NaCl solution for 48 h. (**a1**, **b1**, **c1**, and **d1**) are LEIS spectra for 1 h, (**a2**, **b2**, **c2**, and **d2**) are LEIS spectra for 24 h, and (**a3**, **b3**, **c3**, and **d3**) are LEIS spectra for 48 h.

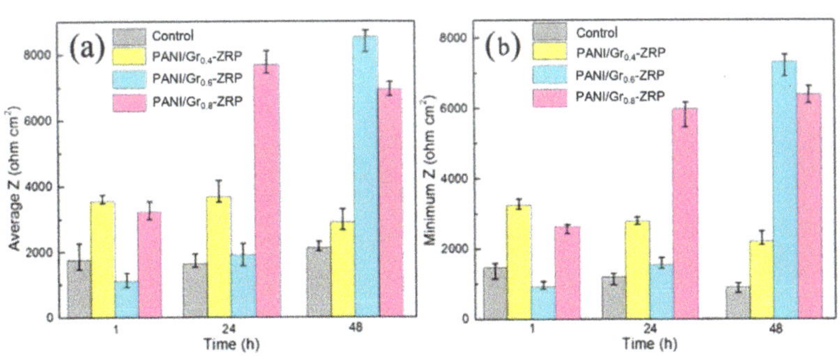

Figure 7. Changes in the (**a**) average impedance and (**b**) minimum impedance in the LEIS maps of Figure 6 for different immersion periods.

Generally, for the ZRP-coated substrate, and once the coating was damaged, the zinc particles around the defect in the coating sacrificially dissolved, providing cathodic protection to the steel substrate. The result of OCP in Figure 5 demonstrates that the substrate exposed in the artificial defect was initially under the cathodic protection of the ZRP coating. However, with prolonged immersion, the sacrificial dissolution of zinc could no longer inhibit the dissolution of the substrate. The results of the LEIS in Figure 6 show the low impedance of the ZRP, which indicated the dissolution of the substrate. For the PANI/Gr-modified ZRP coatings, an increase in the local impedance was observed during the 48-h immersion, especially for the PANI/$Gr_{0.6}$-ZRP coating. The increase in the local impedances could be attributed to the physical barrier effect of scarified zinc products, which precipitated at the defect area with the oxidization of surrounding zinc particles in the ZRP coating. It has been well documented previously in the literature that the PANI pigments in the coating on steel were found to effectively protect the substrate by forming a passive film [29–31]. Furthermore, it has been reported that in the presence of oxygen in the corrosive electrolyte, the PANI-EB could be converted into the PANI-ES, completing the autocatalytic cycle [32,33], which could result in the stabilization of Fe in the passive region. Thus, the flake-shape PANI in ZRP around the defects is another factor of the increase in local impedance by the formation of a protective passive film.

Surface morphologies as well as the compositions of the coatings with artificial defects were investigated by SEM-EDS. Figure 8 compares the surface morphologies of the samples with different PANI contents at artificial defect zones after 30 days of immersion in 3.5% NaCl solution. It was observed that the artificial grooves were almost covered with corrosion products, composed of Zn and/or Fe corrosion products. As mentioned above, the ZRP failed to continuously support the cathodic protection with the extension of immersion, resulting in the dissolution of the substrate. From the EDS results in Figure 8, the presence of Zn elements in the control coating indicated a certain degree of cathodic protection, while the dissolution of the substrate resulted in the enrichment of Fe corrosion products. Interestingly, for the PANI/$Gr_{0.6}$-ZRP and PANI/$Gr_{0.8}$-ZRP samples, almost no Fe, or only a traceable amount, was detected at the defect area, which indicated that the substrate with artificial defects was effectively cathodically protected by Zn sacrificial dissolution.

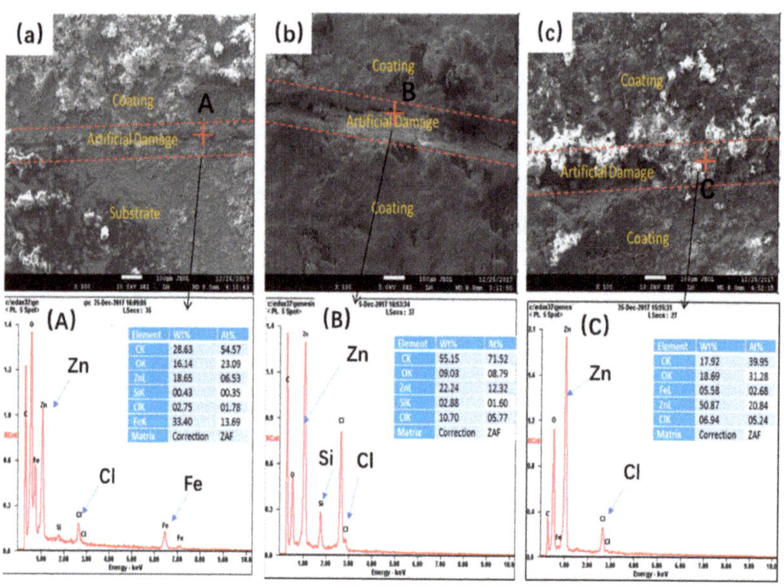

Figure 8. SEM images and EDS spectra for the coated samples: (**a**) PANI/Gr_0-ZRPs, (**b**) PANI/$Gr_{0.6}$-ZRPs, and (**c**) PANI/$Gr_{0.8}$-ZRPs. (**A**) the EDS spectrum of the selected area marked A in (a); (**B**) the EDS spectrum of the selected area marked a in (b); (**C**) the EDS spectrum of the selected area marked C in (c).

In order to investigate the barrier properties of the coatings modified with different contents of PANI/Gr, EIS was measured on the coatings. Figure 9 shows the impedance spectra for the coated samples (PANI/Gr0-ZRPs, PANI/$Gr_{0.4}$-ZRPs, PANI/$Gr_{0.6}$-ZRPs, and PANI/$Gr_{0.8}$-ZRPs) in the 3.5 wt % NaCl solution after 30 days of immersion. The equivalent circuits are also presented in the spectra. It is noted that the impedance of the ZRP coatings could be significantly improved by mixing PANI/Gr, providing the barrier role. Two overlapped capacitive time constants can be distinguished at both low frequency and high frequency during the exposure to NaCl solution due to the penetration of water and oxygen into the coatings, thereby leading to the galvanic corrosion of zinc particles and the corresponding zinc corrosion products [19]. The first capacitance loop is attributed to the coating itself, and the second one is related to the reactions occurring underneath the film [34,35]. Figure 9c indicates the equivalent circuits (Rs(RctQdl)(RcoatQf)) used to simulate the EIS data. Rs refers to the solution resistance. Rct refers to the charge transfer resistance, which denotes the resistance of the electron transfer across the electron double layer from the substrate to the oxidant agent. Rcoat denotes the resistance of the coating, which arises from the kinetic resistance of the related ions or reactant agents through the coating. In order to justify the actual surface conditions of the electrode, a constant phase element (CPE, Q) was used to replace an ideal capacitor, and the impedance of Q is defined as Equation (1), as follows:

$$Z\,(j\omega) = (Y_0)^{-1}(j\omega)^{-n}. \tag{1}$$

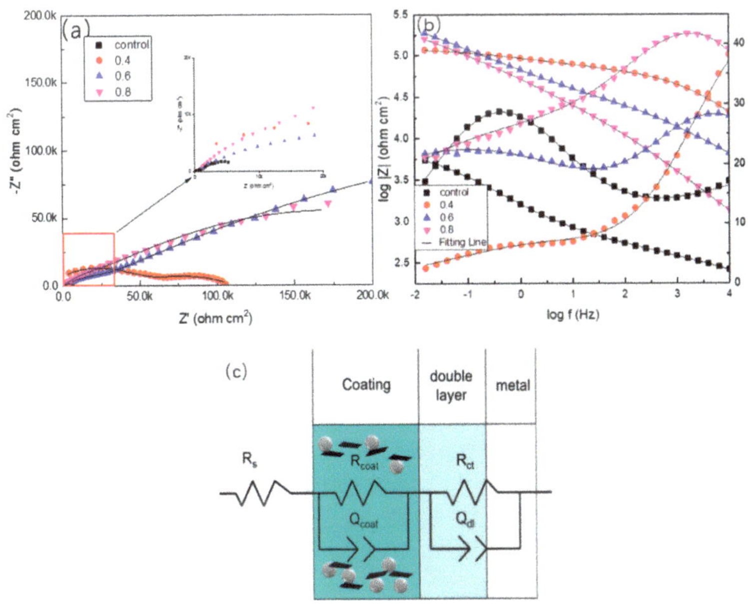

Figure 9. (**a**) Nyquist and (**b**) Bode plots for the samples (PANI/Gr_0-ZRPs, PANI/$Gr_{0.4}$-ZRPs, PANI/$Gr_{0.6}$-ZRPs, and PANI/$Gr_{0.8}$-ZRPs) after immersion in a 3.5 wt % NaCl solution for 25 days. (**c**) the equivalent circuit and model.

In the equation, the parameters Y_0, j, n ($0 \le n \le 1$), and ω ($\omega = 2\pi f$, where f is the frequency) correspond to the Q constant, the imaginary unit, the Q power, and the angular frequency, respectively. It is indicated that Q is a pure capacitance when the value of n is 1. Heterogenicity of the electrode derives the value of n from the unit [36]. The impedance parameters, which were obtained using ZSimpwin software, are presented in Table 2.

Table 2. Optimum fit parameters of the coatings after 30 days of immersion.

Sample	R_s ($\Omega \cdot cm^2$)	C_{coat} ($\Omega^{-1} \cdot cm^{-2} \cdot s^n$)	R_{coat} ($\Omega \cdot cm^2$)	C_{dl} ($\Omega^{-1} \cdot cm^{-2} \cdot s^n$)	R_{ct} ($\Omega \cdot cm^2$)
PANI/Gr_0-ZRPs	0.253	1.36×10^{-4}	653	3.12×10^{-4}	7790
PANI/$Gr_{0.4}$-ZRPs	0.575	8.38×10^{-4}	7030	7.38×10^{-4}	5960
PANI/$Gr_{0.6}$-ZRPs	0.842	4.85×10^{-5}	79200	9.64×10^{-4}	2280
PANI/$Gr_{0.8}$-ZRPs	0.734	1.13×10^{-3}	31400	4.37×10^{-4}	1820

For the PANI/Gr_0-ZRPs, the film resistance (Rcoat) was only 653 $\Omega \cdot cm^2$ after 30 days. This is due to the porosity of the coating which supplied the permeation path of the electrolyte. For PANI/$Gr_{0.4}$-ZRPs, the coating resistance (Rcoat) was 7030 $\Omega \cdot cm^2$ after 30 days. With the increase of PANI/Gr to 0.6 wt %, the coating resistance (Rcoat) was stable around 79,200 $\Omega \cdot cm^2$. On the contrary, the coating resistance (Rcoat) of PANI/$Gr_{0.8}$-ZRPs was lower than the coating resistance of PANI/$Gr_{0.6}$-ZRPs, at 31,400 $\Omega \cdot cm^2$. In other words, the resistance of the coatings reached a maximum value at 0.6% PANI. Another useful parameter for understanding the behavior of coatings is coating capacitance (C_{coat}). Generally, the coatings with lower C_{coat} exhibit a better barrier protection against the penetration of corrosive ions [37], and water absorption causes an improvement in the coating's dielectric constant, resulting in the increase in coating capacitance. It can be seen that the lowest C_{coat} value was obtained from PANI/$Gr_{0.6}$-ZRP coatings after 30 days of immersion, exhibiting the best barrier properties of the PANI/$Gr_{0.6}$-ZRP coatings. It is apparent that the highest value of coating resistance is consistent with the highest value of coating capacitance and both occurred at 0.6% PANI.

The corrosion protection properties of the ZRP coatings with different PANI/Gr contents were lastly evaluated using the salt spray test. Figure 10 shows the visual observations of the samples. Samples with X-scribes were exposed under salt spray fog for different time periods, namely, 500, 1000 and 2000 h. As shown in Figure 10, red rust formed on the scribed area of the PANI/Gr_0-ZRPs, PANI/$Gr_{0.4}$-ZRPs, and PANI/$Gr_{0.8}$-ZRPs after 1000 h of salt spray testing. More corrosion products were observed at scribed regions of the PANI/Gr_0-ZRP and PANI/$Gr_{0.4}$-ZRP samples as the exposure time increased up to 2000 h, exhibiting poor cathodic protection performance. Compared to the control and PANI/$Gr_{0.4}$-ZRPs, the PANI/$Gr_{0.8}$-ZRPs showed better cathodic protection performance with the relatively low quantity of red corrosion products after 2000 h test. Interestingly, for the PANI/$Gr_{0.6}$-ZRPs sample, no Fe corrosion products were formed after 1000 h of exposure and only a small amount of red rust appeared after 2000 h exposure. The salt spray test results revealed that PANI/$Gr_{0.6}$-ZRPs exhibited the best cathodic protection performance for long time exposure.

Mixing the PANI/Gr into the ZRP primer significantly improved the physical barrier protection due to fact that the mixture effectively restrained the permeation of caustic chloride ions to the coating and prolonged the action time of the sacrificial zinc particles. Based on the previous results, it can be seen the quantity of PANI/Gr additives in the coating has a significant effect on the corrosion behavior of the coating, and the PANI/$Gr_{0.6}$-ZRPs coating exhibits the best anticorrosion performance. Figure 11 schematically explains the effects of PANI/Gr on the corrosion protection of ZRP primers. In the presence of 0.6 wt % PANI/Gr, the sacrificial protection is a predominant mechanism, which relates to the increase in electrical connection between the zinc particles and the steel substrate. During the immersion time, a layer of zinc corrosion products rather than iron corrosion products formed on the steel surface. For the 0.4 wt % PANI/Gr-ZRPs sample, the coating resistance was higher than the control (Figure 9), while the cathodic protection properties were not significantly enhanced (Figure 7). When the content of the PANI/Gr was further increased to 0.6 wt %, the barrier performance enhanced considerably (Figure 9), which was due to the layered structure and oxidation of zinc particles. However, if the content of Zn in the PANI is further increased (e.g., 0.8%), it may lead to too many zinc corrosion products due to the oxidation of PANI and isolate the connection between the zinc particles, resulting in a significant degradation of the cathodic protection.

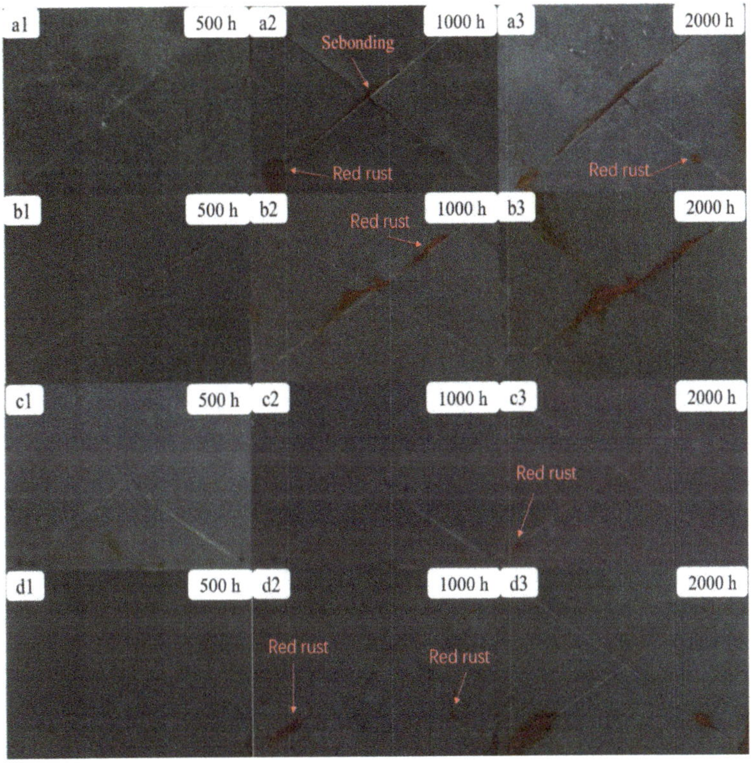

Figure 10. Salt spray test results of the (**a1–a3**) PANI/Gr_0-ZRP, (**b1–b3**) PANI/$Gr_{0.4}$-ZRP, (**c1–c3**) PANI/$Gr_{0.6}$-ZRP, and (**d1–d3**) PANI/$Gr_{0.8}$-ZRP samples during 2000 h exposure.

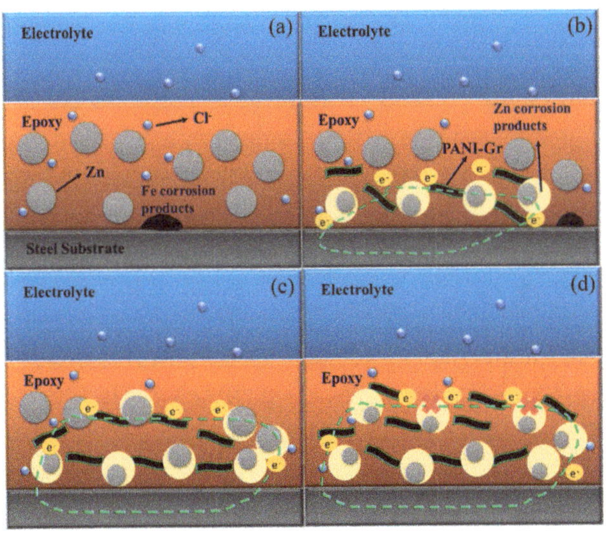

Figure 11. Schematic representation of (**a**) PANI/Gr_0-ZRPs, (**b**) PANI/$Gr_{0.4}$-ZRPs, (**c**) PANI/$Gr_{0.6}$-ZRPs, and (**d**) PANI/$Gr_{0.8}$-ZRPs during immersion in NaCl solution.

4. Conclusions

1. Conductivity measurements indicate that the synthesized PANI/Gr shows higher conductivity compared to the PANI. OCP and LEIS results at the defect region indicate that the mixture of PANI/Gr in this investigation indeed enhances the sacrificially cathodic protection of the ZRP coating by the extension of its active sacrificial duration.
2. Higher values of coating resistance in the EIS results for PANI/Gr-modified ZRP implicate the reduction of the coating porosity, leading to the better barrier properties of the modified ZRP. The ZRP mixed with the PANI/Gr at a content of 0.6% exhibits the best synergetic enhancement of the coating physical barrier as well as enhanced zinc sacrificial cathodic protection across the range of investigated PANI/Gr mixtures.

Author Contributions: Investigation, J.L. and Z.Q.; methodology, T.L.; validation, Z.Q. and D.L.; formal analysis, X.C.; resources, Y.Z.; data curation, N.T.; writing—original draft preparation, J.L.; writing—review and editing, Y.L.; supervision, Y.G.; project administration, Y.L.; funding acquisition, Y.Y.

Funding: This work was supported by the National Key R & D Program of China (No. 2016YFB0300700, 2016YFB0300704) and the National Natural Science Foundation of China (NSFC Grant No. 51602195).

Conflicts of Interest: The authors declare no conflict of interest.

References

1. Akbarinezhad, E.; Ebrahimi, M.; Sharif, F.; Ghanbarzadeh, A. Evaluating protection performance of zinc rich epoxy paints modified with polyaniline and polyaniline-clay nanocomposite. *Prog. Org. Coat.* **2014**, *77*, 1299–1308. [CrossRef]
2. Kowalczyk, K.; Spycha, T. Zinc-free varnishes and zinc-rich paints modified with ionic liquids. *Corros. Sci.* **2014**, *78*, 111–120. [CrossRef]
3. Schaefer, K.; Miszczyk, A. Improvement of electrochemical action of zinc-rich paints by addition of nanoparticulate zinc. *Corros. Sci.* **2013**, *66*, 380–391. [CrossRef]
4. Gergely, A.; Bertóti, I.; Török, T.; Pfeifer, É.; Kálmán, E. Corrosion protection with zinc-rich epoxy paint coatings embedded with various amounts of highly dispersed polypyrrole-deposited alumina monohydrate particles. *Prog. Org. Coat.* **2013**, *76*, 17–32. [CrossRef]
5. Ramezanzadeh, B.; Moghadam, M.H.M.; Shohani, N.; Mahdavian, M. Effects of highly crystalline and conductive polyaniline/graphene oxide composites on the corrosion protection performance of a zinc-rich epoxy coating. *Chem. Eng. J.* **2017**, *320*, 363–375. [CrossRef]
6. Shirehjini, F.T.; Danaee, I.; Eskandari, H. Effect of nano clay on corrosion protection of Zinc-rich epoxy coatings on steel 37. *J. Mater. Sci. Technol.* **2016**, *32*, 1152–1160. [CrossRef]
7. Anandhi, A.; Palraj, S.; Subramanian, G. Corrosion resistance and improved adhesion properties of propargyl alcohol impregnated mesoporous titanium dioxide built-in epoxy zinc rich primer. *Prog. Org. Coat.* **2016**, *97*, 10–18. [CrossRef]
8. Hayatdavoudi, H.; Rahsepar, M. A mechanistic study of the enhanced cathodic protection performance of graphene enforced zinc rich nanocomposite coating for corrosion protection of carbon steel substrate. *J. Alloys Compd.* **2017**, *727*, 1148–1156. [CrossRef]
9. Pan, T.; Yu, Q. Long-term anti-corrosion performance of a conducting polymer-based coating system for steels. *J. Mater. Eng. Perform.* **2016**, *25*, 2384–2394. [CrossRef]
10. Balgude, D.; Sabnis, A.; Ghosh, S.K. Investigation of cardanol-based reactive polyamide as a crosslinker in epoxy zinc-rich primer. *J. Coat. Technol. Res.* **2017**, *14*, 583–595. [CrossRef]
11. Cubides, Y.; Castaneda, H. Corrosion protection mechanisms of carbon nanotube and zinc-rich epoxy primers on carbon steel in simulated concrete pore solutions in the presence of chloride ions. *Corros. Sci.* **2016**, *109*, 145–161. [CrossRef]
12. Meroufel, A.; Touzain, S. EIS characterization of new zinc-rich powder coatings. *Prog. Org. Coat.* **2007**, *59*, 197–205. [CrossRef]

13. Chen, W.B.; Chen, P.; Chen, H.Y.; Wu, J.; Tsai, W.T. Development of Al-containing zinc-rich paints for corrosion resistance. *Appl. Surf. Sci.* **2002**, *187*, 154–164. [CrossRef]
14. Jalili, M.; Rostami, M.; Ramezanzadeh, B. An investigation of the electrochemical action of the epoxy zinc-rich coatings containing surface modified aluminum nanoparticle. *Appl. Surf. Sci.* **2015**, *238*, 95–108. [CrossRef]
15. Arman, S.Y.; Ramezanzadeh, B.; Farghadani, S.; Mehdipour, M.; Rajabi, A. Application of the electrochemical noise to investigate the corrosion resistance of an epoxy zinc-rich coating loaded with lamellar aluminum and micaceous iron oxide particles. *Corros. Sci.* **2013**, *77*, 118–127. [CrossRef]
16. Gharahcheshmeh, M.H.; Sohi, M.H. Pulse electrodeposition of Zn-Co alloy coatings obtained from an alkaline bath. *Mater. Chem. Phys.* **2012**, *134*, 1146–1152. [CrossRef]
17. Liu, J.; Lei, Y.; Qiu, Z.; Li, D.; Chang, X. Insight into the impact of conducting polyaniline/graphene nanosheets on corrosion mechanism of zinc-rich epoxy primers on low alloy DH32 steel in artificial sea water. *J. Electrochem. Soc.* **2018**, *165*, C878–C889. [CrossRef]
18. Stejskal, J.; Gilbert, R.G. Polyaniline. Preparation of a conducting polymer (IUPAC technical report). *Pure Appl. Chem.* **2002**, *74*, 857–867. [CrossRef]
19. Meroufel, A.; Deslouis, C.; Touzain, S. Electrochemical and anticorrosion performances of zinc-rich and polyaniline powder coatings. *Electrochim. Acta* **2008**, *53*, 2331–2338. [CrossRef]
20. Yan, J.; Wei, T.; Shao, B.; Fan, Z.; Qian, W.; Zhang, M.; Wei, F. Preparation of a graphene nanosheet/polyaniline composite with high specific capacitance. *Carbon* **2010**, *48*, 487–493. [CrossRef]
21. Xu, J.; Wang, K.; Zu, S.Z.; Han, B.H.; Wei, Z. Hierarchical nanocomposites of polyaniline nanowire arrays on graphene oxide sheets with synergistic effect for energy storage. *ACS Nano* **2010**, *4*, 5019–5026. [CrossRef]
22. Furukawa, Y.; Ueda, F.; Hyodo, Y.; Harada, I.; Nakajima, T.; Kawagoe, T. Vibrational spectra and structure of polyaniline. *Macromolecules* **1988**, *21*, 1297–1305. [CrossRef]
23. Trchov, M.; Sedenkov, I.; Tobolkov, E.; Stejskal, J. FTIR spectroscopic and conductivity study of the thermal degradation of polyaniline films. *Polym. Degrad. Stab.* **2004**, *86*, 179–185. [CrossRef]
24. Gharahcheshmeh, M.H.; Gleason, K.K. Device fabrication based on oxidative chemical vapor deposition (oCVD) Synthesis of conducting polymers and related conjugated organic materials. *Adv. Mater. Interfaces* **2019**, *6*, 1801564. [CrossRef]
25. Trchova, M.; Stejskal, J. Polyaniline: The infrared spectroscopy of conducting polymer nanotubes (IUPAC Technical Report). *Pure Appl. Chem.* **2011**, *83*, 1803–1817. [CrossRef]
26. Cong, H.P.; Ren, X.C.; Wang, P.; Yu, S.H. Flexible graphene–polyaniline composite paper for high-performance supercapacitor. *Energy Environ. Sci.* **2013**, *6*, 1185–1191. [CrossRef]
27. Al-Mashat, L.; Shin, K.; Kalantar-zadeh, K.; Plessis, J.D.; Han, S.H.; Kojima, R.W.; Wlodarski, W. Graphene/polyaniline nanocomposite for hydrogen sensing. *J. Phys. Chem. C* **2010**, *114*, 16168–16173. [CrossRef]
28. National Association of Corrosion Engineers (NACE). *NACE Standard Practice SP0169: Control of External Corrosion on Underground or Submerged Metallic Piping Systems*; NACE: Houston, TX, USA, 2013.
29. Ohtsuka, T. Corrosion protection of steels by conducting polymer coating. *Int. J. Corros.* **2012**, *907*, 2012. [CrossRef]
30. Kowalski, D.; Ueda, M.; Ohtsuka, T. Corrosion protection of steel by bi-layered polypyrrole doped with molybdophosphate and naphthalenedisulfonate anions. *Corros. Sci.* **2006**, *49*, 1635–1644. [CrossRef]
31. Mrad, M.; Amor, Y.B.; Dhouibi, L.; Montemor, F. Electrochemical study of polyaniline coating electropolymerized onto AA2024-T3 aluminium alloy: Physical properties and anticorrosion performance. *Synth. Met.* **2017**, *234*, 145–153. [CrossRef]
32. Njoku, D.I.; Cui, M.M.; Xiao, H.G.; Shang, B.H.; Li, Y. Understanding the anticorrosive protective mechanisms of modified epoxy coatings with improved barrier, active and self-healing functionalities: EIS and spectroscopic techniques. *Sci. Rep.* **2017**, *7*, 15597. [CrossRef] [PubMed]
33. Martins, J.I.; Reis, T.C.; Bazzaoui, M.; Bazzaoui, E.A.; Martins, L.I. The effect of pH on the pyrrole electropolymerization on iron in malate aqueous solutions. *Prog. Org. Coat.* **2009**, *65*, 62–70. [CrossRef]
34. Margarit, I.C.P.; Mattos, O.R. About coatings and cathodic protection: Properties of the coatings influencing delamination and cathodic protection criteria. *Electrochim. Acta* **1998**, *44*, 363–371. [CrossRef]

35. Magalhaes, A.A.O.; Margarit, I.C.P.; Mattos, O.R. Electrochemical characterization of chromate coatings on galvanized steel. *Electrochim. Acta* **1997**, *44*, 4281–4287. [CrossRef]
36. Sheng, X.; Ting, Y.; Pehkonen, S. The influence of sulphate-reducing bacteria biofilm on the corrosion of stainless steel AISI 316. *Corros. Sci.* **2007**, *49*, 2159–2176. [CrossRef]
37. Zamanizadeh, H.R.; Shishesaz, M.R.; Danaee, I. Investigation of the corrosion protection behavior of natural montmorillonite clay/bitumen nanocomposite coatings. *Prog. Org. Coat.* **2015**, *78*, 256–260. [CrossRef]

polymers

MDPI

Article

Graphene Oxide–Platinum Nanoparticle Nanocomposites: A Suitable Biocompatible Therapeutic Agent for Prostate Cancer

Sangiliyandi Gurunathan *, Muniyandi Jeyaraj, Min-Hee Kang and Jin-Hoi Kim *

Department of Stem Cell and Regenerative Biotechnology, Konkuk University, Seoul 05029, Korea; muniyandij@yahoo.com (M.J.); pocachippo@gmail.com (M.-H.K.)

* Correspondence: gsangiliyandi@yahoo.com (S.G.); jhkim541@konkuk.ac.kr (J.-H.K.); Tel.: +82-2-450-0581 (S.G.)

Received: 5 March 2019; Accepted: 16 April 2019; Published: 23 April 2019

Abstract: Metal nanoparticles and the combination of metal nanoparticles with graphene oxide are widely used in environmental, agriculture, textile, and therapeutic applications. The effect of graphene oxide–green platinum nanoparticles (GO-PtNPs) on human prostate cancer cells (LNCaP) is unclear. Therefore, this study aimed to synthesize a nanocomposite of GO-PtNPs and evaluate their effect on prostate cancer cells. Herein, we synthesized GO-PtNPs using vanillin and characterized GO-PtNPs. GO-PtNP cytotoxicity in LNCaP cells was demonstrated by measuring cell viability and proliferation. Both decreased in a dose-dependent manner compared to that by GO or PtNPs alone. GO-PtNP cytotoxicity was confirmed by increased lactate dehydrogenase release and membrane integrity loss. Oxidative stress induced by GO-PtNPs increased malondialdehyde, nitric oxide, and protein carbonyl contents. The effective reactive oxygen species generation impaired the cellular redox balance and eventually impaired mitochondria by decreasing the membrane potential and ATP level. The cytotoxicity to LNCaP cells was correlated with increased expression of proapoptotic genes (p53, p21, Bax, Bak, caspase 9, and caspase 3) and decreased levels of antiapoptotic genes (Bcl2 and Bcl-xl). Activation of the key regulators p53 and p21 inhibited the cyclin-dependent kinases Cdk2 and Cdk4, suggesting that p53 and p21 activation in GO-PtNP-treated cells caused genotoxic stress and apoptosis. The increased expression of genes involved in cell cycle arrest and DNA damage and repair, and increased levels of 8-oxo-deoxyguanosine and 8-oxoguanine suggested that GO-PtNPs potentially induce oxidative damage to DNA. Thus, GO-PtNPs are both cytotoxic and genotoxic. LNCaP cells appear to be more susceptible to GO-PtNPs than to GO or PtNPs. Therefore, GO-PtNPs have potential as an alternate and effective cancer therapeutic agent. Finally, this work shows that the combination of graphene oxide with platinum nanoparticles opens new perspectives in cancer therapy. However further detailed mechanistic studies are required to elucidate the molecular mechanism of GO-PtNPs induced cytotoxicity in prostate cancer.

Keywords: graphene oxide–platinum nanoparticles nanocomposites; prostate cancer; cytotoxicity; oxidative stress; mitochondrial membrane potential; DNA damage

1. Introduction

The mortality rate of cancer is increased rapidly by both aging and growth of the population and is also associated with socioeconomic development [1]. According to the International Agency for Research on Cancer, it is estimated 18.1 million new cancer cases and 9.6 million cancer deaths in 2018. Prostate cancer is the second most common cancer in men and fourth most commonly occurring cancer overall and most frequently diagnosed cancer, which is occurs one in nine in older man [1,2]. Prostate cancer can be treated by conventional therapies such as radiation, chemo, hormone, cryo and surgery;

however, the treatment is depends on individual cases. Chemo-drugs, such as docetaxel, cabazitaxel, mitoxantrone and estramustine, are used to treat prostate cancer, however they cause undesired side effects. Another major cause of radiotherapy failure is the radioresistance of certain cancers. Therefore, finding alternative, biocompatible treatments is necessary.

Nanomedicine has been proposed as a new tool and alternative for cancer therapy and diagnosis. Recently, biomolecule mediated synthesis of metal nanoparticles shows great interest and rapidly used both academic and medical industry aspects due to their unique properties and promising applications as catalysts, ferrofluids, and semiconductors [3,4]. Several metal nanoparticles, such as silver, gold, and palladium, were synthesized using biomolecules, such as bacterial extracts, fungi extracts, plant extracts, and purified phenolic compounds, and tested for cytotoxicity against various types of cancer cells including human breast cancer cells, lung cancer cells, human ovarian cancer cells, and neuroblastoma cancer cells [5–8]. Small metallic nanoparticles seem to be potential nanodrugs to optimize the performances of radiotherapy. Among other nanomaterial systems, platinum nanoparticles (PtNPs) with radiation are used as radiation dose enhancers and anticancer drug carriers in cancer therapy. For example, Porcel et al. [9] developed a new strategy based on the combination of platinum nanoparticles with irradiation by fast ions effectively used in hadron therapy. The results demonstrated that PtNPs strongly enhance lethal damage in DNA, with an efficiency factor close to 2 for double-strand breaks. Platinum complexes and platinum NPs (PtNPs) have shown excellent properties to amplify radiation effects [9–11].

Since several years, platinum-based drug molecules have received much attention due to their electro-catalytic properties. For instance, platinum-based therapeutic drugs, notably cisplatin and carboplatin, have been exploited in chemotherapy to kill cancer cells [12]. However, these drugs do not have specificity towards cancer cells and have effects on normal cells leading to substantial dose-limiting acute and chronic toxicities. Since undesired toxic side effects and frequent development of drug resistance represent the major challenges in cancer therapy, it is therefore necessary to develop cisplatin analogs or other metal complexes that are able to offer a more acceptable level of toxicity and improved antitumor activity [13].

Graphene-based nanocomposites with metal nanoparticles show immense interest due to their extraordinary physical, chemical, and biological properties [14,15]. The excellent properties of combination of graphene based materials and metal nanoparticles shows promising nanomaterial in many fields of application such as electronic-devices, sensors, nanocomposites, energy storage, and supercapacitors [16–19]. Metal nanoparticles, such as silver and graphene oxide, exhibit significant antibacterial activity against Gram-negative and -positive bacteria and anticancer activity against human ovarian cancer cells and human neuroblastoma cancer cells [8,20,21]. In particular, PtNPs are highly important metallic catalysts for many electrochemical reactions and used as sensors towards biological and drug molecules [22,23]. The combination of rGO and PtNPs exhibits promising electrocatalytic activity and selectivity in the detection of target molecules [24]. The major advantage of using graphene oxide for the preparation of nanocomposites, such as graphene oxide–platinum nanoparticles, is a cheap and accessible nanomaterial with abundant oxygen-containing functional groups, which is indispensable to anchor novel metal ions [25,26]. Furthermore, Pt ions are easily absorbed uniformly by GO due to the presence of abundant hydroxyl and carboxyl functional groups. Wu et al. [27] fabricated reduced graphene oxide (RGO)/metal (oxide) composites using glucose as the reducing agent and the stabilizer. The developed composite nanomaterials show excellent electrode catalyst to simultaneous electrochemical analysis of L-ascorbic acid, dopamine, and uric acid. Ali et al. [28] synthesized variety of composites by simultaneous reduction of variety of nanoparticles, such as palladium, platinum, silver, and gold with graphene oxide using a black pepper extract (BPE) for quantification and kinetic analysis of epidermal growth factor receptor (ErbB2), for application to breast cancer diagnostics. The developed composites exhibited less toxic, biocompatible, and antioxidants properties, and can detect low concentrations of ErbB2. A nanocomposite consisting of combination of various nanomaterials such as reduced graphene oxide combined with manganese-doped zinc sulfide quantum dots and functionalized with folic acid (FA-rGO/ZnS:Mn) and loaded with doxorubicin (DOX). DOX is adsorbed on the surface of graphene sheets and releases efficiently with specificity, against folate-rich breast cancer cells, which is the best platform for targeted cancer treatment [29]. Single-crystal metastable manganese sulfide nanowires (γ-MnS NWs) coated with graphitic carbon exhibited brittle with a Young's modulus of 65 Gpa show the cycling stability of stable microsized α-MnS, with an initial capacity of 1036 mAh g^{-1} and a reversible capacity exceeding 503 mAh g^{-1} after 25 cycles [30]. A novel regorafenib (REG) electrochemical sensor, developed by reduction of a zirconia-nanoparticle and reduced graphene oxide composite (ZrO_2/rGO/GCE) using hydrazine hydrate as reducing and stabilizing agent, shows an excellent electrocatalytic response and determination of REG in the presence of ascorbic acid and uric acid at lower concentration in serum samples [31]. The PtNPs were found to cause DNA strand breaks in a concentration-, time-, and size-dependent manner [32], and inhibit DNA replication, whereas the rich oxygen-containing functional groups of graphene oxide on the surface provide it with the opportunity to be modified by many functional molecules to expand biological applications and reduce toxicity. Combination of graphene oxide with nanoparticles in a single platform could provide the simultaneous administration of two or more active agents that are known to disrupt multiple targets, resulting in a more efficient solution to cancer or any other treatments with two different physical and chemical properties. The combination of graphene oxide and platinum nanoparticles could provide efficient synergistic effect on cytotoxicity compared to its counterpart.

Although several studies have reported that the effect of combination of graphene oxide with various metal nanoparticles, such as silver, gold, palladium, etc., against various type of cancer cells, so far there is no report about the anticancer effect of graphene oxide–platinum nanoparticles nanocomposites. This is the first study aimed to address the following objectives including synthesis of graphene oxide–platinum nanoparticles nanocomposite using vanillin as reducing agent. The second objective is to address the cytotoxic effect of graphene oxide–platinum nanoparticles nanocomposite in prostate cancer cells. The final objective is to address the mechanism of anticancer effect of graphene oxide–platinum nanoparticles nanocomposite in prostate cancer cells using various cellular assays.

2. Materials and Methods

2.1. Synthesis and Characterization of GO, PtNPs, and GO-PtNPs

Graphene sheets were synthesized by Hummers' method with slight modification as required [8,33]. Synthesis and characterization of the GO-PtNP nanocomposite was done as described previously [20]. Detailed materials and methods are given in the Supplementary Materials.

2.2. Cell Viability and Cell Proliferation Assay

The cells were grown to logarithmic growth phase and mixed with various concentrations of GO, GO-PtNPs, and PtNPs for 24 h, followed by determinations of cytotoxicity. The inhibitory concentration (IC50) is defined as the concentration of PtNPs causing 50% inhibition of growth of the cells. Cell proliferation was determined using bromodeoxyuridine/5-bromo-2′-deoxyuridine (BrdU) (Roche, Basel, Switzerland).

2.3. Measurement of Cytotoxicity and Cell Mortality

The membrane integrity of LNCaP cells was evaluated using an LDH Cytotoxicity Detection Kit (Sigma-Aldrich, St. Louis, MS, USA) according to the manufacturer's instructions. Cell mortality was evaluated using the trypan blue assay as described previously [7].

2.4. Determination of Reactive Oxygen Species (ROS), Malondialdehyde (MDA), Nitric Oxide (NO), and Carbonylated Protein Levels

Reactive oxygen species (ROS) were estimated as described previously [34]. Briefly, LNCaP cells were seeded into wells of 24-well-plates at a density of 5×10^4 cells per well and cultured for 24 h. MDA levels were determined using a thiobarbituric acid reactive substances assay as previously described with suitable modifications [34]. NO production was quantified spectrophotometrically using Griess reagent (Sigma-Aldrich, St. Louis, MO, USA). Absorbance was measured at 540 nm and nitrite concentration was determined using a calibration curve prepared with sodium nitrite as the standard [35]. Carbonylated protein content was measured according to Uehara and Rao [36].

2.5. Measurement of Mitochondrial Membrane Potential (MMP) and ATP Level

MMP was measured according to the manufacturer's instructions (Molecular Probes, Eugene, OR, USA) using the cationic fluorescent indicator, JC-1 (Molecular Probes). The ATP level was measured according to the manufacturer's instructions (Catalog Number MAK135; Sigma-Aldrich, St. Louis, MO, USA).

2.6. Measurement of Antioxidative Marker Levels

The expression levels of oxidative and antioxidative stress markers were measured as described previously [37].

2.7. Measurement of 8-oxo-dG and 8-oxo-G

8-oxo-dG was determined as described previously [38] and using the manufacturer's instructions (Trevigen, Gaithersburg, MD, USA).

2.8. Statistical Analysis

All assays were conducted in triplicate, and each experiment was repeated at least three times. The results are presented as the means ± standard deviation. All experimental data were compared using Student's *t*-test. A p-value < 0.05 was considered statistically significant. Results are expressed as mean ± standard deviation of three independent experiments. There was a significant difference in treated cells compared to untreated cells with Student's *t*-test (* $p < 0.05$).

3. Results and Discussion

3.1. Synthesis and Characterization of GO and GO-PtNP by UV-visible Spectroscopy

The ultraviolet–visible spectrum of synthesized GO particles exhibited two characteristic absorption peaks at 230 nm, which can be attributed to the π–π^* transition of aromatic C=C bonds, and a shoulder at 300 nm, corresponding to the n–π^* transition of C=O bonds [39]. The hydrophilic property of the oxygenated graphene layers imparts significant solubility and stability in water. The absorption peak for GO-PtNPs was red-shifted to 267 nm (Figure 1A,B), owing to the restoration of sp^2 carbon atoms. This characteristic red-shift is considered a monitoring tool for the graphene–platinum nanoparticle nanocomposite [8,20].

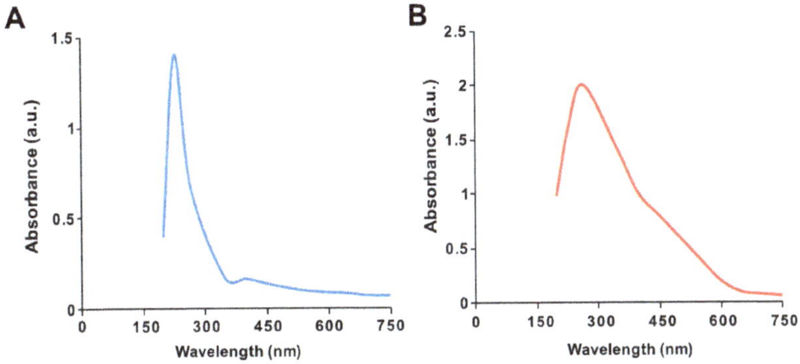

Figure 1. Synthesis and characterization of graphene oxide (GO) and graphene oxide–green platinum nanoparticles (GO-PtNPs). Ultraviolet–visible spectroscopy of GO (**A**) and GO-PtNPs (**B**). At least three independent experiments were performed for each sample and reproducible results were obtained.

3.2. FTIR Analysis of GO and GO-PtNPs

The synthesis of GO from native graphite and its decoration with PtNPs were analyzed by Fourier-transform infrared (FTIR) spectroscopy. The FTIR spectra of GO and the GO-PtNP composite are shown in Figure 2A,B. The spectrum of GO (Figure 2A) showed a strong and broad band at 3300 cm^{-1} due to the –OH stretching vibration. The carbonyl (–C=O) stretching of carboxylic groups present at the edge planes of the GO sheets was observed at 1730 cm^{-1}. The absorption due to –OH bending, epoxide groups, and skeletal ring vibrations were observed at 1600 cm^{-1}. After decoration of PtNPs on the surface of GO, the –OH stretching vibration and carbonyl (–C=O) stretching of carboxylic groups were shifted to 3320 and 1725 cm^{-1}, respectively. Interestingly, the deformation stretching frequency of –OH groups attached to the aromatic ring was 1380 cm^{-1} [40]. The peaks were observed in the spectrum of GO-PtNPs at 1725 and 1650 cm^{-1} corresponding to C=O stretching vibrations of COOH groups, which were attributed to C=O bonds in the carboxylic acid and carbonyl moieties, respectively (Figure 2B), and another strong peak appears at 1150 indicating C–OH stretching. All these data confirmed the formation of GO from native graphite, generation of oxygen-containing functionalities during oxidation process, and decoration of PtNPs on the surface of GO. These observations agreed with those reported in the literature [41,42]. The collect data suggested that the vanillin, aphenolic compound is responsible for synthesis of PtNPs and decoration of PtNPs on the surface of GO.

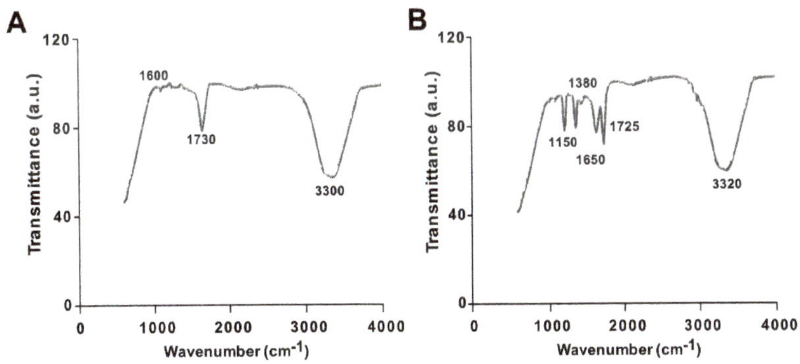

Figure 2. Characterization of GO and GO-PtNPs by Fourier-transform infrared spectroscopy (FTIR). FTIR images of GO (**A**) and GO-PtNPs (**B**). At least three independent experiments were performed for each sample and reproducible results were obtained.

3.3. X-Ray Diffraction Analysis of GO and GO-PtNPs

X-ray diffraction (XRD) was performed to confirm the formation structures of GO and GO-PtNPs. Figure 3A,B display the XRD patterns of GO and GO-PtNPs. The diffraction peak of GO was observed at 11.5, corresponding to the (200) plane and an interlayer distance of 0.76 nm [19,38]. The newly appeared diffraction peaks located at 39.8, 46.5, 55.0, and 70.6 corresponded to the (111), (200), (220), and (311) crystal planes of Pt, respectively (JCPDS No. 01-087-0646), demonstrating that PtNPs were decorated uniformly on the GO surface [19,43], and confirming the presence of platinum particles on the graphene substrate. The average size of the PtNPs was calculated to be 2 nm using the Scherrer equation based on the full width at half maximum of the Pt (111) diffraction peak. The findings strongly agreed with those of previously published reports [19,42,43]. After chemical oxidation, the (002) peak of graphite was shifted to 11.5° with a d spacing of 0.94 nm. This shift might be attributed to the introduction of oxygen-containing functional groups like epoxy, hydroxyl, carbonyl, and carboxylic groups at both the sides and edges of the graphene sheets. This confirmed the formation of GO from natural graphite during the oxidation process and the formation of platinum particles [42].

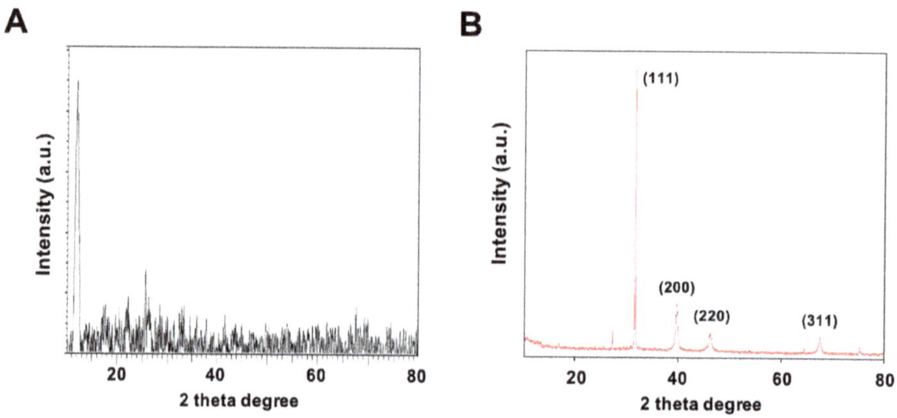

Figure 3. Characterization of GO and GO-PtNPs by XRD. XRD images of GO (**A**) and GO-PtNPs (**B**). At least three independent experiments were performed for each sample and reproducible results were obtained.

3.4. Raman Spectroscopy Analysis of GO and GO-PtNPs

Raman spectroscopy was used to investigate the structure of GO and GO-PtNPs. The main features in the Raman spectra of graphitic carbon-based materials are the G and D peaks and their overtones [44]. Figure 4A,B shows the Raman spectra of the prepared GO and GO-PtNPs. The two most intense peaks were the D and G band at 1360 and 1580 cm^{-1}, respectively. A prominent 2D band at 2690 cm^{-1} was evident, and a defect-activated peak (D + G) was also visible at approximately 2950 cm^{-1}. The D peak represents first order resonance and the breathing mode of aromatic rings arising due to the defects in the sample. The D peak intensity is used to measure of the degree of disorder [45]. The G peak is due to the bond stretching of all pairs of sp^2 atoms in both the rings and chains and corresponds to the optical E2g phonons. The intense D peak along with a large bandwidth suggested the significant structural disorder in GO. The 2D peak at approximately 2690 cm^{-1} is attributed to double resonance transitions resulting in the production of two phonons with opposite momentum. Simultaneously, the ratio of the intensity of the D band to the G band (ID/IG) increased from 1.49 to 1.89 between GO and PtNPs decorated GO.

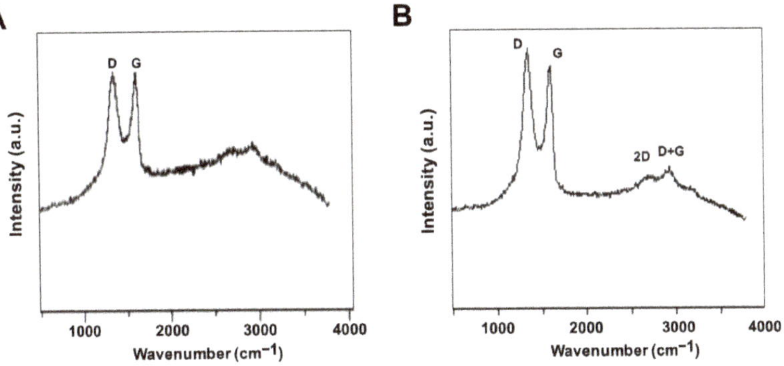

Figure 4. Characterization of GO and GO-PtNPs by Raman spectroscopy. Raman spectroscopy images of GO (**A**) and GO-PtNPs (**B**). At least three independent experiments were performed for each sample and reproducible results were obtained.

3.5. Morphology and Size Analysis of GO and GO-PtNPs using SEM and TEM

The morphology of GO and GO-PtNPs was analyzed by scanning electron microscopy. Micrographs of GO (Figure 5A) revealed a two-dimensional sheet-like structure consisting of multiple lamellar layers. The edges of individual sheets were visible, as has been described [8,46,47]. GO-PtNPs images (Figure 5B) revealed that the PtNPs particles were uniformly distributed throughout the graphene layers without any agglomeration in the substrate on the surface of GO, which agreed with previously published reports of other nanoparticles, such as silver [8]. Field emission scanning electron microscopy images revealed well-decorated PtNPs on the surface of GO nanosheets. Smaller PtNPs were spherical in shape, whereas the bigger particles had an elongated form. This elongated shape was attributed to an agglomeration of the highly concentrated PtNPs. The size of GO and PtNPs decorated GO were characterized by transmission electron microscopy. As shown in Figure 5C,D, transparent and wrinkled layers of GO nanosheets were observed. After successful decoration, PtNPs decorated on the graphene oxide layers were spherical with a diameter < 2 nm, which is consistent with the XRD results, confirming the formation of GO-PtNP nanocomposites. The results indicated that the combination of graphene oxide and PtNPs suggest that graphene oxide facilitates the reshaping and coarsening of PtNPs during simultaneous reduction of graphene oxide and PtNPs. The reducing and stabilizing agents promoted biomolecule induced transformations of nanohybrids (Supplementary Figure S1).

3.6. Effect of GO-PtNPs on Viability of LNCaP Cells

The effect of GO-PtNPs on LNCaP androgen-sensitive human prostate adenocarcinoma cells was explored to evaluate the approach in the treatment of prostate cancer. To evaluate and optimize the dose, and explore a dose-dependent effect, LNCaP cells were treated with various concentrations of GO (20–100 μg/mL), GO-PtNPs (5–25 μg/mL), and PtNPs (10–50 μg/mL) for 24 h. The three compounds displayed dose-dependent cytotoxicity on LNCaP cells. Interestingly, GO-PtNPs showed effective responses on cell viability compared to control, parental GO, and PtNPs. GO-PtNPs nanocomposites exhibited significant cytotoxicity against the human prostate cancer cells and the cytotoxicity was greater as the concentration increased. GO did not affect cell viability as drastically. Even at very high concentration (100 μg/mL), the loss of viability was only 60%. This low cytotoxicity of GO was probably attributed to the enrichment of oxygen atoms on the surface of GO in the form of carboxyl, epoxy, and hydroxyl groups, which reduces cell toxicity [48]. The loss of viability of PtNPs was dose-dependent and slightly better than GO, and less than GO-PtNPs. Even at a very high concentration (50 μg/mL), the loss of viability was only 80%, whereas at a high concentration of GO-PtNPs (25 μg/mL), the loss of viability was 99%. The findings indicated that the GO-PtNPs nanocomposite had enhanced anticancer capability, in contrast to the relatively low toxicity of GO and PtNPs on LNCaP cells. Collectively, the data indicated a dose-dependent inhibition of the cell viability with GO in the range of 20 to 100 µg/mL, with an IC50 of approximately 80 µg/mL (Figure 6A). Using GO-PtNPs, a dose-dependent inhibition of the cell viability was observed in the range of 5 to 25 µg/mL with an IC50 of approximately 10 µg/mL (Figure 6B). With PtNPs, a dose-dependent inhibition of the cell viability was observed in the range of 10–50 µg/mL with an IC50 of approximately 30 µg/mL (Figure 6C). Thus, GO-PtNPs displayed a more pronounced inhibitory effect on cell viability than the other tested nanomaterials and represent a promising candidate for treatment of prostate cancer cells. Similarly, GO-AgNPs nanocomposites effectively inhibit cell viability of a variety of bacteria, human ovarian cancer cells, and cervical cancer cells [8,20,49–51]. Further, the cell morphology of GO, GO-PtNPs, and PtNPs was examined phase-contrast microscope, all the treated cells compromised cell structure, and cells were round in shape and all the dead cells were detached from surface. The effect on cell viability was clearly dose-dependent (Supplementary Figure S2).

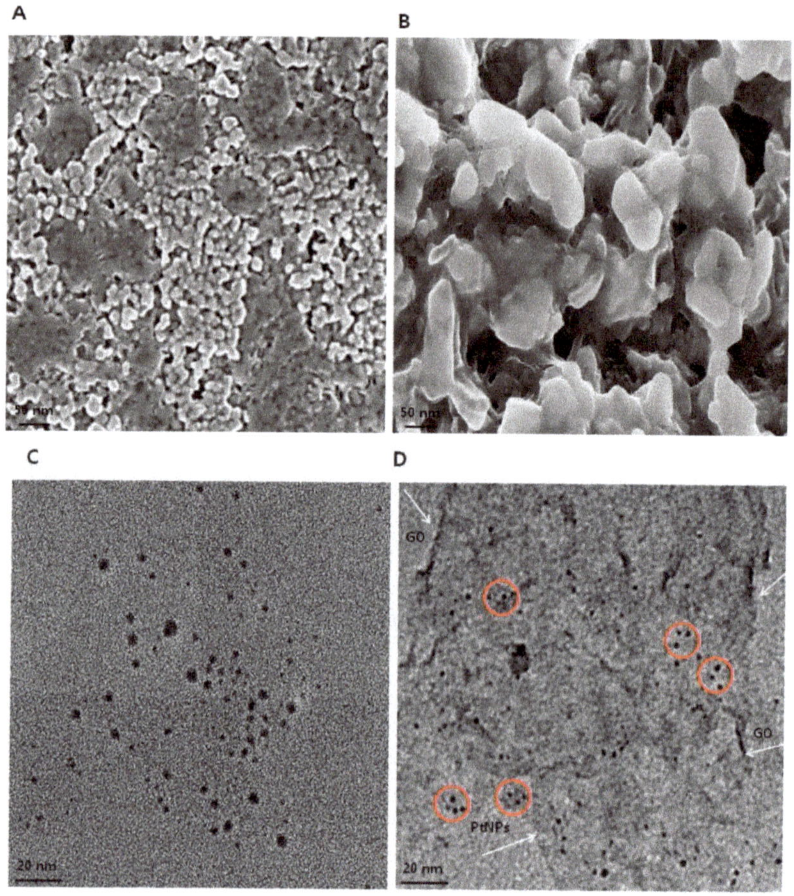

Figure 5. Characterization of GO and GO-PtNPs by SEM and TEM. Morphology of GO (**A**) and GO-PtNPs (**B**), and size of GO (**C**) and GO-PtNPs (**D**) were analyzed by SEM and TEM, respectively. The red circle indicates decoration of PtNPs particles on the surface of graphene sheet (White arrow). The graphene sheet depicted as wrinkled sheet-like structure.

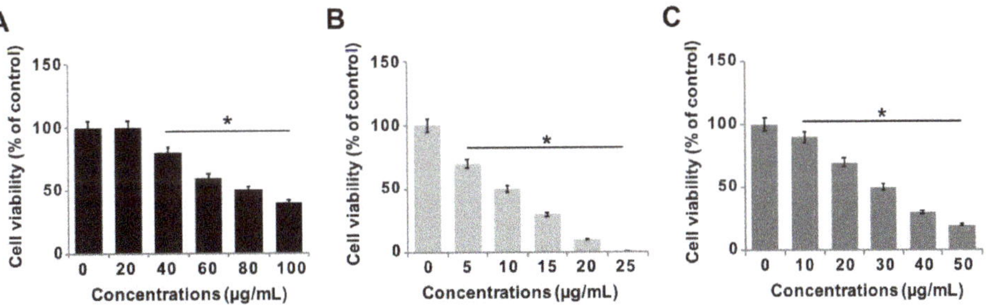

Figure 6. GO, GO-PtNPs, and PtNPs inhibit cell viability of LNCaP cells. The viability of LNCaP cells was determined after 24-h exposure to different concentrations of GO (**A**), GO-PtNPs (**B**), and PtNPs (**C**) using CCK-8. The treated groups showed statistically significant differences from the control group by the Student's *t*-test (* $p < 0.05$).

3.7. GO-PtNPs Inhibit Proliferation of LNCaP Cells

To determine the antiproliferative action of GO-PtNPs on LNCaP cells, the cells were treated with various concentrations of GO, GO-PtNPs, and PtNPs for 24 h, and proliferation was determined using BrdU. The growth rates of prostate cancer cells treated with GO-PtNPs were significantly decreased compared to that of the control, and similar results were observed in LNCaP cells with GO and PtNPs treatment. However, the effective inhibition of proliferation was observed with GO-PtNPs compared to control, GO, and PtNPs (Figure 7A–C). Of note, GO-PtNPs exhibited higher antiproliferative action, which was comparable to the parental GO and PtNPs. The data suggested that GO-PtNPs suppress proliferation. Similarly, goserelin-loaded nanoparticles influence the growth of LNCaP prostate cancer cells by the direct induction of necrosis and apoptosis [52].

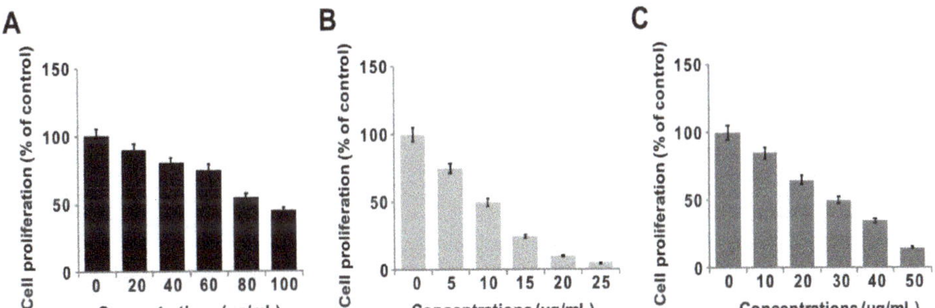

Figure 7. GO, GO-PtNPs, and PtNPs inhibit cell proliferation of LNCaP cells. Cell proliferation of LNCaP cells was determined using BrdU assay after 24-h exposure to different concentrations of GO (**A**), GO-PtNPs (**B**), and PtNPs (**C**). The treated groups showed statistically significant differences from the control group by the Student's *t*-test (* $p < 0.05$).

3.8. GO-PtNPs Induce Cytotoxicity in LNCaP Cells

Membrane integrity determines the fate of cells and is an important factor for cell survival. Membrane integrity can be estimated by measuring leakage of LDH, which is a cytosolic enzyme that aids in the conversion of lactate to pyruvate. When membrane integrity is compromised, the enzyme is secreted. To determine the leakage of LDH, LNCaP cells were treated with IC50 concentrations of GO, GO-PtNPs, and PtNPs for 24 h. GO-PtNPs comparatively significantly increased the leakage of LDH compared to that in the control cells (Figure 8A). GO and PtNPs also increased the leakage of LDH to a greater extent than that in the control group. However, leakage was lower than that resulting from treatment with GO-PtNPs, indicating that the combination of GO and PtNPs acted synergistically to induce cytotoxicity. GO produces moderate leakage of LDH in a variety of cancer cells including human breast cancer cells [47], human ovarian cancer cells [8], and human cervical cancer cells [51]. One study [53] reported that human lung cancer cells treated with silver and platinum nanoparticles released significantly greater amounts of LDH compared to the control. The present and previous data suggest a correlation between cell viability and damage to the membrane caused by the nanoparticles, which ultimately proves lethal [54]. Membrane integrity and cell survival was examined by the trypan blue exclusion assay. Normal healthy cells are able to exclude the dye, but trypan blue will diffuse into cells in which membrane integrity has been lost. LNCaP cells were treated with IC50 concentrations of GO, GO-PtNPs, and PtNPs for 24 h. Significant cytotoxic effect was consistently observed in LNCaP cells. Cytotoxicity was most pronounced for cells treated with GO-PtNPs compared to that for treatments with GO and PtNPs (Figure 8B).

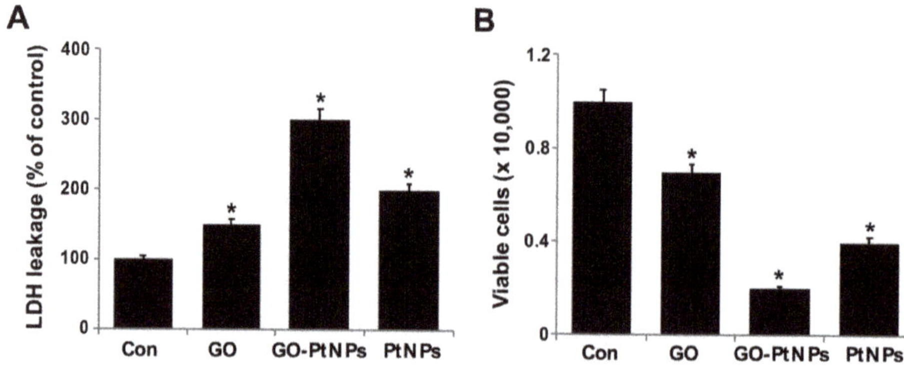

Figure 8. GO, GO-PtNPs, and PtNPs increase the leakage of LDH and cell death. LNCaP cells were treated with respective IC50 concentrations of GO, GO-PtNPs, and PtNPs for 24 h, and the LDH activity was measured at 490 nm using the LDH cytotoxicity kit (**A**). Cell death was determined by trypan blue assay after 24 h of exposure to GO, GO-PtNPs, and PtNPs for 24 h. Cell death was quantified by the ratio of living cells (**B**). The treated groups showed statistically significant differences from the control group by the Student's *t*-test (* $p < 0.05$).

3.9. GO-PtNPs Increase the Level of Oxidative Stress Markers

The effect of GO-PtNPs on ROS generation was evaluated using DCFH2-DA. LNCaP cells were treated with IC50 concentration of GO, GO-PtNPs, and PtNPs for 24 h and then cells were exposed to 40 μM DCFH2-DA for 30 min. First, we measured the distribution of the fluorescence intensity in the presence or absence of GO, GO-PtNPs and PtNPs. The treatment of LNCaP cells with GO-PtNPs led to a marked shift to greater fluorescence peak intensities compared to the untreated control (Figure 9). Moreover, ROS generation induced by GO-PtNPs resulted in high FITC fluorescence intensity, indicating an increased susceptibility to oxidative stress. GO-PtNP-treated cells were more than four times more susceptible than the control samples, while cells treated with GO and PtNPs exhibited susceptible than the control sample (Supplementary Figure S3); these results are consistent with the inhibitions of growth and cell proliferation. Subsequently, we evaluated the level of MDA in GO-, GO-PtNP-, and PtNP-treated cells. The MDA level was significantly higher in GO-PtNP-treated cells than either GO or PtNPs treatments (Figure 9B). Lipid peroxidation is a process of oxidation of polyunsaturated fatty acids due to the presence of several double bonds in their structure and it involves production of peroxides, ROS, and other reactive species, such as MDA. MDA is a reactive byproduct of lipid peroxidation and an end product that interacts with DNA to forming 3-(2-deoxy-β-d-erythro-pentofuranosyl)pyrimido[1,2-α]purin-10(3H)-one adducts [55]. Increased lipid peroxidation and decreased GSH were reported in human embryonic kidney (HEK)293 cells exposed to PtNPs [56].

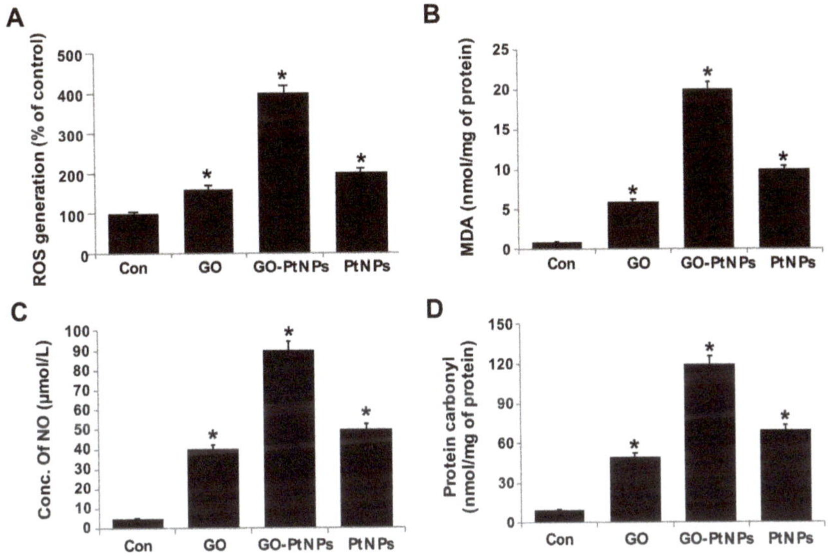

Figure 9. GO, GO-PtNPs, and PtNPs increase ROS generation (**A**), lipid peroxidation (**B**), and nitric oxide (**C**) and carbonylated protein content (**D**) in LNCaP cells. LNCaP cells was exposed to respective IC50 concentrations of GO, GO-PtNPs, and PtNPs for 24 h and then Spectrophotometric analysis of ROS was measured using DCFH-DA (**B**). The concentration of MDA was measured MDA using a thiobarbituric acid reactive substances assay and expressed as nanomoles per milliliter (**C**). NO production was quantified spectrophotometrically using the Griess reagent and expressed as micromoles per milliliter (**D**). Protein carbonyl content was measured and expressed relative to the total protein content. The treated groups showed statistically significant differences from the control group by the Student's *t*-test (* $p < 0.05$).

In general, nanoparticles, such as silver, platinum, and palladium ions, bind to protein disulfide bonds in the cytoplasm, causing deformities in the protein structure. These malformed proteins are then incorporated into the plasma membrane, leading to alterations in cell permeability and cellular death [7,47]. In addition, graphene induces toxicity owing to its distinct physicochemical characteristics such as purity, lateral dimension, size of the sheets, and oxidation state, which may influence its cellular uptake, biodegradation, and toxicity. Once in contact with the cell membrane, graphene sheets can create an impermeable encasement affecting the normal exchange between the cell and the extracellular environment, graphene oxide can also damage the cell membrane through strong electrostatic interactions between the negatively charged oxygen groups on its surface and the positively charged lipids present on cell membranes [57–62].

Nanocytotoxicity may be caused by the induction of oxidative and/or nitro-oxidative stress [63,64]. Therefore, we were interested to determine the fate of GO-PtNPs on the generation of reactive nitrogen species (RNS). LNCaP cells were treated with IC50 concentrations of GO, GO-PtNPs, and PtNPs for 24 h and then the level of nitric oxide (NO) was determined. Cells treated with either GO or PtNPs displayed an 8-fold higher level of NO than control. Interestingly, the level of NO was significantly higher in all tested groups relative to control. The increase in the NO level after 24-h treatment with GO-PtNPs was 18-fold higher than after incubation with the same concentration of GO or PtNPs (Figure 9C). Overproduction of ROS and RNS in cells influences various cellular processes and components, promotes DNA breakage, and impairs the antioxidant potential; it has also been associated with carcinogenesis [65]. An increased level of NO in adenocarcinoma cells treated with silver nanoparticles (AgNPs) alone and in combination with an inhibitor of histone deacetylase (HDAC) was reported [66].

Carbonyl groups appear to be a significant and stable marker of the oxidative stress that results from the oxidation of proteins. Increased oxidative stress during altered homeostasis of prooxidants and antioxidants leads to deleterious effects on cellular components through oxidative damage to proteins, lipids, and nucleic acids. LNCaP cells were treated with IC50 concentrations of GO, GO-PtNPs, and PtNPs for 24 h and then the level of carbonylated protein was determined. Cells treated with GO, PtNPs, and GO-PtNPs showed 5-fold, 8-fold, and 12-fold higher level, respectively, of carbonylated protein than control (Figure 9D).

3.10. GO-PtNPs Decrease MMP

The integrity of the mitochondrial membrane is regulated by its membrane potential, which influences electron transport and oxidative phosphorylation. Alteration of MMP causes cellular apoptosis. Spectrophotometric analysis showed that MMP was significantly compromised in GO-PtNP-treated cells compared with the control, and treatment with GO or PtNPs decreased the MMP, and the decrease was markedly more pronounced using GO-PtNPs. Twenty-four hours of treatment with GO-PtNPs increased the percentage of cells with low MMP ($\Delta\psi m$) compared to that of the untreated cells (Figure 10A). The effect of GO-PtNPs on the MMP of LNCaP cells was evaluated using fluorescence microscopy. LNCaP cells incubated with GO, GO-PtNPs, and PtNPs underwent mitochondrial damage, resulting in changes in the $\Delta\Psi m$. Consequently, the JC-1 aggregate level was significantly decreased by GO-PtNPs compared to GO or PtNPs, which resulted in low FITC fluorescence intensity, indicating an increased susceptibility to oxidative stress (Supplementary Figure S4). Similarly, others [67] reported that a buffalo rat liver cell line treated with silicon oxide nanoparticles displayed increased cytotoxicity and mitochondrial damage accompanied by decreases in mitochondrial dehydrogenase activity, MMP, enzymatic expression in the Krebs cycle, and activity of the mitochondrial respiratory chain complexes I, III, and IV. Similarly, another study [34] reported the decreased level of MMP and ATP in the presence of AgNPs alone and in combination with an HDAC inhibitor.

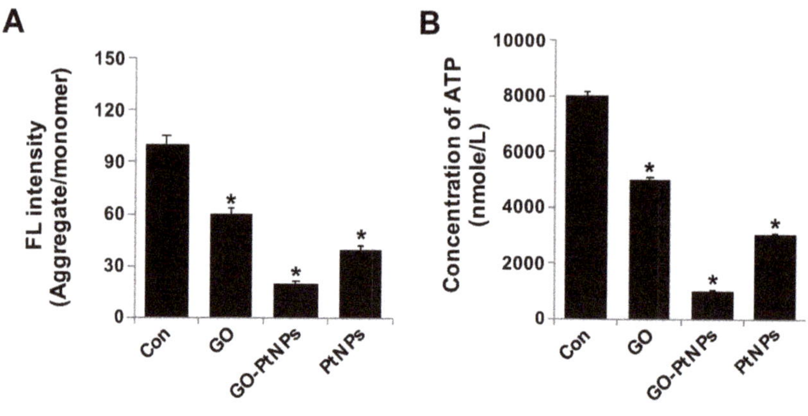

Figure 10. GO, GO-PtNPs, and PtNPs decrease mitochondrial membrane potential and ATP content. LNCaP cells were treated with respective IC50 concentration of GO, GO-PtNPs, and PtNPs for 24 h and spectrophotometric determination of JC-1 monomer/aggregate formation using cationic fluorescent indicator JC-1 (**A**). Intracellular ATP content (**B**). The treated groups showed statistically significant differences from the control group by the Student's *t*-test (* $p < 0.05$).

Next, we examined the impact of GO-PtNPs on ATP synthesis. Synthesis of ATP is highly dependent on the integrity of the mitochondrial membrane, which regulates the pumping of hydrogen ions across the inner membrane during electron transport and oxidative phosphorylation [67]. Therefore, the effect of GO-PtNPs on the ATP synthesis in LNCaP cells was evaluated. As shown in Figure 10B, compared with that in the control, treatment of LNCaP cells with GO, PtNPs, and GO-PtNPs resulted in a decreased level of ATP, but these decreases were highly significant in the latter treatment. These results indicated that ATP synthesis was suppressed by GO-PtNPs. A 24-h treatment with GO-PtNPs resulted in decreased ATP synthesis with low MMP ($\Delta\psi$m), compared to that in the untreated cells. This result suggested that MMP and ATP synthesis are associated.

3.11. GO-PtNPs Impair Antioxidant Systems

The generation of ROS in cells occurs in equilibrium with a wide variety of antioxidant molecules, including SOD, catalase, GPx, and peroxiredoxins, as well as nonenzymatic scavengers, such as vitamin C, vitamin E, GSH, lipoic acid, carotenoids, and iron chelators [68]. Therefore, we evaluated the impact of oxidative stress in GO-PtNP-induced prostate cancer cell death. LNCaP cells were treated with IC50 concentrations of GO, GO-PtNPs, and PtNPs for 24 h and the protein level of selected antioxidant enzymes was determined (GSH, GSH, GSSG, SOD, CAT, GPx, and TRX). A statistically significant reduction in the levels of all enzymes was evident after treatment with GO, PtNPs, and GO-PtNPs. The latter treatment produced the greatest reductions (Figure 11). A previous study reported that AgNPs also reduced the activities of GSH, GSH: GSSG, SOD, CAT, GPx, and TRX [38]. Similarly, a reduced level of SOD was observed in human skin carcinoma and human fibrosarcoma after exposure to 7–20 nm AgNPs [69,70]. Decreased GPX activity in rat pheochromocytoma and mouse neuroblastoma cells by ZrO_2NPs < 100 nm was reportedly related to genotoxic and cytotoxic effects [71]. Others [72] observed that PANC-1 cells treated with 2.6 nm and 18 nm AgNPs displayed decreased levels of SOD1 protein and mRNA, respectively. The lower level of SOD activity in MCF-7 breast cancer cells leads to a drastic alteration in the morphology of the mitochondria associated with increased fragmentation and swelling of the matrix [73]. Consistent with previous results from GO-PtNP-treated LNCaP cells [73], the cell death that occurred following the reduction of SOD level by SOD inhibitor likely occurred through a combination of the regulated mechanism (apoptosis) and unregulated mechanism (oxidative damage to the organelles). The collective data indicate that GO-PtNPs effectively influence the level of antioxidant molecules and eventually compromise the redox balance in LNCaP cells.

3.12. Effect of GO-PtNPs on Expression of Proapoptotic and Antiapoptotic Genes

Reduced ROS is considered an essential regulator of the normal physiological functions of cells. Increased levels of ROS could damage proteins, nucleic acids, lipids, membranes, and organelles, which can lead to activation of cell death processes like apoptosis [74]. ROS play significant roles in the activation of various cellular signaling pathways and transcription factors. One of the most targeted genes with respect to DNA damage are tumor suppressor p53, a gene involved in cell cycle arrest, DNA repair, senescence, and apoptosis [75,76], and p21, a gene involved in cell cycle regulation. Furthermore, ROS induces the activity of genes, such as Bax and Bak, which are involved in mitochondria mediated apoptosis and downregulates Bcl2 and Bcl-xl. To determine the effect of GO-PtNPs on expression of proapoptotic and antiapoptotic genes, cells were treated with IC50 concentrations of GO, GO-PtNPs, and PtNPs for 24 h, and the m-RNA expressions of the target genes were determined. As expected, the levels of p53, p21, Bax, Bak, caspase-9, and caspase-3 were significantly upregulated, whereas the levels of Bcl2 and Bcl-xl were significantly downregulated by up to three-fold (Figure 12). Bcl-2 expression decreased 2.5 times compared to the control. A previous study [77] found that cells treated with PtNPs experienced genotoxic stress due to the activation of p53 and p21, which eventually led to proliferating cell nuclear antigen-mediated growth arrest and apoptosis. Our results were consistent with the recent finding [78] that PtNP-treated DEN animals showed significant increase in liver p53 gene expression level than normal rats. Other [56] reported that exposure of HEK293 cells to PtNPs induced the upregulation of Bax and the downregulation of Bcl2.

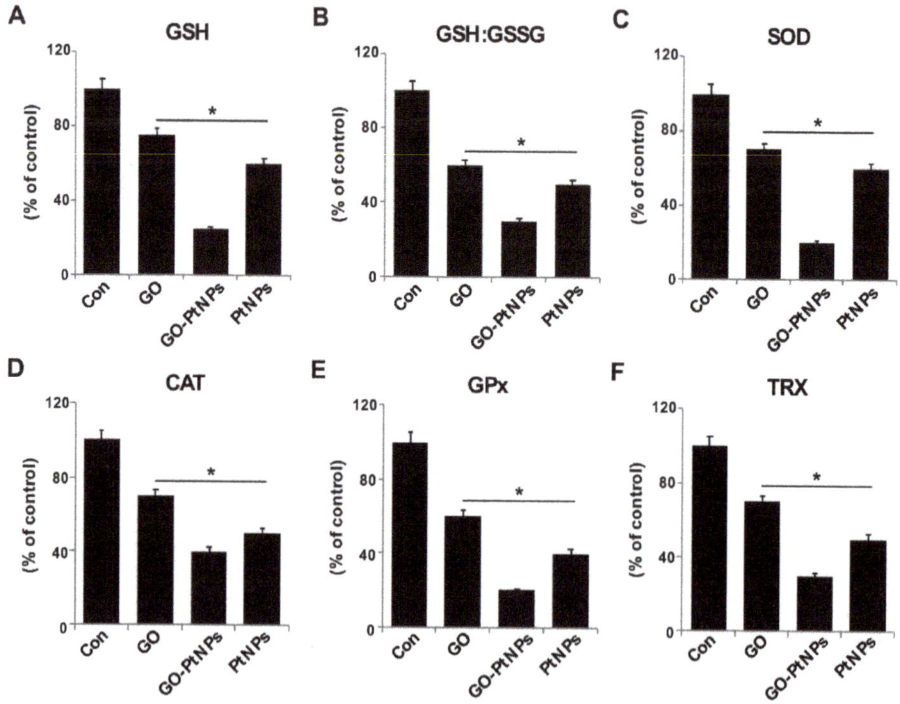

Figure 11. Effect of GO, GO-PtNPs, and PtNPs on antioxidant markers. LNCaP cells were treated with respective IC50 concentration of GO, GO-PtNPs, and PtNPs for 24 h. After incubation, cells were harvested and washed twice with an ice-cold phosphate-buffered saline solution. The cells were collected and disrupted by ultrasonication for 5 min on ice. GSH concentration is expressed as percentage of control (**A**). GSH:GSSG ratio is expressed as percentage of control (**B**). SOD concentration is expressed as percentage of control (**C**). CAT is expressed as percentage of control (**D**). GPx concentration is expressed as percentage of control (**E**). TRX is expressed as percentage of control (**F**). There was a significant difference in treated cells compared to untreated cells with Student's *t*-test (* $p < 0.05$).

The findings confirm that PtNPs induce mitochondria-mediated apoptosis, which is primarily responsible for cisplatin production and ROS generation in cell organelles [79]. Caspases play a significant role in apoptosis. We found that the LNCaP cells treated with IC50 concentrations of GO, GO-PtNPs, and PtNPs for 24 h displayed significantly increased caspase-9 and caspase-3 activities. Caspase-9 plays a significant role in signal transduction by induction of the executioner caspases-3 and -7 [38,80]. Similarly, PtNPs induced caspase-3 activity in a dose-dependent manner in HEK293 cells [56]. Cisplatin-induced DNA damage is related to the ratio of proapoptotic and antiapoptotic proteins, and release of cytochrome c from mitochondria followed by the activation of cysteine caspases selectively degrades the target proteins [81]. The collective data indicate that GO-PtNPs induce the intrinsic pathway of apoptosis, which is mediated by the upregulation of proapoptotic genes and the downregulation of antiapoptotic genes.

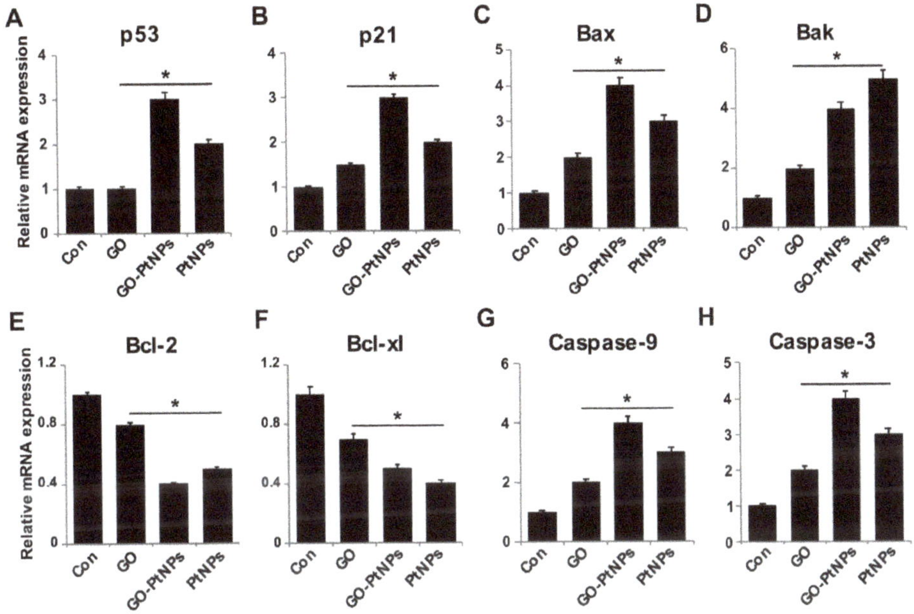

Figure 12. Effect of GO, GO-PtNPs, and PtNPs on the expression of pro- and antiapoptotic genes. LNCaP cells were treated with respective IC50 concentration of GO, GO-PtNPs, and PtNPs for 24 h. The relative messenger RNA(mRNA) expression of P53 (**A**), P21 (**B**), Bax (**C**), Bak (**D**) Bcl-2 (**E**), Bcl-xl (**F**), caspase-9 (**G**), and caspase-3 (**H**) was analyzed by quantitative reverse-transcription polymerase chain reaction in LNCaP cells treated for 24 h. After 24-h treatment, expression fold level was determined as fold changes in reference to expression values against GAPDH. Results are expressed as fold changes. There was a significant difference in treated cells compared to untreated cells with Student's *t*-test (* $p < 0.05$).

3.13. GO-PtNPs Increase the Levels of 8-oxodG and 8-oxo-G by Causing Oxidative Damage to DNA

Overproduction of ROS can induce lipid peroxidation as well as 3-(pyrimido[1,2-α]purin-10(3H)-one and 8-oxodG. Nanoparticles cause several toxic effects, which include chromosomal aberrations, DNA strand breakage, oxidative damage to DNA, and mutations [82,83]. Several assays have been used for the detection of NP-related oxidative damaged to DNA, especially measuring the oxidative level of 8-oxodG. Among several oxidative stress markers, 8-oxodG plays a significant role in DNA damage due to the prevalent oxidative lesions. To measure the levels of 8-oxodG and 8-oxoG, LNCaP cells were treated with IC50 concentrations of GO, GO-PtNPs, and PtNPs for 24 h. Cells treated with GO and PtNPs displayed 4- and 7-fold increases in 8-oxodG and 8-oxoG levels, respectively, compared to those in the control group. Similarly, the cells treated with GO and PtNPs displayed 3-fold and 5-fold increases, respectively, compared to the control group. Cells treated with GO-PtNPs displayed a 12-fold increase compared to the control (Figure 13).

In another study, the exposure of human cells to PtNPs increased DNA damage, accumulation of cells at the S-phase of the cell cycle, and apoptosis [77]. HEK cells exposed to PtNPs reportedly displayed dose-dependent DNA damage that could be severe [56]. In another study [84], Fe_3O_4-NPs induced significant generation of 8-oxodG only at a higher concentration (60 µg/mL) and after more than 24 h of incubation. LNCaP cells exposed to PtNPs are expected to display significantly increased intracellular levels of ROS, which will damage DNA. A recent study suggested that PtNPs inhibit DNA replication and affect the secondary structure of DNA at higher concentrations in human cells and bacteria. In the study, the encapsulation of PtNPs in liposomes (LipoPtNPs) caused approximately

2.4 times higher DNA damage in comparison with CisPt, LipoCisPt, and PtNPs [85]. This was also found in our system, where exposure of LNCaP cells was significantly associated with higher levels of both 8-oxodG and 8-oxoG adducts after 24 h of incubation in comparison to that in the untreated cells. The generation of oxidative DNA lesions seems to be responsible for the induction of apoptotic death in cancer cells.

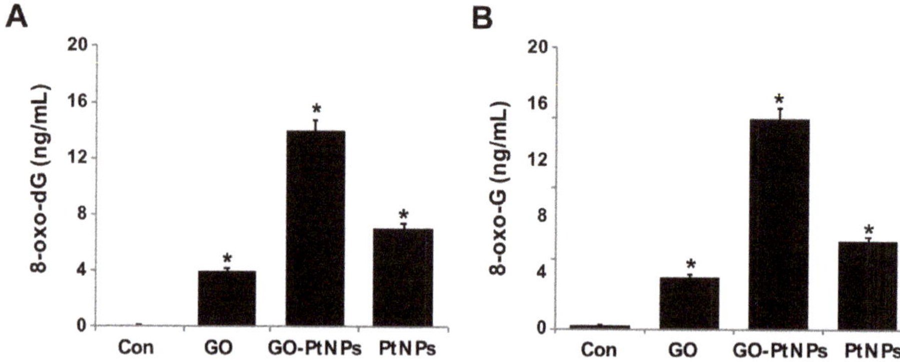

Figure 13. GO, GO-PtNPs, and PtNPs increase DNA damage. LNCaP cells were treated with respective IC50 concentration of GO, GO-PtNPs, and PtNPs for 24 h. 8-oxo-dG and 8-oxo-G were measured after 24 h of exposure of LNCaP cells. There was a significant difference in treated cells compared to untreated cells with Student's *t*-test (* $p < 0.05$).

3.14. Impact of GO-PtNPs on Expression of GENES involved in Cell Cycle and DNA Damage

To further substantiate that oxidative DNA damage induced apoptosis, we evaluated the effect of GO-PtNPs on cell cycle arrest and DNA damage by evaluating the expressions of repair genes, including CDK2, CDK4, GADD45A, OGG1, APEX1, CREB1, UNG, and POLB, by RT-PCR after 24 h exposure to GO, GO-PtNPs, and PtNPs. The genes were significantly upregulated from 1- to 4-fold. CDK2 and CDK4 are tightly regulated by p21, which is controlled by p53. Cdk2 and Cdk4 inhibition by p21 are significant in G_1 arrest upon DNA damage by various stresses and for cellular senescence, and p21 is an essential gene for p53-mediated G1 arrest in human cancer cells [86,87]. p53 is also believed to inhibit Cdk4 activity through p21 and by the repression of Cdk4 synthesis [88,89]. Inhibition of the cyclin-dependent kinases Cdk2 and Cdk4 is initiated by p53 and p21. It was reported [78] that the effect of PtNPs was more potent than that of cisplatin in hepatocellular carcinoma induced in rats. PtNPs inhibit cell proliferation by the induction of apoptotic cell death, which reduces cell viability and causes internucleosomal DNA fragmentation, G2/M cell cycle arrest, and hypodiploid accumulation [90]. The release of platinum ions from PtNPs inhibits cell division by binding to DNA, which causes DNA damage and downregulates the expression of proliferating cell nuclear antigen [77]. Furthermore, the present results suggested that GO, GO-PtNPs, and PtNPs significantly induced the expression of all the tested genes. In particular, the upregulation by GO-PtNPs was 2- to 4-fold higher than that in cells treated with GO or PtNPs (Figure 14).

Graphene quantum dots trigger ROS generation, which triggers upregulation of genes associated with DNA damage [91]. In another study, GO-treated cells showed increased expression of DNA damage genes, including ATM and RAD51 [92]. Thus, GO-PtNPs effectively induce apoptosis compared to GO or PtNPs through inducing oxidative stress in LNCaP cells. These results agree with those of previous studies on the effect of PtNPs on fibroblast, glioblastoma, and A549 lung carcinoma cells [53,77]. Additionally, the increased expressions of all tested genes, 8-oxo-dG, and 8-oxoG is suggestive of oxidative damage to DNA. A hypothetical model demonstrates that the mechanism of GO-PtNPs induced oxidative stress and DNA damage in LnCaP cells (Figure 15).

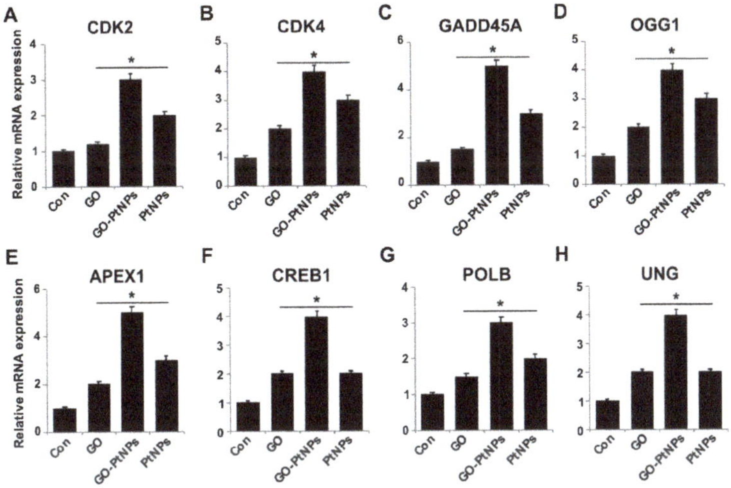

Figure 14. Effect of GO, GO-PtNPs, and PtNPs on expression cell cycle arrest and DNA damage genes. LNCaP cells were treated with respective IC50 concentrations of GO, GO-PtNPs, and PtNPs for 24 h. Relative messenger RNA(mRNA) expression of CDK2 (**A**), CDK4 (**B**), GADD45A (**C**), OGG1 (**D**) APEX1 (**E**), CREB1 (**F**), POLB (**G**), and UNG (**H**) was analyzed by quantitative reverse-transcription polymerase chain reaction in LNCaP cells treated for 24 h. After 24-h treatment, expression fold level was determined as fold changes in reference to expression values against GAPDH. Results are expressed as fold changes. There was a significant difference in treated cells compared to untreated cells with Student's *t*-test (* $p < 0.05$).

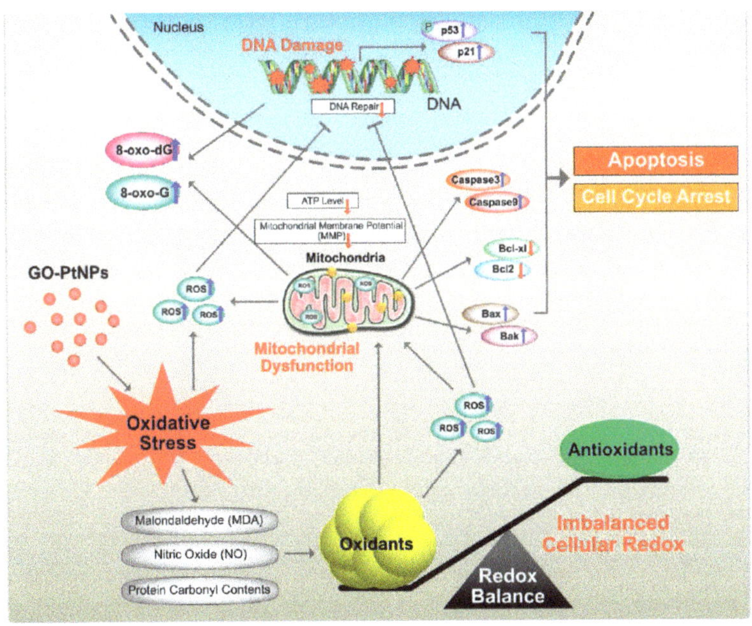

Figure 15. The hypothetical model demonstrates that the impact of GO-PtNPs on oxidative stress induced DNA damage in LnCaP cells.

4. Conclusions

In this study, we have successfully demonstrated a simple, environmentally friendly, and green approach for synthesis of GO-PtNPs using vanillin as a reducing and stabilizing agent. Cytotoxicity of GO-PtNPs was explored by determinations of cell viability, cell proliferation, and series of cellular assays. The results revealed a more effective dose-dependent effect caused by GO-PtNPs compared to either GO or PtNPs. Assessment of cytotoxicity test by examination of lactate dehydrogenase leakage and membrane integrity revealed that GO-PtNPs potentially caused cell death. GO-PtNPs increased the level of reactive oxygen species, malondialdehyde (MDA), nitric oxide (NO), and carbonylated protein levels. The imbalance between pro- and antioxidant levels led to the loss of mitochondrial integrity. In addition, GO-PtNP-treated cells exhibited mitochondrial-mediated apoptosis due to the upregulation of p53, p21, Bax, and Bak, and the downregulation of Bcl-2 and Bcl-xl. GO-PtNPs decreased the MMP and subsequently decreased the level of adenosine triphosphate production. Ultimately, GO-PtNPs caused programmed cell death by the upregulation of proapoptotic genes, including p53, p21, Bax, Bak, caspase 9, and caspase-3, and downregulation of the antiapoptotic marker genes Bcl-2 and Bcl-xl. These results further substantiated that GO-PtNPs nanocomposites can potentially disturb cell viability by inducing DNA cellular damage and genotoxicity by modulating the genes expression responsible for cell cycle arrest, DNA damage, and DNA repair, and increasing the levels of 8-oxo-dG and 8-oxo-G. GO-PtNPs significantly impaired the multiplication of cancer cells compared to GO or PtNPs alone. Hence, we can conclude that GO-PtNPs are potentially valuable as an alternate therapeutic agent for cancer. This piece of work could provide a step forward to improve therapeutic efficiency of biologically synthesized biocompatible and nontoxic agent seems to be one of promising techniques for cancer treatments.

Supplementary Materials: The following are available online at http://www.mdpi.com/2073-4360/11/4/733/s1, Figure S1: Characterization of GO and GO-PtNPs by SEM and TEM, morphology of GO (A), GO-PtNPs (B), and size of GO (C), and GO-PtNPs (D) were analyzed by SEM and TEM, respectively. The red circle indicates decoration of PtNPs particles on the surface of graphene sheet (white arrow). The graphene sheet depicted as wrinkled structure. E. The size distribution of GO. F. Size distribution of GO-PtNPs, Figure S2: The cell morphology was analyzed by phase-contrast microscope. The morphology of LNCaP cells was determined after 24-h exposure to different concentrations of GO (A), GO-PtNPs (B), and PtNPs (C) using light microscope., Figure S3: LNCaP cells was exposed to respective IC50 concentration of GO, GO-PtNPs, and PtNPs for 24 h and then ROS was measured using DCFH-DA-FITC by fluorescence microscopy analysis. Scale bar = 200 μm. Figure S4: LNCaP cells were exposed to respective IC50 concentrations of GO, GO-PtNPs, and PtNPs for 24 h and then MMP was analyzed by fluorescence microscopy analysis. Scale bar = 200 μm.

Author Contributions: S.G. proposed the study idea and was involved in the design, synthesis, and characterization of the GO-PtNP nanocomposites, cellular assays, analysis of the data, and writing of the manuscript. M.J. performed characterization of nanomaterials. M.-H.K. performed references and figure quality. J.-H.K. provided the facility and monitored all experiments. All authors read and approved the final manuscript.

Funding: This work was supported by a grant from the Science Research Center (2015R1A5A1009701) of the National Research Foundation of Korea.

Acknowledgments: Sangiliyandi Gurunathan is the recipient of Konkuk University Research Professor programme.

Conflicts of Interest: The authors declare no conflict of interest.

References

1. Bray, F.; Ferlay, J.; Soerjomataram, I.; Siegel, R.L.; Torre, L.A.; Jemal, A. Global cancer statistics 2018: GLOBOCAN estimates of incidence and mortality worldwide for 36 cancers in 185 countries. *CA Cancer J. Clin.* **2018**, *68*, 394–424. [CrossRef]
2. Hankey, B.F.; Feuer, E.J.; Clegg, L.X.; Hayes, R.B.; Legler, J.M.; Prorok, P.C.; Ries, L.A.; Merrill, R.M.; Kaplan, R.S. Cancer surveillance series: Interpreting trends in prostate cancer—Part I: Evidence of the effects of screening in recent prostate cancer incidence, mortality, and survival rates. *J. Natl. Cancer Inst.* **1999**, *91*, 1017–1024. [CrossRef]
3. Henglein, A. Small-particle research: Physicochemical properties of extremely small colloidal metal and semiconductor particles. *Chem. Rev.* **1989**, *89*, 1861–1873. [CrossRef]

4. Zhang, X.-F.; Shen, W.; Gurunathan, S. Silver Nanoparticle-Mediated Cellular Responses in Various Cell Lines: An in Vitro Model. *Int. J. Mol. Sci.* **2016**, *17*, 1603. [CrossRef]
5. Han, J.W.; Gurunathan, S.; Jeong, J.K.; Choi, Y.J.; Kwon, D.N.; Park, J.K.; Kim, J.H. Oxidative stress mediated cytotoxicity of biologically synthesized silver nanoparticles in human lung epithelial adenocarcinoma cell line. *Nanoscale Res. Lett.* **2014**, *9*, 459. [CrossRef] [PubMed]
6. Park, J.H.; Gurunathan, S.; Choi, Y.-J.; Han, J.W.; Song, H.; Kim, J.-H. Silver nanoparticles suppresses brain-derived neurotrophic factor-induced cell survival in the human neuroblastoma cell line SH-SY5Y. *J. Ind. Eng. Chem.* **2017**, *47*, 62–73. [CrossRef]
7. Gurunathan, S.; Han, J.W.; Eppakayala, V.; Kim, J.H. Green synthesis of graphene and its cytotoxic effects in human breast cancer cells. *Int. J. Nanomed.* **2013**, *8*, 1015–1027. [CrossRef] [PubMed]
8. Gurunathan, S.; Han, J.W.; Park, J.H.; Kim, E.; Choi, Y.J.; Kwon, D.N.; Kim, J.H. Reduced graphene oxide-silver nanoparticle nanocomposite: A potential anticancer nanotherapy. *Int. J. Nanomed.* **2015**, *10*, 6257–6276. [CrossRef]
9. Porcel, E.; Liehn, S.; Remita, H.; Usami, N.; Kobayashi, K.; Furusawa, Y.; Le Sech, C.; Lacombe, S. Platinum nanoparticles: A promising material for future cancer therapy. *Nanotechnology* **2010**, *21*, 85103. [CrossRef]
10. Usami, N.; Furusawa, Y.; Kobayashi, K.; Lacombe, S.; Reynaud-Angelin, A.; Sage, E.; Wu, T.D.; Croisy, A.; Guerquin-Kern, J.L.; Le Sech, C. Mammalian cells loaded with platinum-containing molecules are sensitized to fast atomic ions. *Int. J. Radiat. Biol.* **2008**, *84*, 603–611. [CrossRef]
11. Charest, G.; Paquette, B.; Fortin, D.; Mathieu, D.; Sanche, L. Concomitant treatment of F98 glioma cells with new liposomal platinum compounds and ionizing radiation. *J. Neurooncol.* **2010**, *97*, 187–193. [CrossRef] [PubMed]
12. Kostova, I. Platinum complexes as anticancer agents. *Recent Pat. Anticancer Drug Discov.* **2006**, *1*, 1–22. [CrossRef] [PubMed]
13. Respondek, J.; Engel, J. Organometallics in medicine. *Drug Futur.* **1996**, *21*, 391–408.
14. Geim, A.K.; Novoselov, K.S. The rise of graphene. *Nat. Mater.* **2007**, *6*, 183. [CrossRef] [PubMed]
15. Park, S.; Ruoff, R.S. Chemical methods for the production of graphenes. *Nat. Nanotechnol.* **2009**, *4*, 217. [CrossRef]
16. Zhang, Y.; Rubio, A.; Lay, G.L. Emergent elemental two-dimensional materials beyond graphene. *J. Phys. D Appl. Phys.* **2017**, *50*, 053004. [CrossRef]
17. Liu, Y.; Dong, X.; Chen, P. Biological and chemical sensors based on graphene materials. *Chem. Soc. Rev.* **2012**, *41*, 2283–2307. [CrossRef]
18. Wang, N.; Lin, M.; Dai, H.; Ma, H. Functionalized gold nanoparticles/reduced graphene oxide nanocomposites for ultrasensitive electrochemical sensing of mercury ions based on thymine–mercury–thymine structure. *Biosens. Bioelectron.* **2016**, *79*, 320–326. [CrossRef]
19. Ates, M.; Bayrak, Y.; Yoruk, O.; Caliskan, S. Reduced graphene oxide/Titanium oxide nanocomposite synthesis via microwave-assisted method and supercapacitor behaviors. *J. Alloys Compd.* **2017**, *728*, 541–551. [CrossRef]
20. Gurunathan, S.; Kim, J.H. Synthesis, toxicity, biocompatibility, and biomedical applications of graphene and graphene-related materials. *Int. J. Nanomed.* **2016**, *11*, 1927–1945. [CrossRef]
21. Gurunathan, S.; Kim, J.H. Graphene Oxide-Silver Nanoparticles Nanocomposite Stimulates Differentiation in Human Neuroblastoma Cancer Cells (SH-SY5Y). *Int. J. Mol. Sci.* **2017**, *18*, 2549. [CrossRef] [PubMed]
22. Yan, Y.; Zeitler, E.L.; Gu, J.; Hu, Y.; Bocarsly, A.B. Electrochemistry of Aqueous Pyridinium: Exploration of a Key Aspect of Electrocatalytic Reduction of CO_2 to Methanol. *J. Am. Chem. Soc.* **2013**, *135*, 14020–14023. [CrossRef]
23. Kalambate, P.K.; Sanghavi, B.J.; Karna, S.P.; Srivastava, A.K. Simultaneous voltammetric determination of paracetamol and domperidone based on a graphene/platinum nanoparticles/nafion composite modified glassy carbon electrode. *Sensors Actuators B Chem.* **2015**, *213*, 285–294. [CrossRef]
24. Atta, N.F.; Hassan, H.K.; Galal, A. Rapid and simple electrochemical detection of morphine on graphene–palladium-hybrid-modified glassy carbon electrode. *Anal. Bioanal. Chem.* **2014**, *406*, 6933–6942. [CrossRef] [PubMed]
25. Dreyer, D.R.; Park, S.; Bielawski, C.W.; Ruoff, R.S. The chemistry of graphene oxide. *Chem. Soc. Rev.* **2010**, *39*, 228–240. [CrossRef] [PubMed]
26. Liu, J.; Tang, J.; Gooding, J.J. Strategies for chemical modification of graphene and applications of chemically modified graphene. *J. Mater. Chem.* **2012**, *22*, 12435–12452. [CrossRef]

27. Wu, X.; Xing, Y.; Pierce, D.; Zhao, J.X. One-Pot Synthesis of Reduced Graphene Oxide/Metal (Oxide) Composites. *ACS Appl. Mater. Interfaces* **2017**, *9*, 37962–37971. [CrossRef] [PubMed]
28. Ali, M.A.; Singh, C.; Srivastava, S.; Admane, P.; Agrawal, V.V.; Sumana, G.; John, R.; Panda, A.; Dong, L.; Malhotra, B.D. Graphene oxide–metal nanocomposites for cancer biomarker detection. *RSC Adv.* **2017**, *7*, 35982–35991. [CrossRef]
29. Diaz-Diestra, D.; Thapa, B.; Badillo-Diaz, D.; Beltran-Huarac, J.; Morell, G.; Weiner, B.R. Graphene Oxide/ZnS:Mn Nanocomposite Functionalized with Folic Acid as a Nontoxic and Effective Theranostic Platform for Breast Cancer Treatment. *Nanomaterials* **2018**, *8*, 484. [CrossRef] [PubMed]
30. Beltran-Huarac, J.; Resto, O.; Carpena-Nuñez, J.; Jadwisienczak, W.M.; Fonseca, L.F.; Weiner, B.R.; Morell, G. Single-Crystal γ-MnS Nanowires Conformally Coated with Carbon. *ACS Appl. Mater. Interfaces* **2014**, *6*, 1180–1186. [CrossRef] [PubMed]
31. Venu, M.; Venkateswarlu, S.; Reddy, Y.V.M.; Seshadri Reddy, A.; Gupta, V.K.; Yoon, M.; Madhavi, G. Highly Sensitive Electrochemical Sensor for Anticancer Drug by a Zirconia Nanoparticle-Decorated Reduced Graphene Oxide Nanocomposite. *ACS Omega* **2018**, *3*, 14597–14605. [CrossRef]
32. Gehrke, H.; Pelka, J.; Hartinger, C.G.; Blank, H.; Bleimund, F.; Schneider, R.; Gerthsen, D.; Brase, S.; Crone, M.; Turk, M.; et al. Platinum nanoparticles and their cellular uptake and DNA platination at non-cytotoxic concentrations. *Arch. Toxicol.* **2011**, *85*, 799–812. [CrossRef] [PubMed]
33. Hummers, W.S., Jr.; Offeman, R.E. Preparation of graphitic oxide. *J. Am. Chem. Soc.* **1958**, *80*, 1339. [CrossRef]
34. Gurunathan, S.; Kang, M.H.; Kim, J.H. Combination Effect of Silver Nanoparticles and Histone Deacetylases Inhibitor in Human Alveolar Basal Epithelial Cells. *Molecules* **2018**, *23*, 2046. [CrossRef] [PubMed]
35. Chen, F.; Kuhn, D.C.; Sun, S.-C.; Gaydos, L.J.; Demers, L.M. Dependence and reversal of nitric oxide production on NF-κ-B in silica and lipopolysaccharide induced macrophages. *Biochem. Biophys. Res. Commun.* **1995**, *214*, 839–846. [CrossRef]
36. Uehara, H.; Rao, V.A. Metal-Mediated Protein Oxidation: Applications of a Modified ELISA-Based Carbonyl Detection Assay for Complex Proteins. *Pharm. Res.* **2015**, *32*, 691–701. [CrossRef] [PubMed]
37. Zhang, X.F.; Huang, F.H.; Zhang, G.L.; Bai, D.P.; Massimo, D.F.; Huang, Y.F.; Gurunathan, S. Novel biomolecule lycopene-reduced graphene oxide-silver nanoparticle enhances apoptotic potential of trichostatin A in human ovarian cancer cells (SKOV3). *Int. J. Nanomed.* **2017**, *12*, 7551–7575. [CrossRef]
38. Gurunathan, S.; Qasim, M.; Park, C.; Yoo, H.; Choi, D.Y.; Song, H.; Park, C.; Kim, J.H.; Hong, K. Cytotoxicity and Transcriptomic Analysis of Silver Nanoparticles in Mouse Embryonic Fibroblast Cells. *Int. J. Mol. Sci.* **2018**, *19*, 3618. [CrossRef]
39. Dikin, D.A.; Stankovich, S.; Zimney, E.J.; Piner, R.D.; Dommett, G.H.; Evmenenko, G.; Nguyen, S.T.; Ruoff, R.S. Preparation and characterization of graphene oxide paper. *Nature* **2007**, *448*, 457–460. [CrossRef]
40. Stankovich, S.; Dikin, D.A.; Piner, R.D.; Kohlhaas, K.A.; Kleinhammes, A.; Jia, Y.; Wu, Y.; Nguyen, S.T.; Ruoff, R.S. Synthesis of graphene-based nanosheets via chemical reduction of exfoliated graphite oxide. *Carbon* **2007**, *45*, 1558–1565. [CrossRef]
41. Dey, R.S.; Raj, C.R. Development of an amperometric cholesterol biosensor based on graphene−Pt nanoparticle hybrid material. *J. Phys. Chem. C* **2010**, *114*, 21427–21433. [CrossRef]
42. Kempegowda, R.; Antony, D.; Malingappa, P. Graphene–platinum nanocomposite as a sensitive and selective voltammetric sensor for trace level arsenic quantification. *Int. J. Smart Nano Mater.* **2014**, *5*, 17–32. [CrossRef]
43. Ma, J.; Liu, R.; Wang, X.; Liu, Q.; Chen, Y.; Valle, R.P.; Zuo, Y.Y.; Xia, T.; Liu, S. Crucial role of lateral size for graphene oxide in activating macrophages and stimulating pro-inflammatory responses in cells and animals. *ACS Nano* **2015**, *9*, 10498–10515. [CrossRef]
44. Ferrari, A.C. Raman spectroscopy of graphene and graphite: Disorder, electron–phonon coupling, doping and nonadiabatic effects. *Solid State Commun.* **2007**, *143*, 47–57. [CrossRef]
45. Alam, S.N.; Sharma, N.; Kumar, L. Synthesis of graphene oxide (GO) by modified hummers method and its thermal reduction to obtain reduced graphene oxide (rGO). *Graphene* **2017**, *6*, 1–18. [CrossRef]
46. Gurunathan, S.; Han, J.W.; Kim, J.H. Green chemistry approach for the synthesis of biocompatible graphene. *Int. J. Nanomed.* **2013**, *8*, 2719–2732. [CrossRef]
47. Gurunathan, S.; Han, J.; Park, J.H.; Kim, J.H. An in vitro evaluation of graphene oxide reduced by Ganoderma spp. in human breast cancer cells (MDA-MB-231). *Int. J. Nanomed.* **2014**, *9*, 1783. [CrossRef] [PubMed]
48. Jin, C.; Wang, F.; Tang, Y.; Zhang, X.; Wang, J.; Yang, Y. Distribution of graphene oxide and TiO 2-graphene oxide composite in A549 cells. *Biol. Trace Elem. Res.* **2014**, *159*, 393–398. [CrossRef] [PubMed]

49. Jaworski, S.; Wierzbicki, M.; Sawosz, E.; Jung, A.; Gielerak, G.; Biernat, J.; Jaremek, H.; Łojkowski, W.; Woźniak, B.; Wojnarowicz, J. Graphene oxide-based nanocomposites decorated with silver nanoparticles as an antibacterial agent. *Nanoscale Res. Lett.* **2018**, *13*, 1–17. [CrossRef] [PubMed]
50. Vi, T.T.T.; Rajesh Kumar, S.; Rout, B.; Liu, C.H.; Wong, C.B.; Chang, C.W.; Chen, C.H.; Chen, D.W.; Lue, S.J. The Preparation of Graphene Oxide-Silver Nanocomposites: The Effect of Silver Loads on Gram-Positive and Gram-Negative Antibacterial Activities. *Nanomaterials* **2018**, *8*, 163. [CrossRef]
51. Yuan, Y.G.; Gurunathan, S. Combination of graphene oxide-silver nanoparticle nanocomposites and cisplatin enhances apoptosis and autophagy in human cervical cancer cells. *Int. J. Nanomed.* **2017**, *12*, 6537–6558. [CrossRef] [PubMed]
52. Tomar, P.; Jain, N.; Dixit, V.K. Nanoparticulate delivery of LHRH analogue for the treatment of prostate cancer. *Pharm. Dev. Technol.* **2013**, *18*, 645–652. [CrossRef]
53. Shiny, P.; Mukherjee, A.; Chandrasekaran, N. DNA damage and mitochondria-mediated apoptosis of A549 lung carcinoma cells induced by biosynthesised silver and platinum nanoparticles. *RSC Adv.* **2016**, *6*, 27775–27787. [CrossRef]
54. Horie, M.; Kato, H.; Endoh, S.; Fujita, K.; Nishio, K.; Komaba, L.K.; Fukui, H.; Nakamura, A.; Miyauchi, A.; Nakazato, T.; et al. Evaluation of cellular influences of platinum nanoparticles by stable medium dispersion. *Metallomics* **2011**, *3*, 1244–1252. [CrossRef] [PubMed]
55. Marnett, L.J. Oxy radicals, lipid peroxidation and DNA damage. *Toxicology* **2002**, *181*, 219–222. [CrossRef]
56. Almeer, R.S.; Ali, D.; Alarifi, S.; Alkahtani, S.; Almansour, M. Green Platinum Nanoparticles Interaction With HEK293 Cells: Cellular Toxicity, Apoptosis, and Genetic Damage. *Dose-Response* **2018**, *16*, 1559325818807382. [CrossRef] [PubMed]
57. Yue, H.; Wei, W.; Yue, Z.; Wang, B.; Luo, N.; Gao, Y.; Ma, D.; Ma, G.; Su, Z. The role of the lateral dimension of graphene oxide in the regulation of cellular responses. *Biomaterials* **2012**, *33*, 4013–4021. [CrossRef] [PubMed]
58. Mendes, R.G.; Koch, B.; Bachmatiuk, A.; Ma, X.; Sanchez, S.; Damm, C.; Schmidt, O.G.; Gemming, T.; Eckert, J.; Rümmeli, M.H. A size dependent evaluation of the cytotoxicity and uptake of nanographene oxide. *J. Mater. Chem. B* **2015**, *3*, 2522–2529. [CrossRef]
59. Zhang, W.; Yan, L.; Li, M.; Zhao, R.; Yang, X.; Ji, T.; Gu, Z.; Yin, J.J.; Gao, X.; Nie, G. Deciphering the underlying mechanisms of oxidation-state dependent cytotoxicity of graphene oxide on mammalian cells. *Toxicol. Lett.* **2015**, *237*, 61–71. [CrossRef] [PubMed]
60. Russier, J.; Treossi, E.; Scarsi, A.; Perrozzi, F.; Dumortier, H.; Ottaviano, L.; Meneghetti, M.; Palermo, V.; Bianco, A. Evidencing the mask effect of graphene oxide: A comparative study on primary human and murine phagocytic cells. *Nanoscale* **2013**, *5*, 11234–11247. [CrossRef]
61. Liao, K.H.; Lin, Y.S.; Macosko, C.W.; Haynes, C.L. Cytotoxicity of graphene oxide and graphene in human erythrocytes and skin fibroblasts. *ACS Appl. Mater. Interfaces* **2011**, *3*, 2607–2615. [CrossRef] [PubMed]
62. Linares, J.; Matesanz, M.C.; Vila, M.; Feito, M.J.; Goncalves, G.; Vallet-Regi, M.; Marques, P.A.; Portoles, M.T. Endocytic mechanisms of graphene oxide nanosheets in osteoblasts, hepatocytes and macrophages. *ACS Appl. Mater. Interfaces* **2014**, *6*, 13697–13706. [CrossRef] [PubMed]
63. Kim, S.; Choi, J.E.; Choi, J.; Chung, K.H.; Park, K.; Yi, J.; Ryu, D.Y. Oxidative stress-dependent toxicity of silver nanoparticles in human hepatoma cells. *Toxicol. In Vitro* **2009**, *23*, 1076–1084. [CrossRef] [PubMed]
64. Rahman, M.F.; Wang, J.; Patterson, T.A.; Saini, U.T.; Robinson, B.L.; Newport, G.D.; Murdock, R.C.; Schlager, J.J.; Hussain, S.M.; Ali, S.F. Expression of genes related to oxidative stress in the mouse brain after exposure to silver-25 nanoparticles. *Toxicol. Lett.* **2009**, *187*, 15–21. [CrossRef] [PubMed]
65. Wiseman, H.; Halliwell, B. Damage to DNA by reactive oxygen and nitrogen species: Role in inflammatory disease and progression to cancer. *Biochem. J.* **1996**, *313 Pt 1*, 17–29. [CrossRef]
66. Yuan, Y.-G.; Zhang, S.; Hwang, J.-Y.; Kong, I.-K. Silver Nanoparticles Potentiates Cytotoxicity and Apoptotic Potential of Camptothecin in Human Cervical Cancer Cells. *Oxid. Med. Cell. Longev.* **2018**, *2018*, 6121328. [CrossRef]
67. Xue, Y.; Chen, Q.; Ding, T.; Sun, J. SiO(2) nanoparticle-induced impairment of mitochondrial energy metabolism in hepatocytes directly and through a Kupffer cell-mediated pathway in vitro. *Int. J. Nanomed.* **2014**, *9*, 2891–2903. [CrossRef]
68. Halliwell, B. Free radicals and antioxidants—Quo vadis? *Trends Pharmacol. Sci.* **2011**, *32*, 125–130. [CrossRef] [PubMed]

69. Arora, S.; Jain, J.; Rajwade, J.; Paknikar, K. Cellular responses induced by silver nanoparticles: In vitro studies. *Toxicol. Lett.* **2008**, *179*, 93–100. [CrossRef] [PubMed]
70. Jin, C.Y.; Zhu, B.S.; Wang, X.F.; Lu, Q.H. Cytotoxicity of titanium dioxide nanoparticles in mouse fibroblast cells. *Chem. Res. Toxicol.* **2008**, *21*, 1871–1877. [CrossRef]
71. Asadpour, E.; Sadeghnia, H.R.; Ghorbani, A.; Sedaghat, M.; Boroushaki, M.T. Oxidative stress-mediated cytotoxicity of zirconia nanoparticles on PC12 and N2a cells. *J. Nanopart. Res.* **2016**, *18*, 14. [CrossRef]
72. Barcinska, E.; Wierzbicka, J.; Zauszkiewicz-Pawlak, A.; Jacewicz, D.; Dabrowska, A.; Inkielewicz-Stepniak, I. Role of Oxidative and Nitro-Oxidative Damage in Silver Nanoparticles Cytotoxic Effect against Human Pancreatic Ductal Adenocarcinoma Cells. *Oxid. Med. Cell. Longev.* **2018**, *2018*, 8251961. [CrossRef]
73. Papa, L.; Hahn, M.; Marsh, E.L.; Evans, B.S.; Germain, D. SOD2 to SOD1 switch in breast cancer. *J. Biol. Chem.* **2014**, *289*, 5412–5416. [CrossRef]
74. Redza-Dutordoir, M.; Averill-Bates, D.A. Activation of apoptosis signalling pathways by reactive oxygen species. *Biochim. Biophys. Acta* **2016**, *1863*, 2977–2992. [CrossRef]
75. Bargonetti, J.; Manfredi, J.J. Multiple roles of the tumor suppressor p53. *Curr. Opin. Oncol.* **2002**, *14*, 86–91. [CrossRef]
76. Brancato, B.; Munnia, A.; Cellai, F.; Ceni, E.; Mello, T.; Bianchi, S.; Catarzi, S.; Risso, G.G.; Galli, A.; Peluso, M.E. 8-Oxo-7, 8-dihydro-2′-deoxyguanosine and other lesions along the coding strand of the exon 5 of the tumour suppressor gene P53 in a breast cancer case-control study. *DNA Res.* **2016**, *23*, 395–402. [CrossRef] [PubMed]
77. Asharani, P.V.; Xinyi, N.; Hande, M.P.; Valiyaveettil, S. DNA damage and p53-mediated growth arrest in human cells treated with platinum nanoparticles. *Nanomedicine* **2010**, *5*, 51–64. [CrossRef]
78. Medhat, A.; Mansour, S.; El-Sonbaty, S.; Kandil, E.; Mahmoud, M. Evaluation of the antitumor activity of platinum nanoparticles in the treatment of hepatocellular carcinoma induced in rats. *Tumor Biol.* **2017**, *39*. [CrossRef]
79. Marullo, R.; Werner, E.; Degtyareva, N.; Moore, B.; Altavilla, G.; Ramalingam, S.S.; Doetsch, P.W. Cisplatin induces a mitochondrial-ROS response that contributes to cytotoxicity depending on mitochondrial redox status and bioenergetic functions. *PLoS ONE* **2013**, *8*, e81162. [CrossRef] [PubMed]
80. Govender, R.; Phulukdaree, A.; Gengan, R.M.; Anand, K.; Chuturgoon, A.A. Silver nanoparticles of Albizia adianthifolia: The induction of apoptosis in human lung carcinoma cell line. *J. Nanobiotechnol.* **2013**, *11*, 5. [CrossRef] [PubMed]
81. Konieczny, P.; Goralczyk, A.G.; Szmyd, R.; Skalniak, L.; Koziel, J.; Filon, F.L.; Crosera, M.; Cierniak, A.; Zuba-Surma, E.K.; Borowczyk, J.; et al. Effects triggered by platinum nanoparticles on primary keratinocytes. *Int. J. Nanomed.* **2013**, *8*, 3963–3975. [CrossRef]
82. Watanabe, M.; Yoneda, M.; Morohashi, A.; Hori, Y.; Okamoto, D.; Sato, A.; Kurioka, D.; Nittami, T.; Hirokawa, Y.; Shiraishi, T.; et al. Effects of Fe3O4 Magnetic Nanoparticles on A549 Cells. *Int. J. Mol. Sci.* **2013**, *14*, 15546–15560. [CrossRef] [PubMed]
83. Koedrith, P.; Boonprasert, R.; Kwon, J.Y.; Kim, I.-S.; Seo, Y.R. Recent toxicological investigations of metal or metal oxide nanoparticles in mammalian models in vitro and in vivo: DNA damaging potential, and relevant physicochemical characteristics. *Mol. Cell. Toxicol.* **2014**, *10*, 107–126. [CrossRef]
84. Cellai, F.; Munnia, A.; Viti, J.; Doumett, S.; Ravagli, C.; Ceni, E.; Mello, T.; Polvani, S.; Giese, R.W.; Baldi, G.; et al. Magnetic Hyperthermia and Oxidative Damage to DNA of Human Hepatocarcinoma Cells. *Int. J. Mol. Sci.* **2017**, *18*, 939. [CrossRef]
85. Nejdl, L.; Kudr, J.; Moulick, A.; Hegerova, D.; Ruttkay-Nedecky, B.; Gumulec, J.; Cihalova, K.; Smerkova, K.; Dostalova, S.; Krizkova, S.; et al. Platinum nanoparticles induce damage to DNA and inhibit DNA replication. *PLoS ONE* **2017**, *12*, e0180798. [CrossRef] [PubMed]
86. el-Deiry, W.S.; Tokino, T.; Velculescu, V.E.; Levy, D.B.; Parsons, R.; Trent, J.M.; Lin, D.; Mercer, W.E.; Kinzler, K.W.; Vogelstein, B. WAF1, a potential mediator of p53 tumor suppression. *Cell* **1993**, *75*, 817–825. [CrossRef]
87. Waldman, T.; Kinzler, K.W.; Vogelstein, B. p21 is necessary for the p53-mediated G1 arrest in human cancer cells. *Cancer Res.* **1995**, *55*, 5187–5190. [PubMed]
88. Ewen, M.E.; Oliver, C.J.; Sluss, H.K.; Miller, S.J.; Peeper, D.S. p53-dependent repression of CDK4 translation in TGF-beta-induced G1 cell-cycle arrest. *Genes Dev.* **1995**, *9*, 204–217. [CrossRef]

89. He, G.; Siddik, Z.H.; Huang, Z.; Wang, R.; Koomen, J.; Kobayashi, R.; Khokhar, A.R.; Kuang, J. Induction of p21 by p53 following DNA damage inhibits both Cdk4 and Cdk2 activities. *Oncogene* **2005**, *24*, 2929–2943. [CrossRef] [PubMed]
90. Alshatwi, A.A.; Athinarayanan, J.; Vaiyapuri Subbarayan, P. Green synthesis of platinum nanoparticles that induce cell death and G2/M-phase cell cycle arrest in human cervical cancer cells. *J. Mater. Sci. Mater. Med.* **2015**, *26*, 5330. [CrossRef]
91. Qin, Y.; Zhou, Z.-W.; Pan, S.-T.; He, Z.-X.; Zhang, X.; Qiu, J.-X.; Duan, W.; Yang, T.; Zhou, S.-F. Graphene quantum dots induce apoptosis, autophagy, and inflammatory response via p38 mitogen-activated protein kinase and nuclear factor-κB mediated signaling pathways in activated THP-1 macrophages. *Toxicology* **2015**, *327*, 62–76. [CrossRef]
92. Liu, Y.; Luo, Y.; Wu, J.; Wang, Y.; Yang, X.; Yang, R.; Wang, B.; Yang, J.; Zhang, N. Graphene oxide can induce in vitro and in vivo mutagenesis. *Sci. Rep.* **2013**, *3*, 3469. [CrossRef]

polymers

MDPI

Article

Near-Infrared Radiation-Based Mild Photohyperthermia Therapy of Non-Melanoma Skin Cancer with PEGylated Reduced Nanographene Oxide

Raquel Costa-Almeida [1,2,†], Diana Bogas [3,†], José R. Fernandes [4], Licínia Timochenco [3], Filipa A. L. S. Silva [1,2], João Meneses [3], Inês C. Gonçalves [1,2], Fernão D. Magalhães [3] and Artur M. Pinto [1,2,3,*]

1 i3S—Instituto de Investigação e Inovação em Saúde, Universidade do Porto, 4200-180 Porto, Portugal; rcalmeida@i3s.up.pt (R.C.-A.); flsilva@i3s.up.pt (F.A.L.S.S.); icastro@ineb.up.pt (I.C.G.)

2 INEB—Instituto de Engenharia Biomédica, Universidade do Porto, Rua Alfredo Allen, 208, 4200-180 Porto, Portugal

3 LEPABE, Faculdade de Engenharia, Universidade do Porto, 4200-180 Porto, Portugal; dianabogas@fe.up.pt (D.B.); up201809122@fe.up.pt (L.T.); up201503132@fe.up.pt (J.M.); fdmagalh@fe.up.pt (F.D.M.)

4 CQVR—Centro de Química Vila Real, Departamento de Física, ECT, Universidade de Trás-os-Montes e Alto Douro, 5001-801 Vila Real, Portugal; jraf@utad.pt

* Correspondence: arturp@fe.up.pt

† These author contributed equally to this work.

Received: 17 July 2020; Accepted: 14 August 2020; Published: 17 August 2020

Abstract: Using a one-step thermal reduction and non-covalent chemical functionalization process, PEGylated reduced nanographene oxide (rGOn-PEG) was produced from nanographene oxide (GOn) and characterized in terms of particle size, dispersion stability, chemistry, and photothermal properties, in view of its use for photothermal therapy (PTT) of non-melanoma skin cancer. GOn infrared spectrum presented more intense bands assigned to oxygen containing functional groups than observed for rGOn-PEG. GOn C/O ratio decreased more than 50% comparing with rGOn-PEG and nitrogen was present in the latter (N *at* % = 20.6) due to introduction of PEG-NH_2. Thermogravimetric analysis allowed estimating the amount of PEG in rGOn-PEG to be of about 56.1%. Simultaneous reduction and PEGylation increased the lateral dimensions from 287 ± 139 nm to 521 ± 397 nm, as observed by transmission electron microscopy and dynamic light scattering. rGOn-PEG exhibited ≈13-fold higher absorbance in the near-infrared radiation (NIR) region, as compared to unmodified GOn. Low power (150 mW cm^{-2}) NIR irradiation using LEDs resulted in rGOn-PEG heating up to 47 °C, which is within the mild PTT temperature range. PEGylation strongly enhanced the dispersibility of rGOn in physiological media (phosphate buffered saline, fetal bovine serum, and cell culture medium) and also improved the biocompatibility of rGOn-PEG, in comparison to GOn (25–250 μg mL^{-1}). After a single NIR LED irradiation treatment of 30 min, a decrease of ≈38% in A-431 cells viability was observed for rGOn-PEG (250 μg mL^{-1}). Together, our results demonstrate the potential of irradiating rGOn-PEG using lower energy, cheaper, smaller, and safer LEDs, as alternative to high power lasers, for NIR mild hyperthermia therapy of cancer, namely non-melanoma skin cancer.

Keywords: graphene; light emitting diode; phototherapy; polyethylene glycol; thermal reduction

1. Introduction

Non-melanoma skin cancer (NMSC) has been reported as one of the most common types of cancer worldwide, with an estimation of over 3 million diagnoses each year in the USA [1]. Despite including different malignancies, basal cell and squamous cell carcinomas are the most frequent types of NMSC [1–3]. Depending on disease risk level, surgical excision or radiotherapy are the standard treatment options. Severe cases of high-risk primary, recurrent, or metastatic NMSC frequently require multimodal therapy [1,3–5]. However, functional and cosmetic outcomes are still variable, requiring novel treatment strategies.

Photothermal therapy (PTT) is being increasingly explored as an alternative non-invasive cancer treatment. It relies on tumor irradiation with a near-infrared (NIR) laser, either topically or interstitially through an optical fiber, leading to light energy conversion into heat, ultimately resulting in tumor ablation [6]. Two different mechanisms can be broadly identified. Particularly, hyperthermia results from mild temperature increases leading to the activation of cellular apoptotic pathways, whereas tumor ablation is achieved with rapid temperature increases (> 50 °C) [7]. Nanomaterials hold potential as photothermal absorbers to enhance PTT selectivity within the target tumor tissue toward achieving therapeutic temperatures (> 41 °C) using less total light energy, minimizing damage to the healthy surrounding tissue [6].

Over the past decade, graphene-based materials (GBM) have been widely explored for a huge number of applications, including energy, aerospace, biomedicine, and health [8–15]. Recently, GBM emerged as potential PTT-based cancer therapy platforms owing to their high NIR absorption [16–18]. Similarly to graphene-based materials, other 2D nanomaterials are being increasingly explored for PTT cancer treatment, including black phosphorus (BP) and transition metal dichalcogenides (TMDs), among others [17]. However, BP is highly sensitive to water and oxygen [19] and frequently requires the combination with gold nanoparticles to achieve NIR photothermal transduction efficiency [20,21]. In turn, TMDs are gaining interest, but low-yield and low-quality exfoliation methods still pose enormous challenges to their application in biomedicine [22,23].

Graphene oxide (GO) has been receiving increased attention in the field of biomedicine due to its surface chemistry with abundant oxygen reactive functional groups that enables a wide range of chemical modifications and bioconjugation approaches [13,24–26]. We have previously shown that oxidation of graphene nanoplatelets resulted in enhanced in vitro biocompatibility of these 2D nanomaterials when cultured with human dermal fibroblasts at concentrations up to 100 $\mu g\ mL^{-1}$ [27]. However, while improving water dispersibility and biocompatibility, oxidation of graphene into GO results in diminished electronical and optical conductivity [24,28]. Chemical, thermal, and electro/photochemical reduction methods are widely investigated to restore the aromatic structure of graphene and obtain reduced GO with high NIR absorption capacity and biocompatibility for applications in human health [13]. Over the years, green reduction combined with chemical functionalization methods using biocompatible polymers, such as polyethylene glycol (PEG), to act as surfactants have been reported [29–34]. Such methods yield NIR-absorbing photothermal agents that can strongly absorb NIR light and convert it into cytotoxic temperature increases for local hyperthermia.

In the present work, we envisioned to develop a graphene-based PTT agent to be explored for low power NIR-induced photohyperthermia therapy with prospective application in the treatment of NMSC. To achieve this goal, key requisites were considered: (1) a nano-sized material to enhance skin permeation and retention [35,36]; (2) strong NIR absorption capacity to enable low power NIR-triggered hyperthermia; and (3) biocompatibility in the absence of NIR irradiation. Building up on a previously described protocol for single step thermal reduction and PEGylation of graphene oxide [37], we prepared non-covalently functionalized nanosized reduced graphene oxide (rGOn-PEG) and characterized in detail the impact of this process on the physicochemical properties of developed materials through multiple complementary techniques. Herein, we obtained a stable dispersion of rGOn-PEG through a 'water-only' reduction protocol. This study enabled us to identify the main chemical features and quantify the chemical modification obtained through such a green and facile method. We further explored the developed material as a NIR-absorbing agent for PTT treatment using low power NIR light source to achieve a hyperthermia effect, envisioning application in the treatment of superficial non-melanoma skin cancers, which constitutes a novel approach.

2. Materials and Methods

2.1. Synthesis of Nanographene Oxide (GOn)

Graphene oxide was prepared from graphite powder (size ≤ 20 μm, Sigma Aldrich, St. Louis, MO, USA) by oxidation using the modified Hummer's method, as described before [27,38]. Briefly, a mixture of 320 mL of sulfuric acid (H_2SO_4, VWR, Darmstadt, Germany) and 80 mL of phosphoric acid (H_3PO_4, Chem-Lab, Zedelgem, Belgium) was added to 8 g of graphite while stirring, and the solution was cooled using an ice bath. Then, 48 g of potassium permanganate ($KMnO_4$, JMGS, Odivelas, Portugal) were added gradually and the solution was heated to 35 °C and stirred for 2 h. Subsequently, 1200 mL of H_2O were slowly added under stirring and with temperature control using an ice bath. Finally, hydrogen peroxide (H_2O_2, VWR, Darmstadt, Germany) was added to stop the reaction. After overnight resting, the solution was decanted to separate the solid phase from the acidic solution, centrifuged at 4000 rpm for 20 min, and redispersed in distilled water. The process was repeated until achieving a neutral pH in the supernatant. In order to produce smaller GO flakes, the pellet recovered in the previous step was re-dispersed in distilled water and placed in an ultrasonic bath (Ovan ATM40-3LCD) for 4 h. The sonication was followed by a centrifugation step at 13,000 rpm for 30 min, allowing the separation of two different phases. The upper phase corresponds to nanographene oxide (GOn) particles and was recovered for further use.

2.2. One-Step Reduction and PEGylation of GOn

Reduced nanographene oxide (rGOn) was produced and functionalized through a single step process, as previously described [37]. For this purpose, 10 mL of GOn (500 μg mL^{-1}) were mixed with 50 mg of poly(ethylene glycol) bis(amine) (PEG-NH_2, average M_n = 3350, Sigma, St. Louis, MO, USA). The mixture was then homogenized using an ultrasonic bath (Ovan ATM40-3LCD) for 10 min. Finally, the mixture was left in a water bath for 24 h at 90 °C. In order to remove unstable aggregates and excess PEG-NH_2, rGOn-PEG was washed four times with deionized water and centrifuged at 13,000 rpm for 10 min. After the washing step, the pellet was redispersed in water to obtain purified rGOn-PEG suspensions. As controls, GOn dispersions incubated with PEG were maintained at room temperature for 24 h; and rGOn was prepared following the same thermal reduction protocol in the absence of PEG.

2.3. Physicochemical Characterization of Graphene-Based Materials

2.3.1. Transmission Electron Microscopy

Morphology and lateral dimensions of GOn and rGOn-PEG dispersions were analyzed by transmission electron microscopy (TEM, JEOL JEM 1400 TEM, Tokyo, Japan). Aqueous dispersions were prepared at a concentration of 50 μg mL^{-1}. For each sample, 10 μL were deposited on a carbon coated TEM grid and allowed to sediment for 1 min, followed by excess material removal by capillarity using filter paper. Nanomaterial lateral dimensions were measured on several TEM images using ImageJ software. Results are presented as frequency distribution of lateral dimensions from over 100 or 25 particles for GOn or rGOn-PEG, respectively.

2.3.2. Dynamic Light Scattering

Particle size distributions for GOn and rGOn-PEG were determined by dynamic light scattering (DLS, LS230 particle size analyzer, Beckman Coulter, Brea, CA, USA). The materials were dispersed in water at a concentration of 100 μg mL^{-1} prior to the measurements. Data were collected performing two scans of 60 s, including polarization intensity differential scattering and using Fraunhofer's model. This model assumes spherical shape for particles in suspension. This evaluation of size distributions does not correspond to precise estimations of particle size and must be considered as relative evaluations of deagglomeration of the different materials in water [27].

2.3.3. Zeta Potential Measurements

Prior to zeta potential measurements, GOn and rGOn-PEG dispersions were prepared at a concentration of 50 μg mL^{-1} and pH 7.4. Zeta potentials of aqueous dispersions were determined using a Zetasizer Nano-ZS (Malvern Instruments, Worcestershire, UK) in a disposable Zetasizer cuvette (Malvern Instruments, Worcestershire, UK). Each measurement was performed in triplicate at room temperature and results are reported as mean and standard deviation.

2.3.4. UV/Visible Spectroscopy

Absorption spectra of GOn and rGOn-PEG were acquired using a Lambda 35 UV/vis spectrometer (Perkin-Elmer, Waltham, MA, USA). Samples were transferred to a 160 μL quartz cuvette (Hellma Analytics, Analytica Munich, Germany) with 10 mm light path length and spectra were recorded in the range of 200–850 nm. Measurements were performed at room temperature by averaging three scans with baseline correction based on water as a blank control.

2.3.5. Fourier Transform Infrared (FTIR) Spectroscopy

Infrared spectra of GOn and rGOn-PEG dehydrated samples were recorded using a VERTEX 70 FTIR spectrometer (Bruker, Karlsruhe, Germany) in transmittance mode at room temperature. Samples were measured in ATR mode, with a A225/Q PLATINUM ATR Diamond crystal with single reflection accessory. Spectra were recorded by averaging 64 scans at a resolution of 4 cm^{-1} over the wavenumber range between 4000 and 400 cm^{-1}.

2.3.6. X-ray Photoelectron Spectroscopy

X-ray photoelectron spectroscopy (XPS) analysis was performed at CEMUP (Centro de Materiais da Universidade do Porto, Porto, Portugal) using a Kratos Axis Ultra HSA for data acquisition. For analysis, a monochromator Al X-ray source operating at 15 kV (90 W) was used. Survey XPS spectra were acquired with pass energy of 80 eV, 1 eV step size and 200 ms dwell time. High resolution C1s XPS spectra were acquired with pass energy of 40 eV, 0.1 eV step size, and 1000 ms dwell time. Spectra were processed using CasaXPS software (Casa Software Ltd., Teignmouth, UK). The effect of the electric charge was corrected by calibrating all samples to the reference of the carbon peak (284.6 eV).

2.3.7. Thermogravimetric Analysis

Thermogravimetric analysis (TGA) (Netzsh STA 449 F3 Jupiter, Selb, Germany) was used for comparison of different materials weigh loss under a constant temperature increase. Sample amounts ranged from 4 to 4.5 mg. The thermograms were recorded between 30 and 1000 °C at a heating rate of 10 °C min^{-1} under nitrogen flow. Results are presented as percentage (%) of weight loss.

2.4. *Photothermal Properties of rGOn-PEG*

To evaluate the light-to-heat conversion ability of GOn and rGOn-PEG, 150 μL of GOn, and rGOn-PEG dispersions at different concentrations in a range between 25 and 250 μg mL^{-1} and water (used as control) were placed in a 48-well plate. All samples were irradiated with a LED-based source with a peak emission around 810 nm (NIR region) and irradiance of 150 mW cm^{-2}. The light-induced temperature change on the samples was monitored during 30 min of irradiation, using a type K thermocouple placed centered and half-height in the suspension. Three replicates were used per condition and results are presented as mean and standard deviation of absolute temperature.

2.5. In Vitro Studies

2.5.1. Cell Culture

Biological studies were performed using A-431 human epidermoid carcinoma cells (ATCC, CRL-1555). Cells were cultured in Dulbecco's modified Eagle's medium (DMEM, ATCC) supplemented with 10% (*v/v*) fetal bovine serum (Alfagene, Carcavelos, Portugal) and 1% (*v/v*) penicillin/streptomycin (Biowest, Pays De La Loire, France). Cells were maintained in a humidified atmosphere with 5% CO_2/95% air at 37 °C.

2.5.2. Cytotoxicity Assays

The effect of GOn and rGOn-PEG on cell viability was evaluated using different concentrations in a range between 25 and 250 μg mL^{-1}. Cells were seeded in 48-well plates at a density of 4×10^4 cells/well, incubated at 37 °C and 5% CO_2. Upon sub-confluence (24 h), culture medium was replaced by GOn or rGOn-PEG dispersions in a final volume of 150 μL/well (in complete DMEM) and cells were incubated with the materials for 24 h. Then, cell viability was quantified by resazurin assay. Briefly, material dispersions were removed, cells were washed with PBS and incubated in 10% (*v/v*) resazurin reagent (Sigma-Aldrich, St. Louis, MO, USA) in culture medium at 37 °C and 5% CO_2 for 4 h. The fluorescence ($\lambda_{ex/em}$ = 530/590 nm) of the supernatant was measured using a micro-plate reader spectrophotometer (Synergy Mx, Bio-Tek Instruments, Winooski, VT, USA). Negative and positive controls for cell viability decrease were performed by incubating A-431 cells with complete DMEM and 10% (*v/v*) dimethyl sulfoxide (DMSO) in complete DMEM, respectively. Data for each sample were normalized to the negative control and results are presented as % of control. All assays were performed in triplicate with six replicates for each condition tested.

2.5.3. Photothermal Irradiation Assays

To evaluate the combined effect of GBM and NIR irradiation, A-431 cells were seeded and incubated with rGOn-PEG dispersions at increasing concentrations (25–250 μg mL^{-1}), as described above. After 24 h of incubation with rGOn-PEG, cells were irradiated for 30 min using a LED-based source with peak emission around 810 nm (NIR region) and irradiance of 150 mW cm^{-2}. Immediately after irradiation, the medium containing rGOn-PEG dispersions was removed, cells were washed with PBS, and resazurin assay was performed as described above. For this purpose, negative and positive controls were performed using irradiated A-431 cells with complete DMEM and 10% (*v/v*) dimethyl sulfoxide (DMSO) in complete DMEM, respectively. To compare the effects of increasing concentrations of rGOn-PEG in the presence (LED on) or absence (LED off) of NIR irradiation, results are presented as mean relative fluorescence units (*RFU*) and standard deviation. To compare the effects of NIR irradiation, a negative control in the absence of materials was considered for normalization. Additionally, GOn dispersions at the same concentration were included in NIR irradiation assays as non-absorbing materials for comparison. All assays were performed in triplicate with six replicates for each condition tested.

2.6. Statistical Analyses

Statistical analyses were performed using GraphPad Prism software (version 8.4.2, San Diego, CA, USA). One-way and two-way analysis of variance (ANOVA) with Tukey tests for multiple comparisons were performed. Differences between experimental groups were considered significant with a confidence interval of 95%, whenever $p < 0.05$.

3. Results

3.1. GBM Morphological Properties, Particle Size, and Stability

Graphene oxide (GO) was prepared from graphite using a modified Hummer's method, followed by ultrasonication to obtain nanosized GO (GOn). GOn was then reduced and non-covalently PEGylated through a one-step procedure, as previously reported [37]. For this purpose, GOn was mixed with PEG-NH_2 for 24 h under thermally reducing conditions. Figure 1 shows as-prepared GBM dispersions. GOn presented its typical appearance, as a brownish stable aqueous dispersion. Conjugation of GOn and PEG in aqueous solution at room temperature similarly resulted in a dark brown dispersion (GOn/PEG), as no reducing conditions were present. During thermal reduction of GOn, in absence of PEG, the dispersion progressively changed its color, evidencing the formation of a black precipitate (rGOn). In opposition, non-covalent functionalization of rGOn with PEG (rGOn-PEG) during thermal reduction resulted in a stable aqueous dispersion.

Figure 1. Images of as-prepared GBM dispersions (500 μg mL^{-1}) in glass vials for stability evaluation. From left to right: GOn dispersion; GOn/PEG (GOn mixed with PEG without thermal treatment); rGOn dispersion (after thermal treatment without PEG functionalization); rGOn-PEG (simultaneously reduced and non-covalently PEGylated).

The morphology of GBM nanosheets was studied using transmission electron microscopy (TEM). Figure 2A shows TEM images of few-hundred-nanometer GOn and rGOn-PEG nanosheets. TEM results are in agreement with previously observed good aqueous dispersibility of GOn and rGOn-PEG, as no agglomerates were observed. Lateral dimensions were determined from TEM image analysis and frequency distribution histograms are shown in Figure 2B. GOn nanosheets were obtained with an average size of 287 nm (minimum and maximum values of 99 nm and 848 nm, respectively, Figure 2A,B). A larger size distribution was observed for rGOn-PEG, which exhibited an average size of 521 nm (minimum and maximum values of 162 nm and 2028 nm, respectively, Figure 2A,B). Similarly, particle size determinations by DLS revealed that GOn nanosheets were constituted by considerably smaller particle sizes than rGOn-PEG (Figure 2C, Figure S1). Figure S1 presents the volume distribution of GOn and rGOn-PEG particle size, whereas Figure 2C corresponds to box plot representations of the same results. For GOn, there were two subpopulations with peaks averaging around 93 nm and 195 nm (Figure S1). From both peaks, the determined average for GOn in general was of 135 nm, while the median was of 120 nm (Figure 2C). The minimum particle size was of 48 nm and the maximum size of 284 nm (Figure 2C, Figure S1). For rGOn-PEG, there were four subpopulations with peaks average at around 64 nm, 214 nm, 545 nm, and 2423 nm (Figure S1). From these peaks, the average was determined to be of 928 nm, while the median value corresponded to 413 nm (Figure 2C). The minimum particle size for rGOn-PEG was of 40 nm, whereas the maximum size was of 3206 nm (Figure 2C, Figure S1).

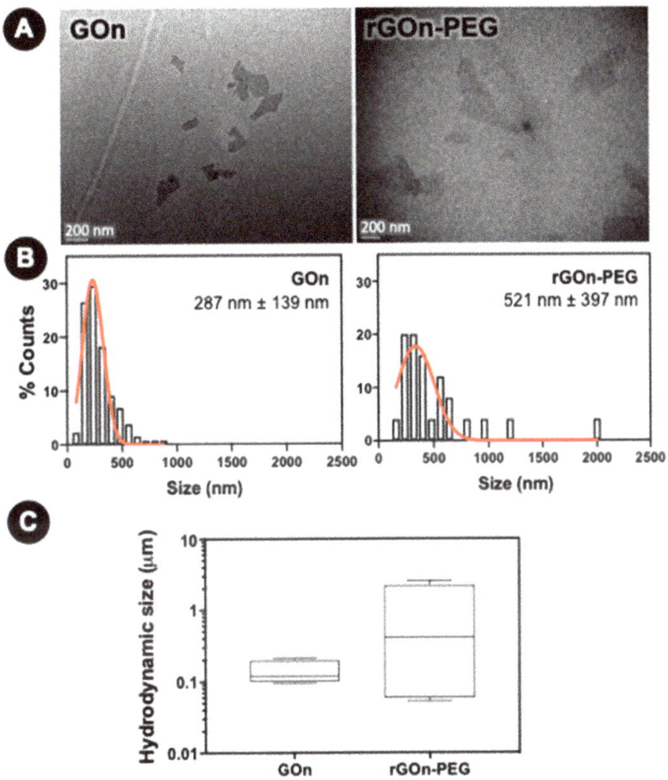

Figure 2. Morphological properties of GOn and rGOn-PEG. (**A**) Representative TEM images of GOn (left) and rGOn-PEG (right) aqueous dispersions and (**B**) respective distribution of particle size and mean and standard deviation, as determined from TEM images. Scale bar, 200 nm. (**C**) Box plot of particle size distributions in volume percentage of GOn and rGOn-PEG dispersed in water at an initial concentration of 250 μg mL^{-1} and determined by light scattering using a Coulter counter.

Both GOn and rGOn-PEG exhibited comparable colloidal stability in aqueous dispersions, according to zeta (ζ)-potential measurements (Table 1). GOn displayed a greater negative surface charge (−25.1 ± 0.8 mV) than rGOn-PEG (−10.2 ± 0.3 mV). This suggests the existence of positive amino-ended branches, resulting in a reduced negative electrostatic charge for non-covalently PEGylated rGOn when compared with GOn [33].

Table 1. Surface charge of GOn and rGOn-PEG aqueous dispersions at an initial concentration of 50 μg mL^{-1} and pH 7.4 (n = 3)

GBM	Surface Charge (mV)
GOn	−25.1 ± 0.8
rGOn-PEG	−10.2 ± 0.3

3.2. Chemical Characterization of GOn and rGOn-PEG Nanosheets

Fourier transform infrared (FTIR) spectra were obtained to confirm the presence of oxygen functionalities on the surface of GOn, as well as its reduction and functionalization into rGOn-PEG (Figure 3). FTIR spectroscopy revealed a broad band in the wavenumber range of 3000 cm^{-1} and

3600 cm^{-1} for GOn, corresponding to O–H stretching vibrations, which are attributed to adsorbed water molecules, hydroxyl, and carboxyl groups [39]. A sharp peak at around 1725 cm^{-1}, which is assigned to C=O stretching vibrations, demonstrated the presence of carbonyl and carboxyl groups [30,39]. An absorption band at ≈1616 cm^{-1} appears owing to the stretching of cyclic alkene (C=C) from unoxidized graphitic domain [30,40,41]. Additionally, the presence of ethers is evidenced by the appearance of strong absorption bands at around 1160 cm^{-1} and 1040 cm^{-1}, which are assigned to C–O stretching vibrations, and through epoxides exhibiting C–O bending vibrations at around 878 cm^{-1} [39].

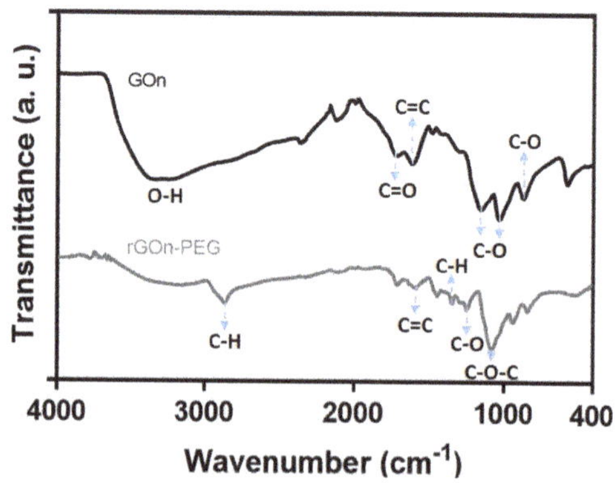

Figure 3. FTIR spectra of GOn and rGOn-PEG. Infrared spectra of GOn (black line) and rGOn-PEG (grey line) describe the contribution of several surface functionalities.

The reduction of GOn was demonstrated in FTIR spectra by a decrease on the intensity of peaks corresponding to the oxygen containing functionalities [41], as compared to the intensities of the peaks of graphene oxide (Figure 3). An absorption band at ≈1590 cm^{-1} appears, which is assigned to C=C stretching vibrations, supporting the restoration of π–π structure of the graphitic domain upon reduction. One significant feature from rGOn-PEG IR spectrum is the presence of a strong absorption band at 2876 cm^{-1} and another peak at 1395 cm^{-1}. These two bands can be assigned to C–H vibrations, attributed to –CH_2– or –C–H groups, strongly supporting the adsorption of PEG molecules onto rGOn-PEG [33]. Additionally, the vibrational C–O stretching at 1248 cm^{-1} corresponds to primary alcohols from PEG molecules. The absorption band at 1055 cm^{-1} could be attributed to C–O–C stretching, which is assigned to ether groups of PEG [33].

Thermogravimetric analysis (TGA) was used to analyze the functionalization degree and thermal stability of GOn and rGOn-PEG. Thermograms for GOn and rGOn-PEG are shown in Figure 4, displaying the weight loss during the heating. The data reveals two main weight loss steps at temperatures above 100 °C for the two materials. The first weight loss occurred between 145 °C and 225 °C for GOn, and between 170 °C and 370 °C for rGOn-PEG. Thermal decomposition during this first step was quantified. It corresponds to the loss of oxygen-containing functional groups, namely carboxyl and epoxy [39]. rGOn-PEG exhibited a lower weight loss when compared with GOn (15.1% and 43.5%, respectively), which suggests fewer oxygen-containing groups are present following reduction and PEGylation [42]. The second weight loss occurred between 225 °C and 630 °C for GOn and 370 °C and 425 °C for rGOn-PEG. In the case of GOn, the weight loss (14.3%) corresponds to the combustion of carbon skeleton and more stable functionalities like carbonyls and residual hydroxyls [43]. The rGOn-PEG substantial weight loss (56.1%) at 425 °C is attributed to the presence of PEG, as at this step pyrolysis of its ether groups occurs [42]. The total percentage weight loss was of 57.8% for GOn and 71.2% for rGOn-PEG. TGA was also performed for non-PEGylated thermally reduced rGOn (Figure S2), which suggests that PEGylation increased the extent of thermal reduction reaction.

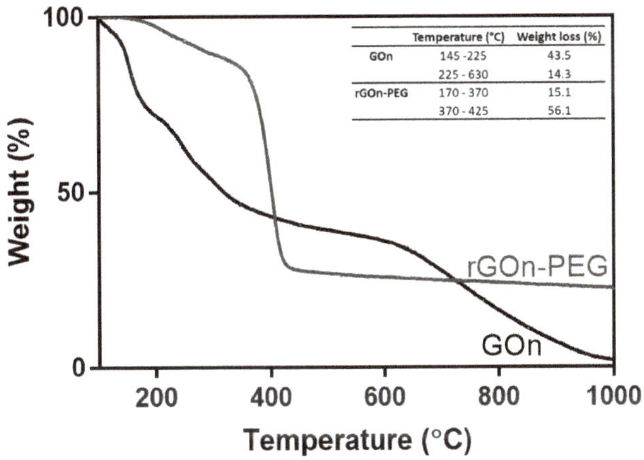

Figure 4. Thermal decomposition of GOn and rGOn-PEG. TGA curves and weight loss values for GOn (black line) and rGOn-PEG (grey line).

X-ray photoelectron spectroscopy (XPS) analyses were performed to characterize the oxidation degree and chemical functional groups at the surface of GOn and rGOn-PEG (Figure 5).

GOn presented a C *at* % of 62.1 and a O *at* % of 32.0 (Figure S3), which demonstrates a successful oxidation and introduction of oxygen functionalities at its surface. Analysis of C1s spectra of GOn revealed two large peaks, which could be further deconvoluted in five peaks (Figure 5A,B). The first binding energy value was attributed to C–C and C=C (284.5 eV, C1s *at* % = 45.5) due to the formation of sp^2 and sp^3 hybridizations of carbon in the graphitic backbone. Single bonds of carbon and oxygen (C–O) in hydroxyls are responsible for the second binding energy value (286.7 eV, C1s *at* % = 44.9). This is the most prevalent carbon bond with oxygen. Carbonyl groups are also present in the form of double bonds between carbon and oxygen (C=O, 287.5 eV, C1s *at* % = 4.2%). The occurrence of carboxyls is responsible for the multiple bonds between carbon and oxygen (O=C–O, 288.5 eV, C1s *at* % = 4.5). Finally, it is possible to observe the π–π^* bond due to the presence of delocalized π electrons in the graphene lattice (290.7 eV, C1s *at* % = 0.93) [39,44].

Regarding the analysis of O1s spectra of GOn, one peak was observed, which could be deconvoluted in three peaks (Figure 5A,C). The first binding energy value was attributed to C=O due to carbonyl groups (531 eV, O1s *at* % = 3.1), C–O bonds from hydroxyls are responsible for the second binding energy (532.3 eV, O1s *at* % = 92.4) and the third peak can be attributed to carboxyls (O=C–O) (533.3 eV, O1s *at* % = 4.5) [39,45]. The relative abundance of chemical bonds found in both C1s and O1s spectra are in accordance and the analysis of the deconvoluted spectra showed that GOn was well oxidized due to the presence of carbon atoms in functional groups (hydroxyl, carbonyl, and carboxyl) with C–O bonds dominating the surface chemistry (Figure 5A).

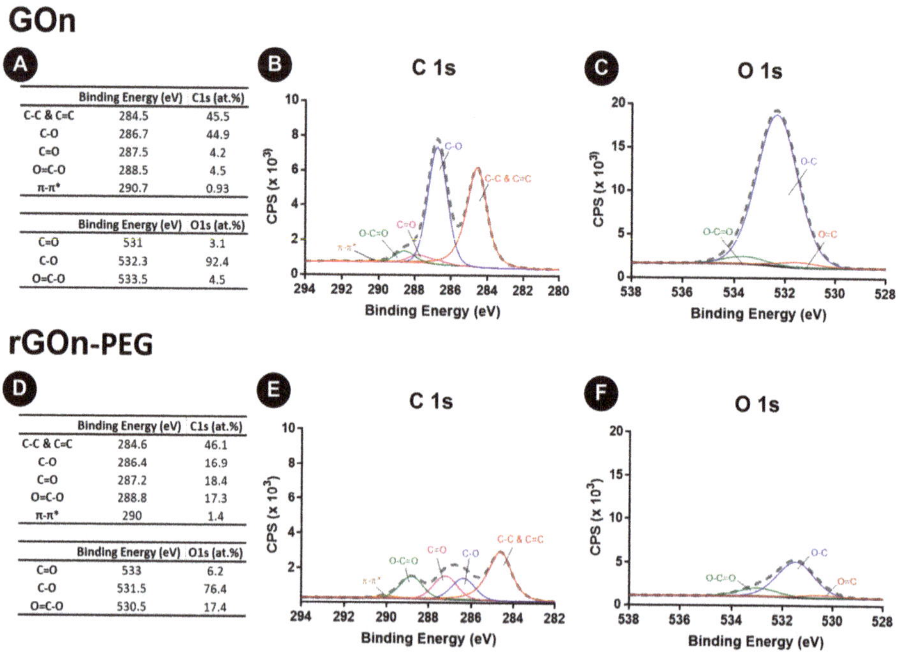

Figure 5. XPS analysis of GOn and rGOn-PEG. (**A**) Atomic composition of GOn and content of C 1s and O 1s chemical functional groups resulting from spectra fitting; (**B**,**C**) Deconvolution of high-resolution (B) C 1s and (C) O 1s XPS spectra for GOn. (**D**) Atomic composition of rGOn-PEG and content of C 1s and O 1s chemical groups resulting from spectra fitting; (**E**,**F**) Deconvolution of high-resolution (E) C 1s and (F) O 1s XPS spectra for rGOn-PEG.

rGOn-PEG presented a C *at* % of 63.5%, an O *at* % of 15.9%, and a N *at* % of 20.6% (Figure S3). These results demonstrated a decrease of the oxygen content comparing with GOn and the presence of N, resulting from the amine-terminated groups from PEG-NH_2. Analysis of C1s and O1s spectra was also performed (Figure 5D–F). Similarly to GOn, C1s spectra was deconvoluted in five peaks. However, different relative abundances in oxygen containing functionalities were found (Figure 5E). The C–C and C=C bond (284.6 eV) continued to largely exist, accounting for 46.1% of carbon bonds (Figure 5D,E). C–O bond (286.4 eV), the most abundant oxygen functionality in GOn, decreased significantly to 16.9% after chemical modification into rGOn-PEG (Figure 5D). The impact of reduction on surface chemistry was also shown by a large increase of C=O bond (287.2 eV) to 18.4% and through the development of a strong peak of O=C–O bond (288.8 eV) with 17.3% of the carbon bonds in rGOn-PEG. In terms of π–π^* bond (290 eV), the results indicate a possible restoration of aromatic structure in rGOn-PEG with an increase from 0.9% in GOn to 1.4% in rGOn-PEG [46,47]. In the case of O1s spectra, it was deconvoluted in three peaks (Figure 5F). An increase of C=O (533 eV, O1s *at* % = 6.2) and O=C–O bonds (530.5 eV, O1s *at* % =17.4) was observed, while there was a decrease on the content of C–O bonds (531.5 eV, O1s *at* % =76.4). These results corroborate those obtained for the C1s spectra. Also, C/O ratio for GOn and rGOn-PEG was of 1.9 and 4, respectively, indicating that rGOn-PEG was successfully reduced. XPS spectra were also obtained for non-PEGylated thermally reduced rGOn (Figure S4), showing a lower extent of reduction in comparison to rGOn-PEG. Successful PEG-NH_2 surface adsorption in rGOn-PEG was confirmed by the presence of a N *at* % of 20.62 (deconvoluted N1s spectra, Figure S5).

3.3. Optical Properties and Photothermal Effect

To assess the potential of the developed materials to be used for photothermal therapy, namely in terms of NIR absorption capacity, the optical properties of both GOn and rGOn-PEG were determined by UV/vis spectroscopy (Figure 6A). Absorbance measurements showed an absorbance peak at λ_{max} = 230 nm for GOn, which is assigned to π–π^* electronic transitions in sp^2 clusters, and a shoulder peak at 300 nm, which is attributed to n–π^* transitions of free electron pairs in oxygen atoms in C=O bonds from carbonyl and carboxyl groups [39]. A red shift of λ_{max} to 263 nm was observed for rGOn-PEG. Additionally, rGOn-PEG exhibited ≈13-fold increment over GOn in NIR absorbance (at 810 nm, Figure 6A inset). These results also support that thermal reduction and non-covalent PEGylation strongly reduced GOn.

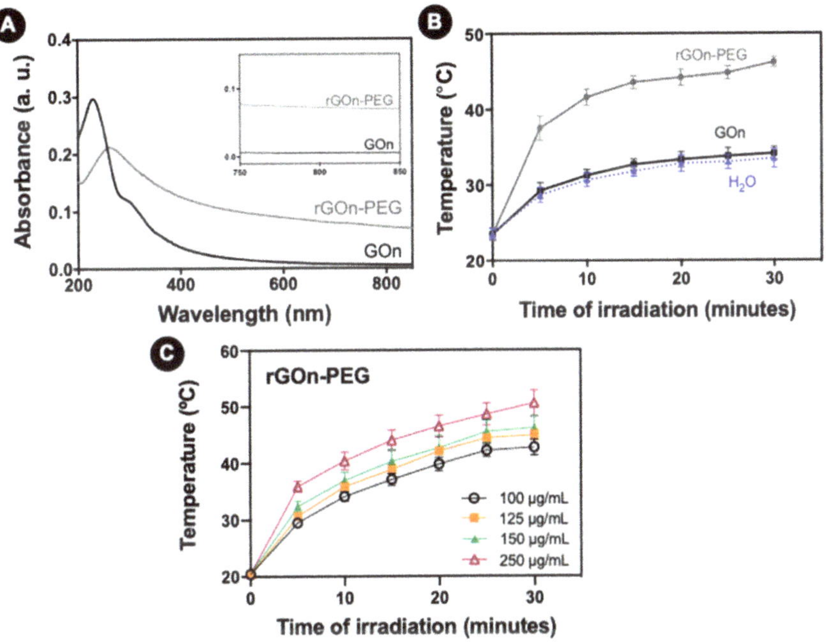

Figure 6. NIR absorption capacity of GBM. (**A**) UV/vis absorption curves of GOn and rGOn-PEG. The inset shows a zoom-in view of the curves in the NIR range from 750 nm to 850 nm. (**B**) Photothermal heating curves of water (dashed line, blue), and GOn (black) and rGOn-PEG (grey) aqueous dispersions at a concentration of 250 μg mL^{-1} in water. (**C**) Concentration-dependent effect on photothermal heating curves for rGOn-PEG.

To further investigate the ability of these nanomaterials to convert NIR light energy into thermal energy, heat generation upon NIR irradiation was evaluated for rGOn-PEG, GOn and water only as control. As seen in Figure 6B, upon 30 min of NIR irradiation, GOn remained at 34 °C, comparable to water. rGOn-PEG displayed concentration-dependent photothermal heating (Figure 6C). In comparison with GOn, aqueous solutions containing 250 μg mL^{-1} of rGOn-PEG showed a higher temperature increase (≈42 °C and ≈47 °C, after 10 and 30 min of NIR irradiation, respectively).

3.4. In Vitro Biocompatibility of GBM Dispersions

To evaluate the biological effect of GOn and rGOn-PEG alone, A-431 cells were used as in vitro model of human skin carcinoma (Figure 7A). Firstly, the dispersibility of GOn and rGOn-PEG in different physiological solutions (PBS, FBS, and complete culture medium) was macroscopically monitored and no precipitates were observed in any of the dispersions up to the highest concentration tested (250 µg mL^{-1}, Figure 7B). Then, cells were incubated with increasing concentrations of developed GBM for 24 h followed by cell viability assessment through resazurin assay (Figure 7A). As control, A-431 cells were cultured in the absence of any material. Increasing concentrations of GOn and rGOn-PEG (25, 50, 100, 125, 150, and 250 µg mL^{-1}) did not affect cell viability, as compared to controls (Figure 7C). Nonetheless, a tendency for a reduction in cell viability was observed for higher concentrations of GOn (150 and 250 µg mL^{-1}, $p < 0.05$). In opposition, rGOn-PEG elicited viability levels above those of control condition. Altogether, these results showed that both GOn and rGOn-PEG do not induce harmful effects on A-431 cells at the concentrations tested, but PEGylation might improve in vitro biocompatibility of these 2D nanomaterials.

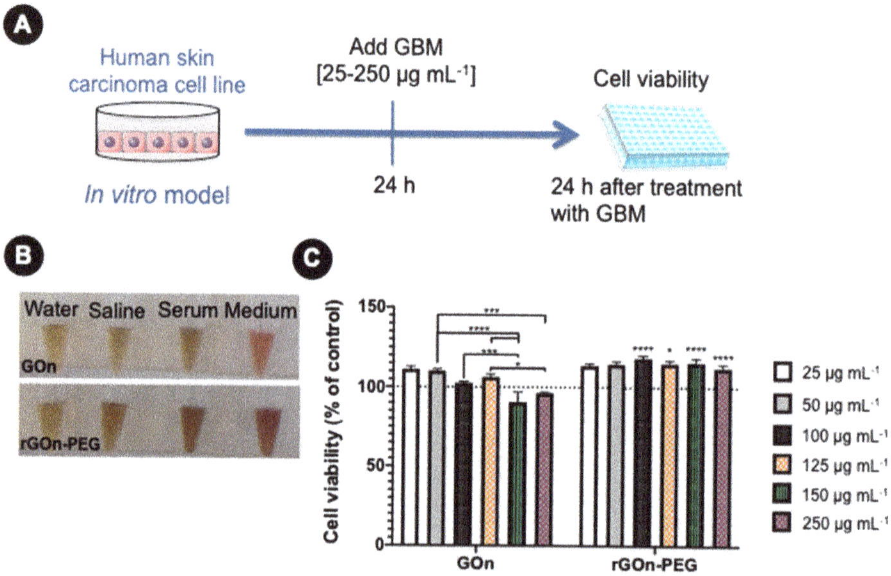

Figure 7. GBM biocompatibility. (**A**) Experimental set-up. One day (24 h) after seeding, human skin carcinoma cells (A-431) were treated with different concentrations of GOn and rGOn-PEG and incubated for an additional 24-h period, prior to resazurin assay. (**B**) GOn and rGOn-PEG (250 µg mL^{-1}) in water and physiological solutions. (**C**) Cellular viability determined using resazurin assay. Results are normalized with respect to values of the control without GBM. Statistically significant differences are shown as * $p < 0.05$, *** $p < 0.001$, **** $p < 0.0001$. Dashed line represents 100% cell viability of the control without GBM.

3.5. In Vitro Photothermal Effect of rGOn-PEG

To determine the cytotoxicity of GOn and rGOn-PEG under NIR irradiation, A-431 cells were incubated with GBM dispersions as described above and then irradiated with a 810 nm LED source for 30 min, followed by cell viability assay (Figure 8A). Increasing concentrations of rGOn-PEG induced a significant decrease in cell viability upon NIR irradiation. Particularly, A-431 cells incubated with > 100 µg mL^{-1} of rGOn-PEG showed significantly lower values in the resazurin assay, in comparison to their non-irradiated counterparts ($p < 0.0001$, Figure 8B,C). Indeed, irradiation of A-431 cells incubated with 250 µg mL^{-1} of rGOn-PEG induced an approximate 38% decrease of cell viability ($p < 0.05$, Figure 8D). These results suggest a potential PTT effect of rGOn-PEG upon low-power (150 mW cm^{-2}) NIR irradiation. As control, no PTT effect was observed for cells incubated with 250 µg mL^{-1} of GOn upon NIR irradiation (Figure 8D).

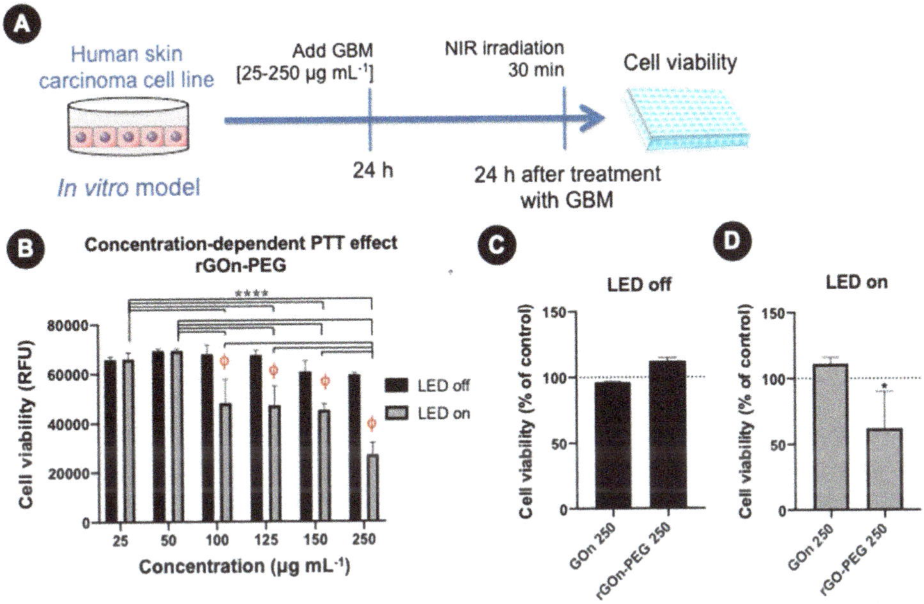

Figure 8. NIR irradiation and in vitro photothermal effect. (**A**) Experimental set-up. One day (24 h) after seeding, human skin carcinoma cells (A-431) were treated with different concentrations of GOn and rGOn-PEG and incubated for an additional 24-h period, prior to NIR irradiation for 30 min and resazurin assay. (**B**) Cellular viability determined using resazurin assay. Results are shown as relative fluorescence units (RFU). Statistically significant differences are shown as ****, $p < 0.0001$, φ, $p < 0.0001$ in comparison to LED off condition. (**C**) Cellular viability without NIR irradiation in the presence of 250 μg mL^{-1} of GOn or rGO-PEG. Results are normalized with respect to values of the control without GBM (**D**) Cellular viability upon NIR irradiation in the presence of 250 μg mL^{-1} of GOn or rGOn-PEG. Results are normalized with respect to values of the control without GBM. Statistically significant differences are shown as *, $p < 0.05$.

4. Discussion

Graphene-based materials have been increasingly investigated for applications in nanomedicine, particularly as PTT platforms to improve the efficacy of cancer treatment strategies. The 2D nanomaterial graphene oxide is commonly obtained from graphite by exfoliation methods, like the modified Hummer's method used here and in our previous work [27]. The exfoliation process results in the introduction of several oxygen functionalities and consequent loss of structural, thermal, and electrical properties [24,28]. In order to restore some of the properties of pristine graphene, over the years different reduction methods have been explored to produce reduced graphene oxide (rGO), including chemical, thermal, and electrochemical reduction pathways [24]. However, the reduction of oxygen functional groups at the surface of GO leads to the formation of unstable colloidal dispersions in aqueous solutions, limiting their potential for biomedical applications and requiring further chemical functionalization toward improving water solubility [25]. Covalent and non-covalent functionalization of GO and rGO with biocompatible polymers like PEG has been widely reported [25,30,31,33,48], using different methods and polymer characteristics (molecular weight, chemistry, etc.). Single step reduction and PEGylation of graphene oxide has been previously described by Chen et al. using a water bath at 90 °C for 24 h and methoxypolyethylene glycol amine (mPEG-NH_2, M_n = 5 kDa) [37]. The authors reported improved water stability of the PEGylated reduced graphene oxide and increased release of resveratrol upon irradiation using a high power 808 nm laser. Following this protocol, we used polyethylene glycol bis(amine) to prepare PEGylated reduced graphene oxide with small sizes for low power NIR-light triggered PTT applications, using cheaper, smaller, and safer LEDs. For this purpose, nano-sized GO (GOn) was obtained by ultrasonication of GO, rendering nanoplatelets with average lateral dimensions below 300 nm. As previously demonstrated, the sonication step has a strong impact on the size of GO flakes, without considerably changing other physicochemical properties [39]. Purified GOn dispersions with controlled lateral dimensions were then reduced and non-covalently functionalized with PEG to attain stable aqueous dispersions with high NIR absorption capacity. The chemical signature of obtained rGOn-PEG colloidal dispersions was carefully investigated through multiple complementary techniques to assess the impact of one-step reduction and PEGylation on the physicochemical properties of graphene oxide.

Upon chemical modification, a significant increase in lateral dimensions from ≈287 nm to ≈521 nm, was found, suggesting the attachment of PEG molecules to GOn. Such an increase in size after PEGylation has been reported by others [33,49]. Smaller sizes have been reported by introducing a sonication step during PEG conjugation reaction [25]. It is relevant to notice that nano-sized particles have been reported to result in improved skin permeation and skin retention, improving treatment outcomes of inflammatory skin diseases [35,36].

The introduction of oxygen functionalities on the surface of GOn, as well as the reduction and non-covalent functionalization of rGOn-PEG were confirmed by FTIR, TGA, and XPS. In comparison with GOn, rGOn-PEG exhibited lower oxygen content given that the majority of oxygen-containing functionalities (carboxyl, hydroxyl, and ketone groups) were removed during the reduction process. Chen and colleagues have previously reported that covalent PEGylation of GOn preserved the aromatic structure of GOn and that a similar green reduction protocol (24 h at 90 °C in a water bath) was able to recover the aromatic structure on rGO-PEG by repairing defects caused during the removal of oxygen moieties from GOn [32]. Additionally, comparing with rGO, simultaneous functionalization with PEG was demonstrated, not only to improve water solubility, but also to increase the extent of the reduction reaction, as confirmed by complementary techniques (TGA and XPS). Other studies have also attributed a role to PEG in strengthening the reduction extent of GOn [32,37]. Indeed, amine groups, which can be oxidized to nitrite, have been reported to exhibit mild reductive ability, being employed as reducing reagents in the preparation of rGO [29]. Similarly to PEG-NH_2, gelatin has numerous amine groups in its backbone and has been reported to act as a reducing agent in the production of stable reduced graphene oxide nanosheets under mild heating conditions [29,50]. In these cases, gelatin formed covalent bonding with rGO through its amine groups. The chemical reactivity of graphene oxide toward amines has been explained by different routes in the literature. Particularly, different types of amine can react with GO functional groups via amidation reaction of carboxylic acid groups at the edges of GO or through ring-opening of epoxides on the surface of GO [50–52]. Hydrogen bonding between amines and hydroxyls of GO is another possibility [32,53]. Nonetheless, given the complex structure of graphene-based materials, the nature of such chemical reactions is still far from being fully understood.

According to TGA measurements, the ratio of grafted PEG was estimated to be 56.1%, supporting the efficiency of single step reduction and PEGylation. Previous reports on rGOn-PEG prepared through a similar method does not provide quantitative data regarding the chemical modification of the material [37], but other non-covalent methods have shown around 78% modification through PEGylation of previously reduced GO [33]. It is worth noting that such non-covalent functionalization methods rely on polymer physisorption onto graphene basal planes via π–π stacking and van der Waals interactions and few examples of covalent functionalization of rGO exist owing to the removal of the majority of reactive functional groups upon reduction [24]. Notwithstanding, non-covalent functionalization methods exhibit strong advantages over covalent functionalization, including the preservation of extended π conjugation and the aromatic structure of GOn, whereas covalent modification creates sp^3 defects on the graphene ring [54].

PEGylation was successful in terms of overcoming the hydrophobicity associated to rGO and rendering a nanomaterial that is stable in different physiological media (PBS, FBS, and culture medium). Herein, we demonstrated that rGOn-PEG had largely restored its aromatic structure showing strongly enhanced absorbance in the NIR region (≈13-fold increase, in comparison to GOn) and light-to-heat energy conversion capacity upon NIR irradiation, comparable to the material previously obtained using a similar protocol with a different PEG [37]. On the other hand, other studies reported approximately 6- to 8-fold increment in NIR absorption by rGOn-PEG produced through covalent PEGylation followed by thermal reduction [32,49]. Several studies have shown a very rapid heating of graphene-based nanomaterials using more powerful irradiation systems (>3 W cm^{-2}, compared to 150 mW cm^{-2} used herein) and reporting heating up to 60–70 °C [32,37,55,56]. Such high temperatures are above those needed for tumor ablation treatments (> 50 °C) [7], and are also likely to induce damage in healthy cells of surrounding tissues, particularly considering the small sizes of superficial non-melanoma skin cancers, such as basal cell carcinoma (≈20 mm in size). For instance, protein-functionalized rGO nanosheets (40 μg mL^{-1}) and polyethylenimine-PEG-rGO (8 μg mL^{-1}) induced a reduction of 35% in cell viability without irradiation [55,57]. Herein, we demonstrated that both GOn and rGOn-PEG alone and below 250 μg mL^{-1} were noncyotoxic to A-431 epidermoid carcinoma cells. PEGylation is an effective chemical functionalization commonly used to improve the biocompatibility of nanomaterials [25,30,34]. Although a tendency for a decrease in cell viability was observed with increasing concentrations of unmodified GOn, the same was not observed for rGOn-PEG, for which concentrations > 100 μg mL^{-1} resulted in higher fluorescence measurements using resazurin assay, in comparison to control. This result supports the use of biocompatible polymers as PEG to enhance the biological effects of developed nanomaterials.

To further determine the in vitro photothermal effect of rGOn-PEG, we irradiated A-431 cells in the presence of prepared GBM using low power (150 mW cm^{-2}) NIR LEDs. Consistently with other studies [32,49,58], NIR irradiation alone was not sufficient to induce cell death, but the combination of rGOn-PEG with NIR irradiation resulted in ≈38% decrease of cell viability after a single treatment during 30 min. As described above, rGOn-PEG temperature increased up to 47 °C upon NIR irradiation, which falls within the hyperthermia range of temperatures (41–50 °C) [6,7]. Hyperthermia triggers apoptotic pathways, interfering with normal cell functions, possibly leading to enhanced membrane permeability, metabolic signaling disruption, dysfunctional membrane transport, and activation of heat shock proteins, among other cellular and molecular changes [7,59,60]. Nonetheless, the effects of hyperthermia, and particularly their combination with graphene-based nanomaterials, are still far from being fully understood. To the best of our knowledge, our study is the first to report the effect of reduced graphene oxide on epidermoid carcinoma cells, which are commonly used as in vitro models of non-melanoma skin cancer. The effect of nanocomposites of gold nanorod-assembled PEGylated graphene oxide has been reported to result in similar levels of cytotoxicity (≈40% decrease in cell viability) upon irradiation with high power (60 W cm^{-2}) Xe-lamp light [58]. Others have shown that comparable PEGylated graphene-based nanomaterials elicited ≈80% cell viability decrease using 4T1 breast cancer cell line [37]; and up to ≈90% decrease of the viability of A549 adenocarcinomic human alveolar basal epithelial cells using covalent chemical functionalization methods [32]. Notwithstanding, major differences in the irradiation time and power of the light source limit the comparison between studies.

5. Conclusions

In this work, we characterized the impact of a single step thermal reduction and PEGylation process on the physicochemical properties of graphene oxide using multiple complementary techniques. Chemical modification with PEG not only resulted in improved water dispersibility of reduced graphene oxide, but also contributed to enhance the extent of the thermal reduction reaction. PEGylation yielded single-layer rGOn-PEG sheets with average nano-sized lateral dimensions of ≈521 nm. The single step process resulted in the restoration of the aromatic structure of graphene, evidenced by the appearance of C=C bonds in rGOn-PEG infrared spectrum, increased C/O ratio, and changes in the optical properties through a red shift of λ_{max}. The obtained rGOn-PEG exhibited a ≈13-fold increase in NIR absorbance and reached 42 °C after 10 min of NIR LED irradiation, exhibiting a continuous heating up to 47 °C after 30 min, whereas unmodified GOn remained at 34 °C even under NIR irradiation. Temperatures registered for rGOn-PEG were within the hyperthermia range. PEGylation of rGOn resulted in improved in vitro biocompatibility, compared to unmodified GOn, which seemed to induce a reduction of A-431 cell viability for concentrations above 150 μg mL^{-1}. Combining NIR irradiation with rGOn-PEG in concentrations above 100 μg.mL^{-1} resulted in a cytotoxic effect. After a single irradiation with a low power NIR LED system, a 38% decrease of cell viability was found, showing the in vitro photothermal effect of rGOn-PEG.

Altogether, our results further support the use of a simple and facile method to obtain functionalized rGOn as a promising photoabsorbing agent for PTT applications in non-melanoma skin cancer treatment. The combination of this functionalized nanomaterial with NIR irradiation using a safer LED-based NIR light source opens new possibilities toward exploring lower power and cheaper systems for mild hyperthermia cancer therapy, enabling better control over nanomaterial heating.

Supplementary Materials: The following are available online at http://www.mdpi.com/2073-4360/12/8/1840/s1. Figure S1. Volume distribution of particle size of GOn and rGOn-PEG dispersed in water at an initial concentration of 250 μg mL^{-1} and determined by light scattering using a Coulter counter. A boxplot, which is a standardized way of displaying the dataset based on a five-number summary (the minimum, the maximum, the sample median, and the first and third quartiles) is presented in Figure 2C. Figure S2. TGA curve of rGOn and weight loss. Figure S3. XPS survey spectra and atomic composition for (A) GOn and (B) rGOn-PEG. Figure S4. Surface chemical properties of rGOn. (A) Atomic composition of rGOn and content of C 1s and O 1s chemical groups resulting from spectra fitting; (B, C) Deconvolution of high-resolution (B) C 1s and (C) O 1s XPS spectra. Figure S5. Quantification of functionalization of rGOn-PEG. (A) Content of N 1s chemical groups resulting from spectra fitting; (B) Deconvolution of high-resolution N 1s XPS spectra.

Author Contributions: Conceptualization, R.C.-A., D.B., I.C.G., F.D.M., and A.M.P.; Methodology, R.C.-A., D.B., J.R.F., L.T., F.A.L.S.S., J.M., and A.M.P.; Validation, R.C.-A., D.B., J.R.F., and A.M.P; Formal analysis, R.C.-A., D.B., and A.M.P.; Investigation, R.C.-A. and D.B.; Resources, J.R.F., I.C.G., F.D.M., and A.M.P.; Data curation, R.C.-A., D.B., J.R.F., L.T., F.A.L.S.S., J.M., and A.M.P.; Writing—original draft preparation, R.C-A., D.B., and A.M.P.; Writing—review and editing, R.C.-A., I.C.G., F.D.M., and A.M.P.; Supervision, I.C.G., F.D.M., and A.M.P.; Project administration, A.M.P.; Funding acquisition, A.M.P. All authors have read and agreed to the published version of the manuscript.

Funding: This work was financed by FEDER funds through the COMPETE 2020 – Operacional Programme for Competitiveness and Internationalisation (POCI), Portugal 2020, and by national funds (PIDDAC) through FCT/MCTES in the framework of the project POCI-01-0145-FEDER-031143, and Base Funding - UIDB/00511/2020 of the Laboratory for Process Engineering, Environment, Biotechnology and Energy – LEPABE. Authors would also like to thank the support of i3S Scientific Platforms and respective funding: HEMS, member of the national infrastructure PPBI – Portuguese Platform of Bioimaging: POCI-01-0145-FEDER-022122; and Biointerfaces and Nanotechnology (BN) Laboratory, Portuguese Funds through FCT, UID/BIM/04293/2019.

Acknowledgments: Authors would like to acknowledge Rui Fernandes from the Histology and Electron Microscopy Service (HEMS) of i3S, Porto, for the assistance with TEM studies; and Daniela Silva from Centro de Materiais da Universidade do Porto (CEMUP) for her support with XPS spectra analyses.

Conflicts of Interest: The authors declare no conflict of interest.

References

1. Rembielak, A.; Ajithkumar, T. Non-melanoma skin cancer—An underestimated global health threat? *Clin. Oncol. (R. Coll. Radiol.)* **2019**, *31*, 735–737. [CrossRef] [PubMed]
2. Rubin, A.I.; Chen, E.H.; Ratner, D. Basal-cell carcinoma. *N. Eng. J. Med.* **2005**, *353*, 2262–2269. [CrossRef] [PubMed]
3. Bander, T.S.; Nehal, K.S.; Lee, E.H. Cutaneous squamous cell carcinoma: Updates in staging and management. *Dermatol. Clin.* **2019**, *37*, 241–251. [CrossRef] [PubMed]
4. Marzuka, A.; Book, S.E. Basal cell carcinoma: Pathogenesis, epidemiology, clinical features, diagnosis, histopathology, and management. *Yale J. Boil. Med.* **2015**, *88*, 167–179.

5. Firnhaber, J.M. Diagnosis and treatment of basal cell and squamous cell carcinoma. *Am. Fam. Physician* **2012**, *86*, 161–168.
6. Doughty, A.C.; Hoover, A.R.; Layton, E.; Murray, C.K.; Howard, E.W.; Chen, W.R. Nanomaterial applications in photothermal therapy for cancer. *Materials* **2019**, *12*, 779. [CrossRef]
7. Zhang, B.; Wang, Y.; Liu, J.; Zhai, G. Recent developments of phototherapy based on graphene family nanomaterials. *Curr. Med. Chem.* **2017**, *24*, 268–291. [CrossRef]
8. Kumar, R.; Sahoo, S.; Joanni, E.; Singh, R.K.; Yadav, T.; Verma, R.K.; Singh, D.P.; Tan, W.K.; Del Pino, A.P.; Moshkalev, S.A.; et al. A review on synthesis of graphene, h-BN and MoS2 for energy storage applications: Recent progress and perspectives. *Nano Res.* **2019**, *12*, 2655–2694. [CrossRef]
9. Kumar, R.; Sahoo, S.; Joanni, E.; Singh, R.K.; Tan, W.K.; Kar, K.K.; Matsuda, A. Recent progress in the synthesis of graphene and derived materials for next generation electrodes of high performance lithium ion batteries. *Prog. Energy Combust. Sci.* **2019**, *75*, 100786. [CrossRef]
10. Kumar, R.; Sahoo, S.; Joanni, E.; Singh, R.K.; Maegawa, K.; Tan, W.K.; Kawamura, G.; Kar, K.K.; Matsuda, A. Heteroatom doped graphene engineering for energy storage and conversion. *Mater. Today* **2020**. [CrossRef]
11. Tiwari, S.K.; Sahoo, S.; Wang, N.; Huczko, A. Graphene research and their outputs: Status and prospect. *J. Sci. Adv. Mater. Devices* **2020**, *5*, 10–29. [CrossRef]
12. Pinto, A.M.; Gonçalves, I.C.; Magalhães, F.D. Graphene-based materials biocompatibility: A review. *Colloids Surfaces B Biointerfaces* **2013**, *111*, 188–202. [CrossRef] [PubMed]
13. Fadeel, B.; Bussy, C.; Merino, S.; Vázquez, E.; Flahaut, E.; Mouchet, F.; Evariste, L.; Gauthier, L.; Koivisto, A.J.; Vogel, U.; et al. Safety assessment of graphene-based materials: Focus on human health and the environment. *ACS Nano* **2018**, *12*, 10582–10620. [CrossRef] [PubMed]
14. Pinto, A.M.; Moreira, J.A.; Magalhães, F.D.; Gonçalves, I.C. Polymer surface adsorption as a strategy to improve the biocompatibility of graphene nanoplatelets. *Colloids Surfaces B Biointerfaces* **2016**, *146*, 818–824. [CrossRef]
15. Gomes, R.N.; Borges, I.; Pereira, A.; Maia, A.F.; Magalhães, F.D.; Magalhães, F.D.; Pinto, A.M.; Gonçalves, I.C. Antimicrobial graphene nanoplatelets coatings for silicone catheters. *Carbon* **2018**, *139*, 635–647. [CrossRef]
16. Chen, Y.-W.; Su, Y.-L.; Hu, S.-H.; Chen, S.-Y. Functionalized graphene nanocomposites for enhancing photothermal therapy in tumor treatment. *Adv. Drug Deliv. Rev.* **2016**, *105*, 190–204. [CrossRef]
17. Liu, S.; Pan, X.; Liu, H. Two-dimensional nanomaterials for photothermal therapy. *Angew. Chem. Int. Ed.* **2020**, *59*, 5890–5900. [CrossRef]
18. Pinto, A.M.; Pereira, A.T.; Gonçalves, I.C. *Carbon Biomaterials in Biomaterials Science*; Wagner, W.R., Sakiyama-Elbert, S.E., Zhang, G., Yaszemski, M.J., Eds.; Elsevier: Amsterdam, The Netherlands, 2020; pp. 327–360.
19. Wan, S.; Zhang, B.; Li, S.; He, B.; Pu, Y. Combination of PEG-decorated black phosphorus nanosheets and immunoadjuvant for photoimmunotherapy of melanoma. *J. Mater. Chem. B* **2020**, *8*, 2805–2813. [CrossRef]
20. Liu, Z.; Chen, H.; Jia, Y.; Zhang, W.; Zhao, H.; Fan, W.; Zhang, W.; Zhong, H.; Ni, Y.; Guo, Z. A two-dimensional fingerprint nanoprobe based on black phosphorus for bio-SERS analysis and chemo-photothermal therapy. *Nanoscale* **2018**, *10*, 18795–18804. [CrossRef]
21. Yang, G.; Liu, Z.; Li, Y.; Hou, Y.; Fei, X.; Su, C.; Wang, S.; Zhuang, Z.; Guo, Z. Facile synthesis of black phosphorus–Au nanocomposites for enhanced photothermal cancer therapy and surface-enhanced raman scattering analysis. *Biomater. Sci.* **2017**, *5*, 2048–2055. [CrossRef]
22. Gong, L.; Yan, L.; Zhou, R.; Xie, J.; Wu, W.; Gu, Z. Two-dimensional transition metal dichalcogenide nanomaterials for combination cancer therapy. *J. Mater. Chem. B* **2017**, *5*, 1873–1895. [CrossRef] [PubMed]
23. Zhang, C.; Hu, D.-F.; Xu, J.-W.; Ma, M.-Q.; Xing, H.; Yao, K.; Ji, J.; Xu, Z.-K. Polyphenol-assisted exfoliation of transition metal dichalcogenides into nanosheets as photothermal nanocarriers for enhanced antibiofilm activity. *ACS Nano* **2018**, *12*, 12347–12356. [CrossRef] [PubMed]
24. Dreyer, D.R.; Park, S.; Bielawski, C.W.; Ruoff, R.S. The chemistry of graphene oxide. *Chem. Soc. Rev.* **2010**, *39*, 228–240. [CrossRef] [PubMed]
25. Yang, K.; Feng, L.; Hong, H.; Cai, W.; Liu, Z. Preparation and functionalization of graphene nanocomposites for biomedical applications. *Nat. Protoc.* **2013**, *8*, 2392–2403. [CrossRef]
26. Zhang, B.; Yu, Q.; Zhang, Y.M.; Liu, Y. Two-dimensional supramolecular assemblies based on beta-cyclodextrin-grafted graphene oxide for mitochondrial dysfunction and photothermal therapy. *Chem. Comm.* **2019**, *55*, 12200–12203. [CrossRef]

27. Pinto, A.M.; Gonçalves, C.; Sousa, D.M.; Ferreira, A.R.; Moreira, J.A.; Gonçalves, I.C.; Magalhães, F.D. Smaller particle size and higher oxidation improves biocompatibility of graphene-based materials. *Carbon* **2016**, *99*, 318–329. [CrossRef]
28. Mohan, V.B.; Lau, K.-T.; Hui, D.; Bhattacharyya, D. Graphene-based materials and their composites: A review on production, applications and product limitations. *Compos. Part B Eng.* **2018**, *142*, 200–220. [CrossRef]
29. Liu, K.; Zhang, J.-J.; Cheng, F.-F.; Zheng, T.; Wang, C.; Zhu, J.-J. Green and facile synthesis of highly biocompatible graphene nanosheets and its application for cellular imaging and drug delivery. *J. Mater. Chem.* **2011**, *21*, 12034. [CrossRef]
30. Zhang, S.; Xiong, P.; Yang, X.; Wang, X. Novel PEG functionalized graphene nanosheets: Enhancement of dispersibility and thermal stability. *Nanoscale* **2011**, *3*, 2169–2174. [CrossRef]
31. Zhang, W.; Guo, Z.; Huang, D.; Liu, Z.; Guo, X.; Zhong, H. Synergistic effect of chemo-photothermal therapy using PEGylated graphene oxide. *Biomaterials* **2011**, *32*, 8555–8561. [CrossRef]
32. Chen, J.; Wang, X.; Chen, T. Facile and green reduction of covalently PEGylated nanographene oxide via a 'water-only' route for high-efficiency photothermal therapy. *Nanoscale Res. Lett.* **2014**, *9*, 86. [CrossRef] [PubMed]
33. Mendonça, M.C.P.; Soares, E.S.; De Jesus, M.B.; Ceragioli, H.J.; Batista, Â.G.; Nyúl-Tóth, Á.; Molnár, J.; Wilhelm, I.; Maróstica, J.M.R.; Krizbai, I.; et al. PEGylation of reduced graphene oxide induces toxicity in cells of the blood–brain barrier: An in vitro and in vivo study. *Mol. Pharm.* **2016**, *13*, 3913–3924. [CrossRef] [PubMed]
34. Syama, S.; Aby, C.P.; Maekawa, T.; Kumar, D.S.; Mohanan, P. Nano-bio compatibility of PEGylated reduced graphene oxide on mesenchymal stem cells. *2D Mater.* **2017**, *4*, 025066. [CrossRef]
35. Try, C.; Moulari, B.; Béduneau, A.; Fantini, O.; Pin, D.; Pellequer, Y.; Lamprecht, A. Size dependent skin penetration of nanoparticles in murine and porcine dermatitis models. *Eur. J. Pharm. Biopharm.* **2016**, *100*, 101–108. [CrossRef] [PubMed]
36. Yokota, J.; Kyotani, S. Influence of nanoparticle size on the skin penetration, skin retention and anti-inflammatory activity of non-steroidal anti-inflammatory drugs. *J. Chin. Med Assoc.* **2018**, *81*, 511–519. [CrossRef]
37. Chen, J.; Liu, H.; Zhao, C.; Qin, G.; Xi, G.; Li, T.; Wang, X.; Chen, T. One-step reduction and PEGylation of graphene oxide for photothermally controlled drug delivery. *Biomaterials* **2014**, *35*, 4986–4995. [CrossRef]
38. Marcano, D.C.; Kosynkin, D.V.; Berlin, J.M.; Sinitskii, A.; Sun, Z.; Slesarev, A.; Alemany, L.B.; Lu, W.; Tour, J.M. Improved synthesis of graphene oxide. *ACS Nano* **2010**, *4*, 4806–4814. [CrossRef]
39. Rodrigues, A.F.; Newman, L.; Lozano, N.; Mukherjee, S.P.; Fadeel, B.; Bussy, C.; Kostarelos, K. A blueprint for the synthesis and characterisation of thin graphene oxide with controlled lateral dimensions for biomedicine. *2D Mater.* **2018**, *5*, 035020. [CrossRef]
40. Ain, Q.T.; Haq, S.H.; Alshammari, A.; Al-Mutlaq, M.A.; Anjum, M.N. The systemic effect of PEG-nGO-induced oxidative stress in vivo in a rodent model. *Beilstein J. Nanotechnol.* **2019**, *10*, 901–911. [CrossRef]
41. Emiru, T.F.; Ayele, D.W. Controlled synthesis, characterization and reduction of graphene oxide: A convenient method for large scale production. *Egypt. J. Basic Appl. Sci.* **2017**, *4*, 74–79. [CrossRef]
42. Zhang, L.; Wang, Z.; Lu, Z.; Shen, H.; Huang, J.; Zhao, Q.; Liu, M.; He, N.; Zhang, Z. PEGylated reduced graphene oxide as a superior ssRNA delivery system. *J. Mater. Chem. B* **2012**, *1*, 749–755. [CrossRef] [PubMed]
43. Wojtoniszak, M.; Chen, X.; Kalenczuk, R.J.; Wajda, A.; Lapczuk, J.; Kurzewski, M.; Drozdzik, M.; Chu, P.K.; Borowiak-Palen, E. Synthesis, dispersion, and cytocompatibility of graphene oxide and reduced graphene oxide. *Colloids Surfaces B Biointerfaces* **2012**, *89*, 79–85. [CrossRef] [PubMed]
44. Ganguly, A.; Sharma, S.; Papakonstantinou, P.; Hamilton, J.W.J. Probing the thermal deoxygenation of graphene oxide using high-resolution in situ x-ray-based spectroscopies. *J. Phys. Chem. C* **2011**, *115*, 17009–17019. [CrossRef]
45. Araújo, M.; Soares, O.S.G.P.; Fernandes, A.J.S.; Pereira, M.F.R.; Freire, C. Tuning the surface chemistry of graphene flakes: New strategies for selective oxidation. *RSC Adv.* **2017**, *7*, 14290–14301. [CrossRef]
46. Chang, H.; Sun, Z.; Saito, M.; Yuan, Q.; Zhang, H.; Li, J.; Wang, Z.; Fujita, T.; Ding, F.; Zheng, Z.; et al. Regulating infrared photoresponses in reduced graphene oxide phototransistors by defect and atomic structure control. *ACS Nano* **2013**, *7*, 6310–6320. [CrossRef]

47. Kellici, S.; Acord, J.; Ball, J.; Reehal, H.S.; Morgan, D.; Saha, B. A single rapid route for the synthesis of reduced graphene oxide with antibacterial activities. *RSC Adv.* **2014**, *4*, 14858. [CrossRef]
48. Kim, H.; Lee, D.; Kim, J.; Kim, T.-I.; Kim, W.J. Photothermally triggered cytosolic drug delivery via endosome disruption using a functionalized reduced graphene oxide. *ACS Nano* **2013**, *7*, 6735–6746. [CrossRef]
49. Robinson, J.T.; Tabakman, S.M.; Liang, Y.; Wang, H.; Casalongue, H.S.; Vinh, D.; Dai, H. Ultrasmall reduced graphene oxide with high near-infrared absorbance for photothermal therapy. *J. Am. Chem. Soc.* **2011**, *133*, 6825–6831. [CrossRef]
50. Piao, Y.; Chen, B. One-pot synthesis and characterization of reduced graphene oxide–gelatin nanocomposite hydrogels. *RSC Adv.* **2016**, *6*, 6171–6181. [CrossRef]
51. Vacchi, I.A.; Spinato, C.; Raya, J.; Bianco, A.; Ménard-Moyon, C. Chemical reactivity of graphene oxide towards amines elucidated by solid-state NMR. *Nanoscale* **2016**, *8*, 13714–13721. [CrossRef]
52. Bourlinos, A.B.; Gournis, D.; Petridis, D.; Szabó, T.; Szeri, A.; Dékány, I. Graphite oxide: Chemical reduction to graphite and surface modification with primary aliphatic amines and amino acids. *Langmuir* **2003**, *19*, 6050–6055. [CrossRef]
53. Matsuo, Y.; Miyabe, T.; Fukutsuka, T.; Sugie, Y. Preparation and characterization of alkylamine-intercalated graphite oxides. *Carbon* **2007**, *45*, 1005–1012. [CrossRef]
54. Layek, R.K.; Nandi, A.K. A review on synthesis and properties of polymer functionalized graphene. *Polymer* **2013**, *54*, 5087–5103. [CrossRef]
55. Cheon, Y.A.; Bae, J.H.; Chung, B.G. Reduced graphene oxide nanosheet for chemo-photothermal therapy. *Langmuir* **2016**, *32*, 2731–2736. [CrossRef] [PubMed]
56. Hu, S.-H.; Chen, Y.-W.; Hung, W.-T.; Chen, I.-W.; Chen, S.-Y. Quantum-dot-tagged reduced graphene oxide nanocomposites for bright fluorescence bioimaging and photothermal therapy monitored in situ. *Adv. Mater.* **2012**, *24*, 1748–1754. [CrossRef] [PubMed]
57. Tong, C.; Zhang, X.; Fan, J.; Li, B.; Liu, B.; Daniyal, M.; Wangb, W. PEGylated mBPEI-rGO nanocomposites facilitate hepotocarcinoma treatment combining photothermal therapy and chemotherapy. *Sci. Bull.* **2018**, *63*, 935–946. [CrossRef]
58. Dembereldorj, U.; Choi, S.Y.; Ganbold, E.-O.; Song, N.W.; Kim, D.; Choo, J.; Lee, S.Y.; Kim, S.; Joo, S.-W. Gold nanorod-assembled PEGylated graphene-oxide nanocomposites for photothermal cancer therapy. *Photochem. Photobiol.* **2013**, *90*, 659–666. [CrossRef] [PubMed]
59. Hegyi, G.; Szigeti, G.P.; Szasz, A. Hyperthermia versus oncothermia: Cellular effects in complementary cancer therapy. *Evid. Based Complement. Altern. Med.* **2013**, *2013*, 1–12. [CrossRef]
60. Ren, G.; Jv, H.; Tian, Z.; Gvetadze, S.R.; Hu, J.; Zhao, M.; Zhou, G.; Zhang, C.; Zhang, Z. Ultrasound hyperthermia induces apoptosis in head and neck squamous cell carcinoma: An in vitro study. *Med. Oral Patol. Oral y Cir. Bucal* **2017**, *22*, 289–296. [CrossRef]

MDPI
St. Alban-Anlage 66
4052 Basel
Switzerland
Tel. +41 61 683 77 34
Fax +41 61 302 89 18
www.mdpi.com

Polymers Editorial Office
E-mail: polymers@mdpi.com
www.mdpi.com/journal/polymers

www.ingramcontent.com/pod-product-compliance
Lightning Source LLC
Chambersburg PA
CBHW061448210926
51850CB00008B/1649